FREE RADICALS: CHEMISTRY, PATHOLOGY AND MEDICINE

Free Radicals: Chemistry, Pathology and Medicine

Edited by
Catherine Rice-Evans
Royal Free Hospital School of Medicine
University of London
and
Thomas Dormandy
Whittington Hospital, University of London

THE RICHELIEU PRESS
LONDON

Published by
THE RICHELIEU PRESS LIMITED
30 Saint Mark's Crescent, London, NW1 7TU

ISBN 0 903840 03 0

Printed in Great Britain at the University Press, Cambridge

PREFACE

Twenty-five years ago free radicals, their importance already recognised in inorganic, organic and industrial chemistry, were regarded by all but a few as irrelevant to biology and medicine. The change since then has been spectacular and it is still accelerating. This is good news to devotees but it has also created problems. A clinical audience today, for example, can include a few outstanding experts among many with only the haziest idea of what free radicals are.

In medicine free radicals are fast acquiring a sinister reputation: they are perceived as an unredeemed menace against which we need constant and elaborate protection. The image is reinforced by a torrent of promotional literature promising to improve on Nature's flagging efforts and inadequate performance. Yet our survival as complex organisms depends on the continuous destruction of our constituent units (which alone governs the rate of replacement) and failure of the disposal mechanisms is probably a more important cause of death and disease than is excessive disposal. In short, free radicals are still a wide open field with few firm guidelines and plenty of pitfalls but also full of promise. Forty years ago they helped to solve one of the oldest and seemingly intractable problems of organic and industrial chemistry - the rancidification of fats and oils. There are many such problems awaiting solution in biology and medicine. It is not surprising perhaps that interest in them follows not a period of slow and cautious rapprochement but sudden illumination, an almost Pauline conversion. Nor is it surprising that scientific meetings devoted to free radicals can be good or bad but are almost never dull.

As free radical concepts advance in new directions - cancer screening, parasitic infection, malnutrition, infertility, to name but a few - an interdisciplinary approach becomes more rather than less important. If war is too serious to be left to the generals, the biology and medicine of free radicals are too treacherous to be left entirely to the biologists and the doctors. The problem again is communication. One of the main objectives of the Society for Free Radical Research is to try and maintain the links between disciplines; all the speakers and ´poster artists´ who were invited to contribute to its 1987 Summer Meeting were also asked to make themselves comprehensible across a wide spectrum of specialities. The Editors of this volume, a written record of the Meeting, are grateful to them for rising to the challenge. They are also deeply grateful to Professor F D Gunstone, our most generous host, and his team of local organisers. Many a first-time visitor to the ancient city of St Andrews commented in surprise on how ´small´ the place is. How true: and what a shining illustration that in scientific achievement small is still usually beautiful. The organisers of the Meeting had also received generous finacial assistance from Lilly Industries, Pfizer, Syntex Pharamaceuticals, Shandon Southern Products, Schering Health Care, Laboratory Data Control (UK), to whom we should like to renew our thanks. For the research on the founding of the University of St Andrews we are indebted to Professors FD Gunstone and RH Burdon.

Thomas Dormandy Catherine Rice-Evans

November 1987

You are a radical. You do not obey the rules.
You are trying to change the system .

Prince Metternich to

Count Istvan Szecheny,

1842.

CONTENTS

RADICAL INTERMEDIATES & THEIR CONSEQUENCES IN BIOLOGICAL SYSTEMS

FREE RADICAL SPECIES & THEIR PATHOLOGICAL IMPLICATIONS

St Salvator´s Colledge
founded by the esteemed and venerable Bishop James Kennedy in the mid-fifteenth century as the second college of the University of St Andrew´s, Scotland

CHEMISTRY OF FREE RADICAL MECHANISMS

FREE RADICALS: CHEMISTRY, PATHOLOGY AND MEDICINE
edited by Catherine Rice-Evans & Thomas Dormandy,

Free Radicals in Solution - 50 Years On

M.J. Perkins
Department of Chemistry, Royal Holloway
and Bedford New College, Egham, U.K. TW20 OEX

In 1937 Hey and Waters published their "landmark" review of the chemistry of Free Radicals in Solution. This brief anniversary review attempts to highlight some key developments in the succeeding 50 years.

This year sees the 50th anniversary of the publication in Chemical Reviews, by two young British scientists, of a seminal article[1] on Free Radicals.

SOME ORGANIC REACTIONS INVOLVING THE OCCURRENCE OF FREE RADICALS IN SOLUTION

DONALD H. HEY

The Victoria University of Manchester, Manchester, England

AND

WILLIAM A. WATERS

The University of Durham (Durham Division), Durham, England

Received May 25, 1937

Supplemented October 6, 1937

In the preface to his 1946 book on Radicals,[2] Alec Waters credits Hey with having produced the first clear evidence for the participation of reactive free radicals in solution. Donald Hey, working with his first research student, W.S.M. Grieve, on substitution reactions of biphenyls, had first to synthesise the biphenyls. This he did using decomposition of diazonium compounds in aromatic solvents to couple together two aromatic nuclei. Irrespective of whether X in the substrate was nitro (a

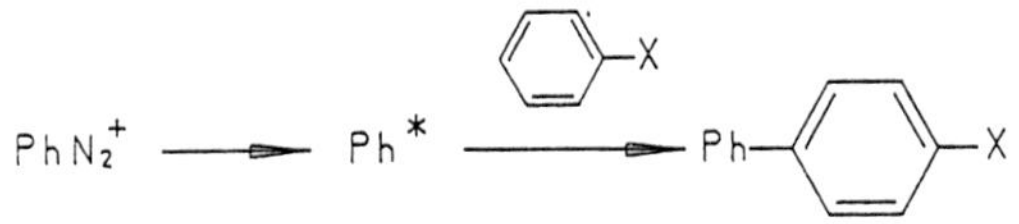

typical meta-directing substituent) or chlorine (which is ortho-/para-directing), the para-substituted product could be isolated. The normal rules of aromatic substitution were being disobeyed; the reactive species drawn here as Ph* was not responding to normal electrical effects of the substituents. Although Grieve and Hey hid behind a non-provocative title[3]

397. *Amphoteric Aromatic Substitution. Part I. Reactions of Sodium Benzenediazoate and Nitrosoacetanilide.*

By William S. M. Grieve and Donald H. Hey.

the text of their 1934 paper left no doubt that a reactive neutral phenyl radical containing trivalent carbon was believed to be involved. Indeed, the same paper discusses participation of divalent carbon species - what we would now call carbenes - in related reactions.

Waters, who, in Durham, had already been speculating along similar lines, though with less direct evidence and in print only in the little-known "Proceedings of the University of Durham Philosophical Society",[4] wrote to Hey and the two agreed to collaborate.

It had been known from the turn of the century that relatively stable carbon radicals such as triphenylmethyl could exist in solution [5], and whilst stable nitroxide radicals had even been

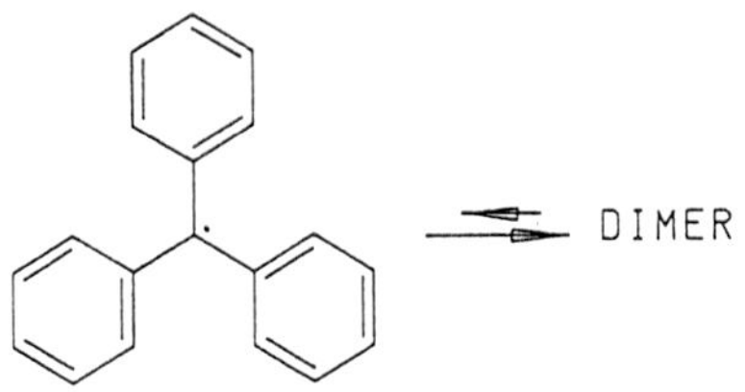

Triphenylmethyl

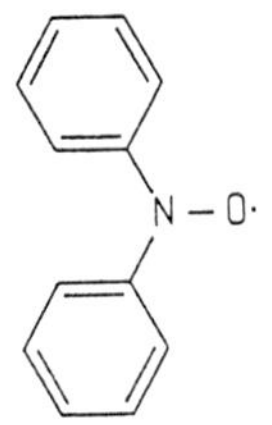

Diphenyl Nitroxide

isolated,[6] in the early 1930's reactive radicals were very much in the domain of the gas-phase physical chemists. Very quickly Hey and Waters found that many of the gas-phase reactions were mirrored in solution by the behaviour of Ph$^{\bullet}$.

$$Ph\cdot + C_6H_{14} \longrightarrow PhH$$
$$Ph\cdot + CCl_4 \longrightarrow PhCl$$
$$Ph\cdot + CS_2 \longrightarrow PhSSPh$$

It is often suggested that proponents of ionic theory, including Christopher Ingold, were obstructive to these new ideas of free radicals. In fact it was Ingold himself who engineered an invitation to Hey and Waters to summarise the field in the Chemical Review whose publication we are recalling in this brief talk.

The 1937 review by Hey and Waters pulled together the then recent literature in which radical chemistry had been discussed. It also examined several processes for which radical mechanisms had not previously been suggested, not least of which was the peroxide effect in Morris Kharasch's work on addition of HBr to alkenes, although it was Kharasch himself who, almost simultaneously, explained in detail the radical chain mechanism for this reaction.[7]

In the subsequent decade - which of course included the war years - Waters's work in particular ranged over many reaction types involving radicals: he recognised the importance of electron transfer in peroxide decomposition, following the earlier suggestions of Haber and Weiss concerning the decompositon of hydrogen peroxide,[8] and he made significant

$$Fe^{++} + H_2O_2 \longrightarrow Fe^{+++} + HO^- + HO\cdot$$

contributions to early work on mechanisms of autoxidation - though another British chemist, E.H. Farmer, first explained the high reactivity in autoxidation of allylic methylene groups adjacent to double bonds (and more particularly of the doubly allylic methylene in drying oils) in terms of bond delocalisation in the derived radical.[9] The same principle is of paramount importance in the biotransformations of arachidonic acid; e.g.

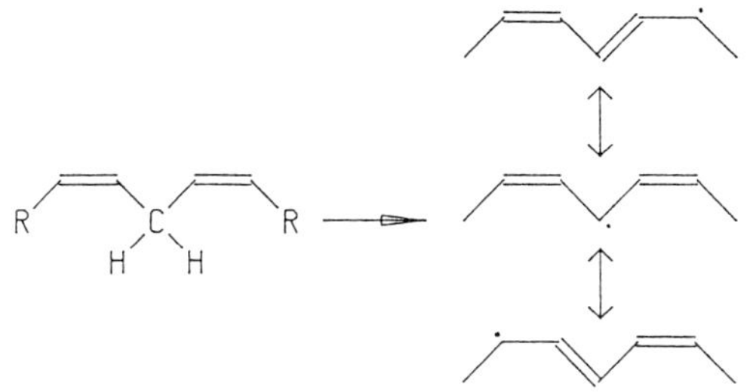

Waters's 1946 treatise on Free Radicals[2] even contains a substantial, if speculative, chapter on the importance of radicals in some enzyme reactions. By the time the fourth edition of another text by Waters, "Physical Aspects of Organic Chemistry" was published in 1949,[10] the foundations of modern free radical chemistry had effectively been laid. But the most rapid progress at that time had been in polymer chemistry, including synthetic rubbers created by the war effort. Extensive industrial research in this field built up a catalogue of quantitative information. However, for a long period there was an attitude towards radical reactions which seemed to inhibit other developments. It was as though the following quotation

> In general, therefore, *the reactions in solution of the very active free radicals are not specific, and take place mainly with solvent molecules.*

from Waters's book had been misinterpreted to imply that all radicals other than the triarylmethyls react in a non-specific way. One has only to consider polymerisation of styrene, in which monomer is converted almost quantitatively into polymer, to recognise that this is not so - even if polystyrene is not the organic chemist's ideal of a "nice crystalline product"!

Waters's research ranged widely over radical reactions, although his later work embraced in particular the complexities of organic oxidation mechanisms. In contrast, Hey's interests focused on the reactions of aryl radicals in aromatic solvents, recognising on the one hand the importance of intramolecular analogues in the synthesis of complex aromatic structures, e.g.[11]

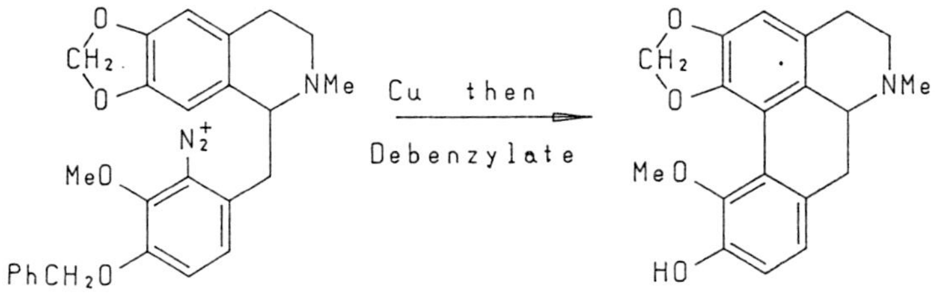

Isobulbocapnine

and exploring on the other the superposition of polar influences on reactions of, formally neutral, free radicals. A more recent perspective on polar effects involves frontier orbital considerations.

Another recent development is the accumulation of absolute rate constants for radical-radical and radical-molecule reactions - much of it from work in Keith Ingold's laboratory in Canada. These have provided a quantitative base for much modern radical chemistry and a firm lead in the planning of synthetic sequences which incorporate radical reactions as key steps. Examples of what can be achieved include new functional group transformations and multistep cyclisation processes,[12,13] as well as the

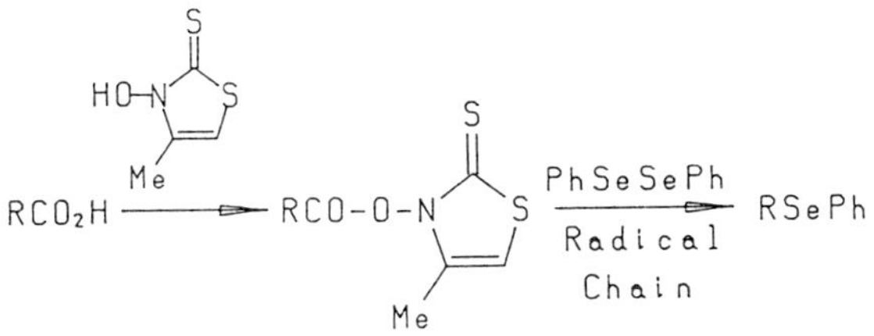

Ref 12

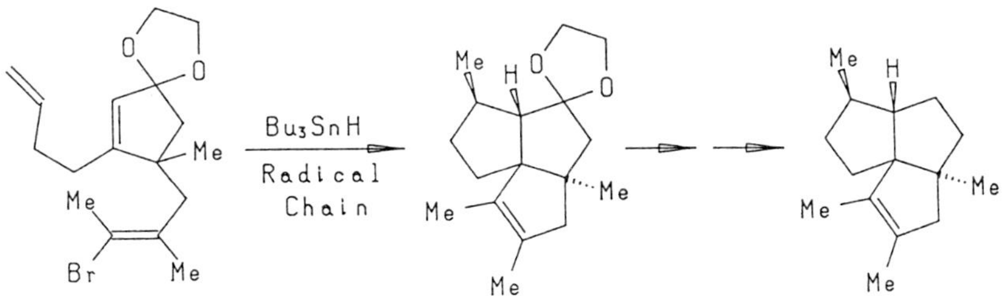

Ref 13

following highly efficient "witches brew" reaction due to Minisci and his colleagues:

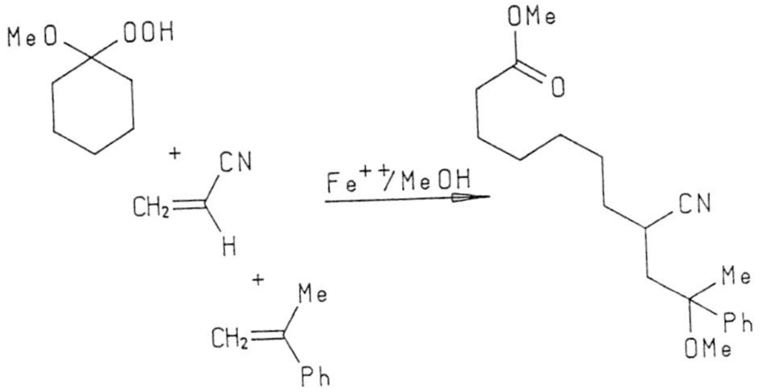

Ref 14

Of course, nature already knew all about radical cyclisation reactions. As we now know, these underlie, for example, the biosynthesis of e.g. prostaglandins from arachidonic acid.

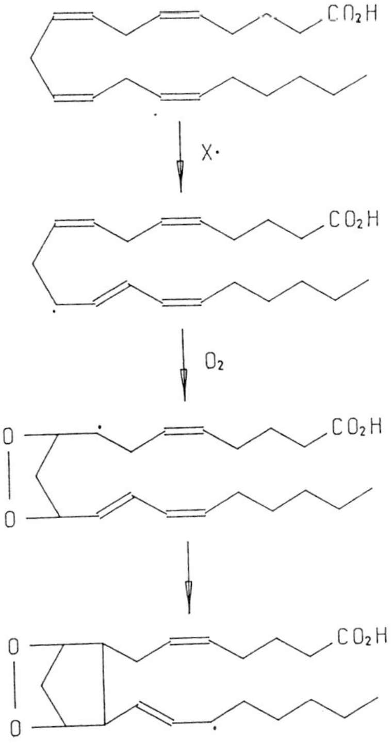

Related chemistry is of current importance in the pharmaceutical industry, but probably much greater activity is prevalent in investigations of the harmful effects in biological systems of oxidising radicals, including hydroxyl, hydroperoxyl and superoxide. Of these, hydroxyl is particularly indiscriminate in the manner described in my quotation from Waters, and only relatively recently have techniques such as spin trapping been developed for investigation of the not-quite random molecular damage that can be inflicted by this species.

Hey and Waters laid down many of the ground rules 50 years ago, but they never worked together. Hey began his academic career in Manchester from where his pioneering papers were published. Waters started in Durham. After the war years Hey accepted a Chair at King's College in London, and Waters moved to Oxford.

The work of both was recognised by their election to the Royal Society in the mid 1950's. Waters retired in 1970, Hey in 1971.

In the later years of his career, Hey had become heavily involved in University administration and, as Vice-Principal of King's College in London he was effectively at the helm of the College for a time whilst the then Principal, Peter Noble, served as Vice-Chancellor of London University. Hey had maintained many other interests throughout his career and into retirement. Notable amongst these was his Welshman's love of music; for many years he was deputy organist at the parish church in Tadworth in Surrey where he lived.

Like Hey, Waters was born in Wales, but, again like Hey, his parents were not Welsh. He, too, loved music, and was a competent, if shyly secretive pianist. His later career in

Oxford was spent as Research Professor, and after retirement he retained space in the Dyson-Perrins laboratory and continued his own researches until well into his seventies.

As recently as three years ago, Waters was himself reminiscing to a conference audience at this University on the development of the field of free radical chemistry. Sadly, however, both men are recently dead. During their lifetimes free-radical chemistry had grown so rapidly that workers in one area can now easily lose sight of even the basic principles of another. We owe it to Hey and Waters to avoid these pitfalls, and one role of our Society, of which both men were Honorary Members, is to bring together academic chemists, industrial chemists, biochemists and those engaged in medical research, where their common interest is the importance of free radicals.

In conclusion I should like to thank Dr. Dormandy and the meeting organisers for allowing me the opportunity to read this opening paper, and you for your patient forbearance with my little history lesson!

1. D.H. Hey and W.A. Waters, Chem. Revs. (1937), 21, 169.
2. W.A. Waters "The Chemistry of Free Radicals", Oxford (1946).
3. W.S.M. Grieve and D.H. Hey, J. Chem. Soc. (1934), 1797.
4. W.A. Waters, Proc. Univ. Durham Phil. Soc., (1932), 8, 459.
5. M. Gomberg, J. Amer. Chem. Soc. (1900), 22, 757.
6. e.g. H. Wieland and M. Offenbacher, Ber., (1914), 47, 2111.
7. M.S. Kharasch, F. Engelmann and F. Mayo, J. Org. Chem. (1937), 2, 288.
8. F. Haber and J. Weiss, Proc. Roy. Soc. (1934), A147, 333.
9. E.H. Farmer, J. Chem Soc. (1943), 541; (1944), 242.

10. W.A. Waters, "Physical Aspects of Organic Chemistry", 4th Edn., Routledge, London (1950).

11. e.g. D.H. Hey and L.C. Lobo, J. Chem. Soc. (1954), 2246.

12. D.H.R. Barton, D. Brydon and S.Z. Zard, Tetrahedron Letters (1984), 25, 5777.

13. D.P. Curran and S.-C. Kuo, J. Amer. Chem. Soc. (1986), 108, 1106.

14. F. Minisci, M. Cecere, R. Galli and R. Bernadi, Tetrahedron (1969), 25, 2667.

FREE RADICALS: CHEMISTRY, PATHOLOGY AND MEDICINE
edited by Catherine Rice-Evans & Thomas Dormandy, © 1988 Richelieu Press, London

Oxidative Damage and Radical Repair: One-Electron Transfer Reactions Involving Metal Ions, Peroxides and Free Radicals.

B. C. Gilbert* and M. Jeff

Department of Chemistry, University of York,
Heslington, York, YO1 5DD.

E.s.r. experiments, with a rapid-flow system, have been employed to investigate the formation of radicals in the metal-catalysed decomposition of hydrogen peroxide and related compounds. Evidence is presented that the free hydroxyl radical is the reactive intermediate generated in the reaction between Fe^{II} and H_2O_2, a process which is dramatically accelerated by certain chelating agents (e.g. EDTA at pH ca.7). Kinetic analysis is presented for this and related radical-initiation processes and for one-electron transfer reactions of first-formed radicals with Fe^{III} and Fe^{II} (as well as other ions, e.g., Cu^{II} and Cu^{I}).

Of all the intermediates which may be generated *in vivo* by the biological activation of dioxygen (e.g. $O_2^{\cdot-}$, $\cdot O_2H$, $\cdot OH$, and $^1O_2$), it is widely recognized that the most potent and damaging to biomolecules is the hydroxyl radical.[1] This is known to be formed in a variety of processes including radiation damage and electron-transfer reactions involving oxygen and hydrogen peroxide; it is believed to participate in a number of potentially damaging processes including lipid peroxidation, degradation of carbohydrates (e.g., hyaluronic acid) and strand-breakage in DNA (see e.g. ref.1).

Despite the convincing body of circumstantial evidence for the formation of ·OH _in vivo_, direct evidence for its formation in realistic model systems is relatively scarce. For example while it seems likely that the decomposition of H_2O_2 by Fe^{II} in aqueous solution (the Fenton reaction) proceeds via the hydroxyl radical [reaction (1), as proposed by Haber and Weiss[2]], an alternative mechanism involving a ferryl species as oxidant [e.g. reaction (2)] remains a possibility:[3] while kinetic analysis of the use of this couple for the oxidation of organic substrates appears to support the former interpretation (see, e.g., refs. 4 and 5), the results of an earlier e.s.r. study by Shiga led to the suggestion[6] that a metal complex is responsible for the oxidation. Although it was subsequently suggested[7,8] that the apparent discrepancies may reflect the reaction of organic radicals (from ·OH) with Fe^{III}, it has since been proposed[9] that the hydroxyl radical may not be produced as the primary intermediate. Further controversy also centres on the role of the complexing agent, e.g., as to whether the ligand DTPA, diethylenetriaminepentaacetic acid [$(HO_2CCH_2)_2NCH_2CH_2N(CH_2CO_2H)CH_2CH_2N(CH_2CO_2H)_2$], compared with

EDTA [ethylenediaminetetraacetic acid, $(HO_2CCH_2)_2NCH_2CH_2N(CH_2CO_2H)_2$] is capable of retarding[10] or accelerating[11] the reaction.

We here describe the results of e.s.r. experiments utilizing a flow system to study directly the reaction between H_2O_2 and Fe^{II} and, in particular, to compare these results with those obtained for the comparable Ti^{III}–H_2O_2 reaction, which is also believed to generate ·OH [reaction (3)]. Our work has then been extended to monitor the further reactions of first-formed free radicals (detected using e.s.r) with transition-metal ions, notably those of iron and copper

$$Fe^{II} + H_2O_2 \longrightarrow Fe^{III} + \cdot OH + OH^- \qquad (1)$$

$$Fe^{II} + H_2O_2 \longrightarrow (Fe^{IV}O)^{2+} + H_2O \qquad (2)$$

$$Ti^{III} + H_2O_2 \longrightarrow Ti^{IV} + \cdot OH + OH^- \qquad (3)$$

Results and Discussion.

The e.s.r.experiments described here are based on previous use of the biomimetic Ti^{III}–H_2O_2 couple as a source of ·OH [reaction (3)] and utilized a continuous mixing system in which three deoxygenated streams, containing the metal ion (typically *ca*. 10^{-3} mol dm^{-3}, after mixing), peroxide (*ca*.10^{-2} mol dm^{-3}) and the substrate (up to 0·5 mol dm^{-3}), are mixed a short time before their passage through the sample cell.
With Ti^{III}–H_2O_2 it is believed that the radicals observed are

formed by reaction of the first-formed hydroxyl radical with added substrate:[8,12,13] under conditions where [RH] is sufficiently high to prevent reaction of ·OH with either metal or peroxide, then the reaction scheme should be represented by reactions (3)-(5). Further it can also be shown that a pseudo steady-state is achieved in the cavity so that the radical concentration detected there is a function of k_3 and k_5 (as well as concentrations of reagents after the time elapsed between mixing and observation), as indicated in equations (6) and (7) (since $[H_2O_2]$ is normally chosen to be in considerable excess over $[Ti^{III}]$). It can also be shown that the concentration of [R·] as a function of $[H_2O_2]_o$ (for a given t and $[Ti^{III}]_o$) should have a maximum given by equation (8) (so that measurement of $[H_2O_2]_{max}$ and t should give k_3).

$$Ti^{III} + H_2O_2 \longrightarrow Ti^{IV} + \cdot OH + OH^- \qquad (3)$$

$$\cdot OH + RH \longrightarrow R\cdot + H_2O \qquad (4)$$

$$2R\cdot \longrightarrow \text{molecular products} \qquad (5)$$

$$k_3[Ti^{III}]_t[H_2O_2]_t = 2k_5[R\cdot]^2_{ss} \qquad (6)$$

$$[R\cdot]^2_{ss} = \frac{k_3[Ti^{III}]_o[H_2O_2]_o e^{-k_1[H_2O_2]_o t}}{2k_5} \qquad (7)$$

$$[H_2O_2]_o = \frac{1}{k_3 t} \qquad (8)$$

(a) <u>The Ti^{III}-H_2O_2 reaction. (i) Generation of ·OH</u>.

Initial experiments were carried out with Bu^tOH since the radical formed by reaction with ·OH (·CH_2CMe_2OH) is not believed to react rapidly with Ti^{III}, Ti^{IV}, or H_2O_2 (in contrast to the radical ·CHMeOH from ethanol, for example).*

As predicted on the basis of equation (7), [·CH_2CMe_2OH] is linearly dependent upon $[Ti^{III}]_0^{½}$. Further, the expected linear relationship is confirmed between ln[·CH_2CMe_2OH] (obtained from peak-heights since the line-width did not vary) and the mixing-time t (obtained from the flow-rate, calibrated[15,16] using a spectrophotometric study of the reaction between Fe^{III} and SCN^-) (see Figure 1a). The predicted maximum in [R·] as a function of $[H_2O_2]$ [see equation (8) and Figure 1b] leads to the determination of the initiation rate constant k_3: with a flow-rate of 2·70 cm^3 s^{-1} (equivalent to a mixing time of 28 ms), as illustrated, the maximum in [R·] occurs for $[H_2O_2]_0$ 1·72 x 10^{-2} mol dm^{-3}, giving k_3 2·08 (± 0·2) x 10^3 dm^3 mol^{-1} s^{-1} (at 20 °C and pH 1·95). A value of 2·15 (± 0·21) x 10^3 dm^3 mol^{-1} s^{-1} was obtained by monitoring ln[·CH_2CMe_2OH] as a function of flow-rate (see above). This value is consistent with the

---

* E.s.r. parameters for radicals from reactions of ·OH with organic substrates have been previously reported[14] and will not be repeated here.

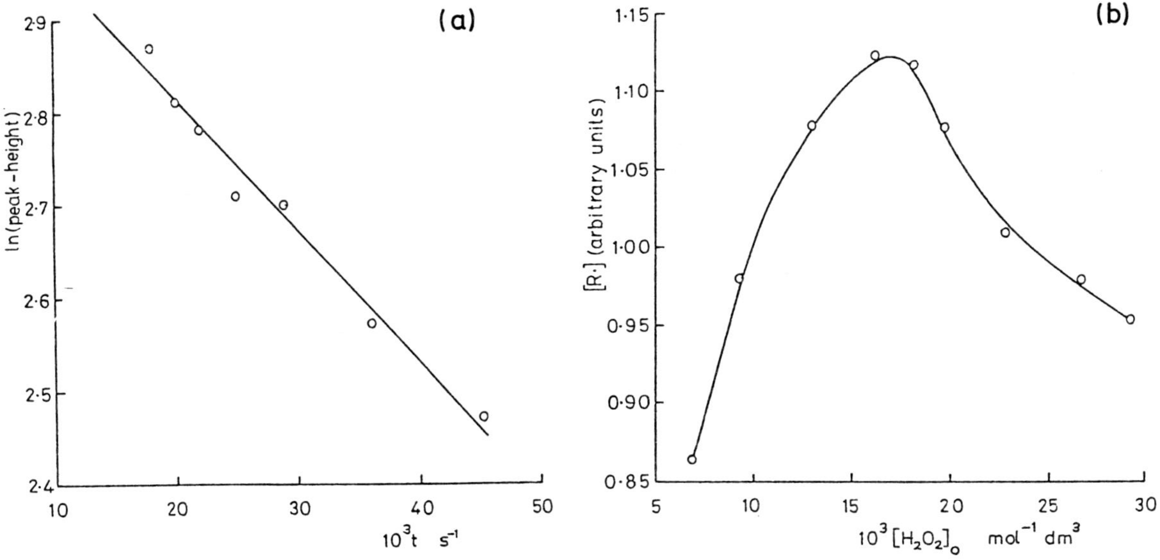

FIGURE 1: Variation of [$\cdot CH_2CMe_2OH$] with (a) mixing time, (b) $[H_2O_2]_o$, for Ti^{III}–H_2O_2–Bu^tOH (pH 1·9, 20 °C); (a) $[Ti^{III}]_o$ 1·7x10^{-4}, $[H_2O_2]_o$ 1·3x10^{-2}, [Bu^tOH] 0·355 (b) $[Ti^{III}]_o$ 8·3x10^{-4} mol dm^{-3}, t 28 ms.

results of Berdnikov and coworkers[17] but significantly larger than that reported by Czapski et al.[18] The kinetic behaviour of $\cdot CH_2CMe_2OH$ was found to be characteristic of a variety of radicals in which le-oxidation and reduction do not compete with bimolecular termination.

The effect of some complexing agents on the initiation rate constant k_3 were also investigated. For EDTA in the pH range 3–7 a small but significant increase was observed, to $4{\cdot}5\ (\pm\ 0{\cdot}5) \times 10^3\ dm^3\ mol^{-1}\ s^{-1}$. For ethanedioate (oxalate) at pH 2 it was found that strong e.s.r. signals could only be detected at very low values of $[H_2O_2]_o$ ($\leqslant 3 \times 10^{-3}\ mol\ dm^{-3}$, for t ca. 0·03s): that this is a result of a significant increase in k_3 was confirmed by the variation of $\ln[\cdot CH_2CMe_2OH]$ with t, which gave $k_3\ 2{\cdot}7(\pm\ 0{\cdot}2) \times 10^4\ dm^3\ mol^{-1}\ s^{-1}$.

No radicals could be detected in the presence of DTPA, but instead a strong absorption was observed from the Ti^{III}-DTPA complex (g 1·9656, with satellite lines from ^{47}Ti and ^{49}Ti). The very slow initiation rate constant ($5\ dm^3\ mol^{-1}\ s^{-1}$) was determined by monitoring the decay of this signal when the flow was stopped.

(ii) Oxidation and Reduction of Radicals.

In experiments with Ti^{III}, H_2O_2, and ethanol as substrate (in the absence of ligands) increase in $[H_2O_2]_o$ led to the expected increase - but then a rapid decrease - in the concentration of the major radical, but a steady increase in the concentration of $\cdot CH_2CH_2OH$, the minor radical (up to $[H_2O_2]$ ca. $0{\cdot}1$ mol dm^{-3}) (see Figure 2).

Comparable results were obtained for the radicals $\cdot CMe_2OH$ and $\cdot CH_2CHMeOH$, respectively, from propan-2-ol. This behaviour is believed to be due to the rapid oxidation of the α-hydroxyalkyl radicals by H_2O_2 (a result of the +M effect of the α-oxygen substituent) so that the hydroxyl radical thus regenerated serves to increase the β-radical concentrations at their expense.* Computer simulations were carried out on the basis of the occurrence of reactions (3),(4), and (9)-(14), with k_3 2×10^3 dm^3 mol^{-1} s^{-1}, k_9 $1{\cdot}9 \times 10^9$ dm^3 mol^{-1} s^{-1},[20] k_9/k_{10} 13 and $2k_{11} = 2k_{12} = k_{13} = 2 \times 10^9$ dm^3 mol^{-1} s^{-1}, together with different values of k_{14}. Good agreement between observed and simulated behaviour (at least up to $[H_2O_2]$ ca. 6×10^{-2} mol dm^3) was obtained for k_{14} $2{\cdot}0 \times 10^5$ dm^3 mol^{-1} s^{-1} (see Figure 2). Similarly oxidation of $\cdot CMe_2OH$ (from

- -

* This effect is more marked in the presence of EDTA, since Ti^{IV}-EDTA is also an effective oxidant for $\cdot CHMeOH$.[19]

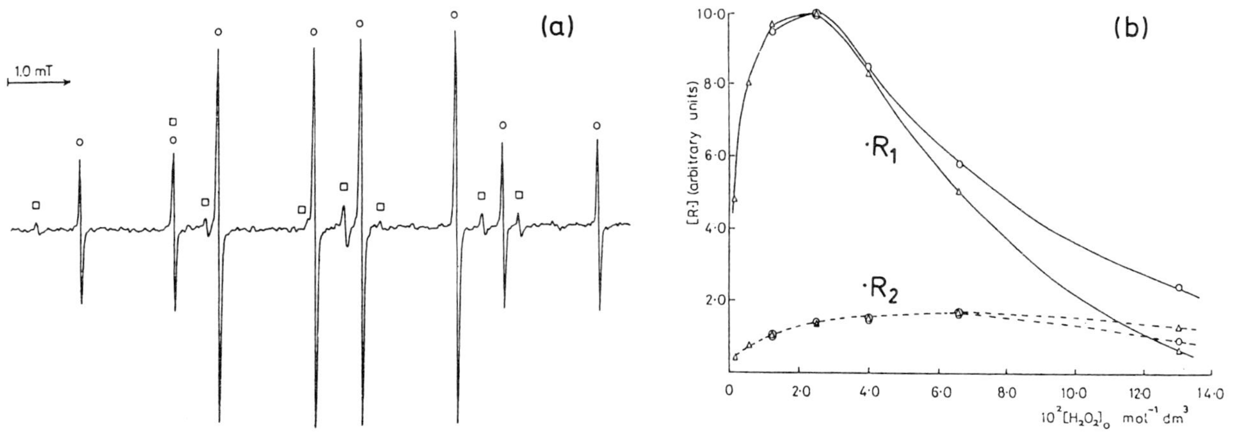

FIGURE 2: (a) E.s.r. spectra of [·CHMeOH](·R_1,o) and ·CH_2CH_2OH (·R_2,□) from Ti^{III}–H_2O_2–EtOH. (b) Variation (o) of [R_1·] and [R_2·] with [H_2O_2] for $[Ti^{III}]_o$ 1·46x10^{-3}, [EtOH] 0·6 mol dm^{-3}, t 27 ms. Simulated variation (Δ) with k_{14} 2 x 10^5 dm^3 mol^{-1} s^{-1}.

propan-2-ol) is calculated to have k 7 x 10^5 dm^3 mol^{-1} s^{-1}, the further increase in k compared with unconjugated radicals suggesting the importance of stabilization of the incipient carbonium ion by the extra alkyl group at the radical centre.

$$HO\cdot + EtOH \longrightarrow \cdot CHMeOH(R_1\cdot) + H_2O \qquad (9)$$

$$HO\cdot + EtOH \longrightarrow \cdot CH_2CH_2OH(R_2\cdot) + H_2O \qquad (10)$$

$$\left.\begin{array}{l} R_1\cdot + R_1\cdot \longrightarrow \quad (11) \\ R_2\cdot + R_2\cdot \longrightarrow \quad (12) \\ R_1\cdot + R_2\cdot \longrightarrow \quad (13) \end{array}\right\} \text{molecular products}$$

$$\cdot CHMeOH + H_2O_2 \longrightarrow \cdot OH + OH^- + {}^+CHMeOH \qquad (14)$$

$${}^+CHMeOH \longrightarrow H^+ + MeCHO \qquad (15)$$

$$\cdot CHMeOH + S_2O_8^{2-} \longrightarrow {}^+CHMeOH + SO_4^{\cdot -} + SO_4^{2-} \qquad (16)$$

Related experiments[21] in which the decrease in, e.g. [·CHMeOH] (from reaction of ·OH) was monitored as a function of [$S_2O_8^{2-}$], from added peroxydisulphate ions, yield even faster rate constants for le-transfer (e.g. with a value for ·CHMeOH [reaction(16)] of 6·9 x 10^5 dm^3 mol^{-1} s^{-1}). This is as expected on the basis of the Marcus theory of electron-transfer reactions,[22] the rate of the process being dependent upon $E^{\ominus}$ for the oxidant.

The predicted dependence of [R·] on $[Ti^{III}]_0^{1/2}$ may also break down for some α-carbonyl-conjugated radicals,[12] as a result of the occurrence of le-reduction of the radicals by the low-valent metal ion [see e.g. reaction (17)]. Further evidence for the efficiency of this reaction is provided by the detection of signals from a 'dimeric' species in the reaction between Ti^{III} and $\cdot CH_2COCH_3$ (e.g. from acetone or via the acid-catalysed dehydration of the 1,2-diol radical $\cdot CMe(OH)CH_2OH$). This novel species, which builds up (as $[\cdot CH_2COCH_3]$ diminishes) when $[Ti^{III}]$ is increased, is attributed the structure $\cdot CMe(OH)CH_2CH_2C(O)Me$: its mechanism of formation evidently involves generation and reaction of the enol or enolate of acetone [reactions (17) and (18)]. Kinetic simulation leads to rate constants for these processes of $8{\cdot}5 \times 10^6$ and $7{\cdot}5 \times 10^4$ $dm^3\ mol^{-1}\ s^{-1}$ respectively. The overall process is significantly catalysed by the addition of low concentrations of Cu^{II}, probably via rapid one-electron reduction of $\cdot CH_2COCH_3$ by Cu^I (k _ca_. 10^{10} $dm^3\ mol^{-1}\ s^{-1}$: see later)

$$Ti^{III} + \cdot CH_2COCH_3 \longrightarrow Ti^{IV} + CH_2{=}\overset{\displaystyle O^-}{\overset{|}{C}}{-}CH_3 \qquad (17)$$

$$\cdot CH_2COCH_3 + CH_2{=}\overset{\displaystyle OH}{\overset{|}{C}}CH_3 \longrightarrow CH_3C(O)CH_2CH_2\dot{C}(OH)CH_3 \qquad (18)$$

b) The Fe^{II}- H_2O_2 reaction.

Initial flow experiments employed ethanol (typically ca. 0·6 mol dm^{-3}) together with Fe^{II} and EDTA (both 10^{-3} mol dm^{-3} overall) and, in the third stream H_2O_2 [ca. 6 x 10^{-3} mol dm^{-3} overall], all in phosphate buffer at pH 7 (with a mixing time of ca. 0·03s). Under these conditions, which resemble those employed by Shiga,[6] spectra were weaker than obtained from Ti^{III}-H_2O_2 and, in further contrast, had [$\cdot CH_2CH_2OH$] > [$\cdot$CHMeOH]. Other substrates gave weak signals from carbon-centred radicals. Similar signals were obtained at pH 7 in the absence of buffer, but a significant improvement in signal intensity was obtained by careful removal of dissolved oxygen before the metal-containing solutions were prepared.

With Bu^tOH, strong signals were obtained from $\cdot CH_2CMe_2OH$ (which is not readily oxidized) in the presence of EDTA in the pH range 3-7, with $[Fe^{II}]_o$ = $[EDTA]_o$ = 2 x 10^{-3} mol dm^{-3} and $[H_2O_2]_o$ 2 x 10^{-3} - 7 x 10^{-2} mol dm^{-3}, with $[R\cdot]_{max}$ for $[H_2O_2]_o$ ca. 10^{-3} mol dm^{-3}; with $[H_2O_2]_o$ 1.5 x 10^{-2} mol dm^{-3}, [$\cdot CH_2CMe_2OH$] increased as $[Fe^{II}]_o$ was increased from 3 x 10^{-4} to 3 x 10^{-3} mol dm^{-3}. DTPA behaved similarly.

In the absence of EDTA and at pH $\leq$ 2, only weak signals were obtained with $[Fe^{II}]_o$ 2 x 10^{-3} and $[H_2O_2]_o$ 7 x 10^{-3} mol dm^{-3}. However, when much higher concentrations of peroxide were employed (up to 0.3 mol dm^{-3}) considerably enhanced intensities were observed: this evidently reflects the much slower reaction of uncomplexed Fe^{II} with H_2O_2 (though still to give $\cdot OH$), so that high $[H_2O_2]_o$ is required to give sufficiently rapid radical generation in the sample cell.

Kinetic experiments and analysis (as for Ti^{III}) established that, for example, in the presence of EDTA, $[R\cdot]$ is proportional to $[Fe^{II}]_o^{1/2}$ [at least up to $[Fe^{II}]_o$ 2.5 x 10^{-3} mol dm^{-3} with $[H_2O_2]_o$ 1.5 x 10^{-2} mol dm^{-3}: cf. equation (7)], that the variation of $[R\cdot]$ with $[H_2O_2]_o$ shows the expected maximum [cf. equation (8)], and that ln$[\cdot R]$ varies linearly with mixing time [equation (7)] (see Figure 3). The value derived for k(Fe^{II}-EDTA-H_2O_2) of 1.0 x 10^4 dm^3 mol^{-1} s^{-1} (at 20° C) is in good accord with that (9.1 x 10^3 dm^3 mol^{-1} s^{-1}) reported by Borggard and coworkers.[23]

Similar experiments with the much slower reaction of Fe^{II} (uncomplexed) with H_2O_2 at pH 2 give k ca. 2 x 10^2 dm^3 mol^{-1} s^{-1}, though the much weaker signals under these conditions imply a considerably greater uncertainty in the result (cf. a value of 53 dm^3 mol^{-1} s^{-1} previously reported for this reaction[24]). The initiation rate constant for DTPA (1·35 x 10^4 dm^3 mol^{-1} s^{-1}), closely similar to that for EDTA, is in marked contrast to the Ti^{III} system and suggests that an outer-sphere mechanism is involved for iron.

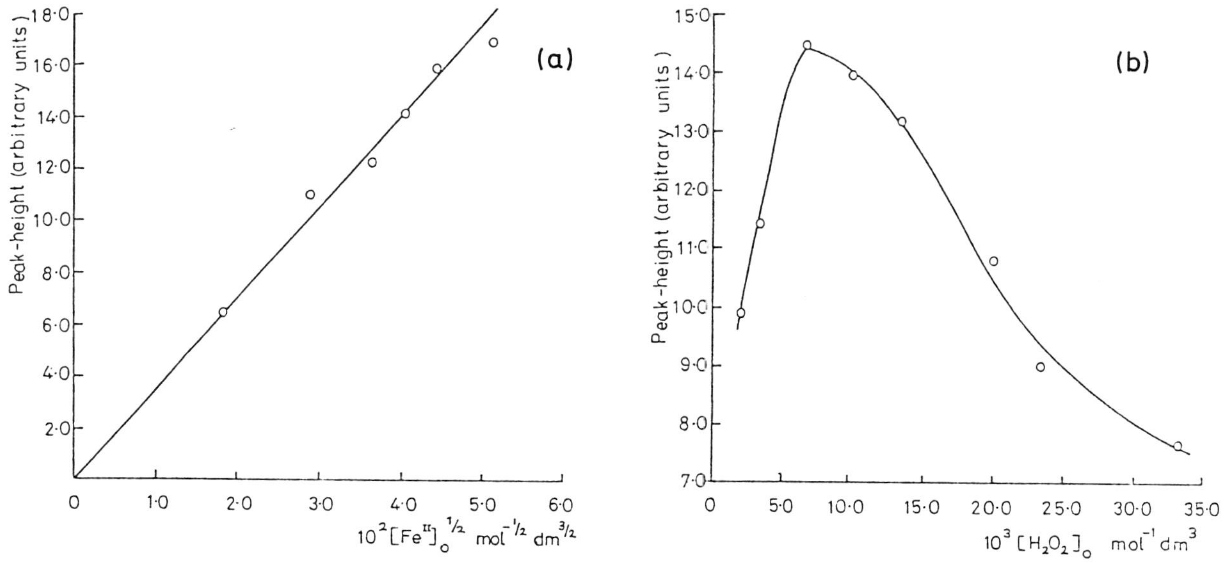

FIGURE 3: Variation of $[\cdot CH_2CMe_2OH]$ with $[H_2O_2]$ and $[Fe^{II}]$ in the Fe^{II}-EDTA-H_2O_2-Bu^tOH system at pH 7. (a) $[Fe^{II}]$ = [EDTA], $[H_2O_2]$ $1{\cdot}5x10^{-2}$, $[Bu^tOH]$ 0·35 mol dm^{-3} (b) $[Fe^{II}]$ = [EDTA] = $2x10^{-3}$ mol dm^{-3}.

Our results are in reasonable agreement with earlier kinetic studies on the Fenton system and strongly support the assertion that free hydroxyl radicals are generated (even in the presence of polyaminocarboxylate ligands such as DTPA and EDTA). To verify this, we examined the selectivity of the intermediate formed in its reaction with a variety of substrates.

For those substrates whose reaction with Ti^{III}–H_2O_2 produces a single detectable radical, precisely the same spectra were obtained from Fe^{II}–EDTA/H_2O_2 (but with slightly lower concentrations). Examples included t-butyl alcohol, ethane-1,2-diol, dimethyl sulphoxide (to give ·Me) and several alkenes [e.g. methallyl alcohol, to give $\cdot CMe(CH_2OH)_2$]. Where two or more radicals were obtained, the spectra (splittings and relative intensities) in most cases were essentially identical in the two systems: examples include 2-methylpropene (to give $\cdot CMe_2CH_2OH$ and the minor radical $\cdot CH_2CMe{=}CH_2$), allyl alcohol [$\cdot CH(CH_2OH)_2$ and $\cdot CH_2CH(OH)CH_2OH$, with traces of $\cdot CH(OH)CH{=}CH_2$] and diethyl ketone (·CHMeCOEt and $\cdot CH_2CH_2COEt$ in the ratio 2:1). This suggests that there is essentially no difference in the nature of the reactive intermediate in the Fenton and Ti^{III}–H_2O_2 systems.

On the other hand, with ethanol the Fenton reaction gives a ratio of ca. 0·3:1 for [·CHMeOH]:[$\cdot CH_2CH_2OH$], despite the established preference for attack at α C–H by ·OH. This observation is completely consistent with the rapid oxidation of the α-radical by Fe^{III} [reaction (19)], both in the presence and absence of EDTA. Simulations of the α/β ratio as a function of k_{19} suggest that the rate constant is $5{\cdot}7 \times 10^8$ $dm^3\ mol^{-1}\ s^{-1}$ at 20° and pH 7 (for Fe^{III}-EDTA). Closely similar results were obtained for $\cdot CH_2OH$ and $\cdot CMe_2OH$. The observed α/β ratio of ca. 0·1 for EtOH in the presence of Fe^{III}–DTPA leads to an oxidation rate constant of $1{\cdot}8 \times 10^9$ $dm^3\ mol^{-1}\ s^{-1}$ in this case.

$$\cdot CHMeOH + Fe^{III} \longrightarrow {}^{+}CHMeOH + Fe^{II} \qquad (19)$$

Small differences between the relative concentrations of $\cdot CHMeCO_2^-$ and $\cdot CH_2CH_2CO_2^-$ from propionic acid at pH 7 in the Ti^{III}-H_2O_2 and Fe^{II}-H_2O_2 systems may either reflect differential changes in line-width (which are somewhat greater in the presence of iron), a change in selectivity associated with the generation of ·OH in the vicinity of a chelated carboxylate (to give an increase in the α-radical concentration with iron), or the occurrence of preferential oxidation and/or reduction (to reduce the β-radical concentration) in the Fenton system.

Since linear plots of $[\cdot CH_2CMe_2OH]$ versus $[Fe^{II}]_0^{1/2}$ were obtained for the Fe^{II}-EDTA (and DTPA)-H_2O_2 reactions at pH 7, it appears that further reaction of $\cdot CH_2CMe_2OH$ with either Fe^{II} or Fe^{III} is unimportant under the conditions employed here (i.e. with $[H_2O_2] \gg Fe^{II}$). However, Eberhardt has reported[25] the formation of 2-methylpropene in the reaction of Bu^tOH with Fe^{II}-EDTA and H_2O_2 (with the metal in excess), which may reflect one-electron reduction of first-formed $\cdot CH_2CMe_2OH$: in view of this observation and the claim[26] that the Fenton system in the presence of excess iron does not simply proceed via $\cdot OH$ we extended our investigation to higher concentrations of Fe^{II}. Monitoring $[\cdot CH_2CMe_2OH]$ as a function of $[Fe^{II}]$, up to $[Fe^{II}]$ 10^{-2} mol dm^{-3} (for $[H_2O_2]_o$ $1{\cdot}3 \times 10^{-3}$ dm^3 mol^{-1}) we satisfactorily simulated the increase in $[\cdot CH_2CMe_2OH]$ and maximum for $[Fe^{II}]$ 3×10^{-3} mol dm^{-3} with a straightforward mechanism involving generation of $\cdot OH$ (Fenton), reaction of $\cdot OH$ with Bu^tOH, Fe^{II}, and H_2O_2, and removal of $\cdot CH_2CMe_2OH$ by dimerization and also reaction (20): the best fit to experimental behaviour was obtained for k_{20} 7 $\times 10^5$ dm^3 mol^{-1} s^{-1} . No evidence for metal- or radical-complexes as intermediates was obtained.

$$\cdot CH_2CMe_2OH + Fe^{II} \longrightarrow CH_2{=}CMe_2 + OH^- + Fe^{III} \qquad (20)$$

(c) Radical Reactions of Cu^I and Cu^{II}.

Our approach has been extended in an attempt to investigate the possible roles of Cu^{II} and Cu^I in biological systems involving radicals.

(i) $Ti^{III}/Cu^{II}-H_2O_2$.

Initial experiments with $Ti^{III}-H_2O_2$ and Me_2SO, Bu^tOH and MeOH as substrates showed that the $[R\cdot]_{obs}$ decreased significantly on introduction of Cu^{II} (in the substrate stream, up to a concentration after mixing of 2×10^{-5} mol dm^{-3}). The rate of decrease is in the order Me· > $\cdot CH_2CMe_2OH$ > $\cdot CH_2OH$. With ethanol there was a marked reduction in the intensity of the signal from $\cdot CH_2CH_2OH$ (cf. Me·) but only a small reduction in the dominant radical ·CHMeOH. These contrasts suggest that the observed behaviour is not due to changes in the rate of overall hydroxyl-radical generation (similar effects on different radicals would then have been expected): the decreases appear to be due to alternative decay pathways via reactions of radicals with a copper species.[27,28] The observed order for the decrease in [R·] suggests that the reactive species is Cu^I (formed via reaction (21)*] rather than Cu^{II} since oxidation in the latter case

---

* A reduction in [R·] was also observed when Cu^{II} was premixed with Ti^{III} before flowing: spectra recorded ca. 10 minutes after mixing showed virtually no detectable signals (and metallic copper was deposited, presumably from disproportionation, or further reduction of Cu^I, to give Cu^0). Spectrophotometric experiments also showed rapid removal of Cu^{II} (800 nm) and Ti^{III}(500 nm). Copper(II) chloride catalyses the oxidation of $TiCl_3$ with O_2, probably also via reaction (21): K_{21} is reported to be $4{\cdot}2 \times 10^6$ at 25°C, consistent with the redox potentials of the Ti^{IV}/Ti^{III} and Cu^{II}/Cu^I couples.[29]

would be expected to be faster for radicals of the type $\cdot CR^1R^2OH$ [oxidations of $\cdot CHMeOH$ and $\cdot CH_2CH_2OH$ with Cu^{II} are reported[27,28] to have k 9×10^7 and $2{\cdot}2 \times 10^7$ $dm^3\ mol^{-1}\ s^{-1}$, respectively: reaction (22) is reported[28] to have k $1{\cdot}9 \times 10^{10}$ $dm^3\ mol^{-1}\ s^{-1}$].

$$Ti^{III} + Cu^{II} \rightleftharpoons Ti^{IV} + Cu^{I} \qquad (21)$$

$$Cu^{I} + \cdot CH_2CH_2OH \longrightarrow Cu^{II} + CH_2{=}CH_2 + OH^- \qquad (22)$$

Computer simulations of the predicted variation of $\cdot CH_2CMe_2OH$ with $[Cu^{II}]_o$ in the Ti^{III}–H_2O_2–Bu^tOH system incorporated proposed rate constants for both radical-oxidation by Cu^{II} ($3{\cdot}2 \times 10^6$ $dm^3\ mol^{-1}\ s^{-1}$) and reduction by Cu^I ($2{\cdot}6 \times 10^{10}$ $dm^3\ mol^{-1}\ s^{-1}$ (cf. refs 27 and 28). Good agreement with experimental behaviour was then obtained for k_{21} 1×10^6 $dm^3\ mol^{-1}\ s^{-1}$.

The rate of decrease in $[\cdot CH_2CH_2OH]$ with increase in added Cu^{II} (and hence Cu^I) was significantly greater than for $\cdot CH_2CMe_2OH$, which is somewhat unexpected in terms of the similar rate constants reported[28] for reduction by Cu^I ($\cdot CH_2CMe_2OH$ is claimed to be reduced slightly faster than $\cdot CH_2CH_2OH$). A possible explanation lies in the fact that in the reaction of ethanol, the major radical would also be expected to regenerate Cu^I via one-electron transfer. To investigate this possibility further we generated a mixture of $\cdot CH_2CMe_2OH$ (readily reducible) and $\cdot CH_2OH$ (readily oxidizable) from a mixture of Bu^tOH and MeOH with $\cdot OH$ (from Ti^{III}–H_2O_2) in the presence of copper. As expected on the basis of reactions (23)-(28), the rate of decrease of $[\cdot CH_2CMe_2OH]$ with Cu^{II} was much greater in the presence, rather than the absence, of $\cdot CH_2OH$ (and similar to that of $\cdot CH_2CH_2OH$ from EtOH), as was also that of $\cdot CH_2OH$.

$$\cdot OH + MeOH \longrightarrow \cdot CH_2OH + H_2O \quad (23)$$

$$\cdot OH + Bu^tOH \longrightarrow \cdot CH_2Me_2OH + H_2O \quad (24)$$

$$Ti^{III} + Cu^{II} \rightleftharpoons Ti^{IV} + Cu^{I} \quad (25)$$

$$Cu^{I} + \cdot CH_2CMe_2OH \longrightarrow Cu^{II} + CH_2{=}CMe_2 + OH^- \quad (26)$$

$$Cu^{II} + \cdot CH_2OH \longrightarrow Cu^{I} + H_2C{=}O + H^+ \quad (27)$$

$$\cdot CH_2OH, \cdot CH_2CMe_2OH \longrightarrow \text{molecular products} \quad (28)$$

No direct evidence regarding the feasibility or rate of reaction (29) could be obtained from our experiments (except that an upper limit of _ca_. 10^3 dm^3 mol^{-1} s^{-1} can be placed on the rate constant). However we observed that addition of Cu^{II} to the Ti^{III}–EDTA–$S_2O_8^{2-}$ system caused considerable enhancement of signal intensity from some aliphatic radicals (e.g. $\cdot CH_2CMe_2OH$ from Bu^tOH), evidently as a result of an extremely rapid reaction between Cu^I and the peroxide [_cf_. reactions (30) and (31): in contrast the reaction between Ti^{III} and $S_2O_8^{2-}$ is relatively slow]. An estimate of ca. 6×10^5 dm^3 mol^{-1} s^{-1} can be made for reaction (30).

$$Cu^{I} + H_2O_2 \longrightarrow Cu^{II} + HO\cdot + OH^- \quad (29)$$

$$Cu^{I} + S_2O_8^{2-} \longrightarrow Cu^{II} + SO_4^{\cdot-} + SO_4^{2-} \quad (30)$$

$$SO_4^{\cdot-} + Bu^tOH \longrightarrow HSO_4^- + \cdot CH_2CMe_2OH \quad (31)$$

(ii) Addition of Cu^{II} to the Fenton system.

With the Fe^{II}-EDTA-H_2O_2 system at pH ca.2 (with $[Fe^{II}]_o$ = $[EDTA]_o$ ca. 10^{-3} mol dm^{-3}, $[H_2O_2]_o$ 3×10^{-3}-10^{-2} mol dm^{-3}, $[Cu^{II}]_o$ was varied in the range 10^{-5}-10^{-4} mol dm^{-3}. Little decrease was observed in the signals from Me· or $\cdot CH_2CMe_2OH$ (from DMSO and Bu^tOH), in contrast to the Ti^{III} system. This is as anticipated if copper remains as Cu^{II} in solution (despite the reduction in $E^\ominus$ for Fe^{III}/Fe^{II} in the presence of EDTA). The failure of even α-hydroxyalkyl radicals (e.g. $\cdot CH_2OH$) to be significantly reduced in intensity presumably reflects their much faster rate of removal by Fe^{III} (as noted earlier, with k ca. 10^9 dm^3 mol^{-1} s^{-1}) than Cu^{II} (k $1{\cdot}6 \times 10^8$ dm^3 mol^{-1} s^{-1}).

Conclusions

Our results establish directly that ·OH is indeed the reactive intermediate involved in the oxidation of organic substrates by Fenton's reagent under certain biomimetic conditions (e.g. at physiological pH and in the presence of polyaminocarboxylate ligands); differences between the Ti^{III}-H_2O_2 and Fe^{II}-H_2O_2 systems largely reflect the extreme ease of oxidation of electron-rich radicals by Fe^{III}.

The choice of complexing ligand and pH dramatically affects the (initiation) rate of the Fe^{II}-H_2O_2 reaction and hence of the reagent concentrations needed to observe strong spectra in such flow-system studies. The decrease in redox potential of the Fe^{III}/Fe^{II} couple on complexation of iron with EDTA and DTPA (0·12 and 0·03v respectively, compared with 0·77 V in neutral or acidic solution in the absence of chelating agent[30]) provides an essential thermodynamic rationale for the faster rates of reaction of Fe^{II}-EDTA and Fe^{II}-DTPA ($E^{\ominus}$ values for the reduction of H_2O_2 to ·OH are 0·87 and 0·46 V at pH 0 and 7 respectively[30]). The demonstration that Fe^{II}-DTPA reacts faster with H_2O_2 than Fe^{II}-EDTA at pH 7 (in agreement with the findings of Buettner[11] and Floyd[31] but in contrast to those of Sutton and Winterbourn[10]) is consistent with the lower redox potential of the Fe^{II}-DTPA complex and suggests the occurrence of an outer-sphere process. DTPA has both a greater affinity for Fe^{III} and Fe^{II} than EDTA and an increased polydentate nature[32] such that more effective shielding is provided for the metal's inner coordination sphere. In contrast, the Ti^{III}-DTPA complex presumably reacts via an inner-sphere process.

The ability of Fe^{III} to oxidise radicals formed on reaction of ·OH with a particular substrate governs both the nature of the products (dimers or oxidised species) and the potency of the Fenton system in biological damage: not only is the initial damage (e.g., to carbohydrate or DNA) "fixed" but regeneration of Fe^{II} leads to a process with only a catalytic requirement for metal ion.

Though only radicals giving relatively stable carbonium ions (e.g., $R^1R^2\dot{C}OH$) are very readily oxidised by Fe^{III}, Cu^{II} appears to be much less discriminating and is an excellent oxidant for a range of organic radicals: longer catalytic chains are in principle possible in oxidations induced by Fenton's reagent in the presence of Cu^{II} because of the ease of regeneration of Cu^{II} from Cu^{I} in radical-induced oxidation of the low-valent metal.

Acknowledgements

We thank the SERC and Interox Chemicals Ltd. for a CASE award, and Mark Burkitt, Susan Croft, Jonathan Stell and Adrian Whitwood for helpful discussions.

References

1. B. Halliwell and J. M. C. Gutteridge, "Free Radicals in Biology and Medicine", Clarendon Press, Oxford, 1985.

2. F. Haber and J. Weiss, Naturwiss., 20 (1932), 948; Proc.Roy.Soc.(London), A147 (1934), 332.

3. W. C. Bray and M. H. Gorin, J.Amer.Chem.Soc., 54 (1932), 2124.

4. J. H. Merz and W. A. Waters, Disc.Faraday Soc., 2 (1947), 179; J.Chem.Soc., (1949), 515.

5. J. H. Baxendale, M. G. Evans, and G. S. Park, Trans.Faraday Soc., 42 (1946), 155.

6. T. Shiga, J.Phys.Chem., 69 (1965), 3505; 71 (1967), 3559, 4264.

7. C. Walling, J.Amer.Chem.Soc., 104 (1982), 1185.

8. G. Czapski, J.Phys.Chem., 75 (1971), 2957.

9. See e.g. W. M. Koppenol, J.Free Radicals Biol.Med., 1 (1985), 281.

10. H. C. Sutton and C. C. Winterbourn, Arch.Biochem.Biophys., 235 (1984), 106.

11. G. R. Buettner, T. P. Doherty and L. K. Patterson, FEBS Letters, 158 (1983), 143.

12. B. C. Gilbert, R. O. C. Norman and R. C. Sealy, J.Chem.Soc.Perkin Transactions II., (1973) 2174.

13. B. C. Gilbert, R. O. C. Norman, and R. C. Sealy, J.Chem.Soc.Perkin Transactions II., (1974), 824.

14. "Magnetic Properties of Free Radicals", Group II, vol.9. Part b; Landolt-Börnstein New Series, Springer-Verlag, Berlin, 1977.

15. M. Jeff, D. Phil. Thesis, University of York (1986).

16. J. F. Below, R. E. Connick, and C. P. Coppel, J.Amer.Chem.Soc., 80 (1958), 2961.

17. V. M. Berdnikov, V. K. Federov and V. F. Schuvalov, Kinetica i Kataliz, 15 (1974), 34.

18. A. Samuni, D. Meisel, and G. Czapski, J.Chem.Soc.Dalton Trans., (1972), ·1273.

19. B. C. Gilbert, R. O. C. Norman, P. S. Williams and J. N. Winter, J.Chem.Soc.Perkin Transactions II., (1982), 1439.

20. Farhataziz and A. B. Ross "Selected Specific Rates of Reactions of Transients from Water in Aqueous Solution III. Hydroxyl Radical and Perhydroxyl Radical and their Radical Ions." National Standard Reference Data systems, National Bureau of Standards, Washington, 1977.

21. M. J. Davies, B. C. Gilbert, and R. O. C. Norman, J.Chem.Soc.Perkin Transactions II., (1984), 503.

22. See e.g. L. Eberson, Adv.Phys.Org.Chem., 18 (1982), 79.

23. O.K. Borggard, D. Forver and V. S. Anderson, Acta Chem.Scand., 25 (1971), 3541.

24. W. G. Barb, J. H. Baxendale, P. George and K. R Hargrave, Trans.Faraday Soc., 47 (1951), 462.

25. M. K. Eberhardt, J.Org.Chem., 1984, 49, 3720.

26. J. D. Rush and W. H. Koppenol, J.Biol.Chem., 261, (1986), 6730.

27. M. Freiberg and D. Meyerstein, J.Chem.Soc.Faraday Trans.I, 76 (1980), 1825.

28. G. V. Buxton and J. C. Green, J.Chem.Soc.Faraday Trans.I, 74 (1978), 697.

29. D. P. Fel'dman and B. P. Matseevskii, Kinetica i Kataliz., 15 (1974), 1173.

30. W. H. Koppenol and J. Butler, Adv.Free Radical Biol.Med., 1 (1985), 91.

31. R. A. Floyd, Canad J.Chem., 1982, 60, 1577.

32. J. Vandegaer, S. Chaberek and A. E. Frost, J.Inorg.Nucl.Chem., 11 (1959), 210.

FREE RADICALS: CHEMISTRY, PATHOLOGY AND MEDICINE
edited by Catherine Rice-Evans & Thomas Dormandy, © 1988 Richelieu Press, London

ELECTRON TRANSFER BETWEEN RADICALS AND ORGANIC MOLECULES VIA ADDITION/ELIMINATION. AN INNER-SPHERE PATH.[1]

Steen Steenken

Max-Planck-Institut für Strahlenchemie,
D-4330 Mülheim a.d. Ruhr, Federal Republic of Germany

ABSTRACT

In aqueous solution the electron transfer between (reducing) carbon-centered radicals or (oxidizing) hetero-atom-centered inorganic radicals and organic molecules often proceeds by covalent bond formation between the radical and the molecule followed by heterolysis of the so-formed bond between the carbon and the hetero-atom. It is the heterolysis step in which the actual electron transfer between the radical and the molecule takes place. This makes electron transfer a part of the area of (heterolytic) solvolysis reactions. The rate constant for the heterolysis is sensitive to and therefore indicative of the difference in effective electron density or affinity between the radical and the molecule. Factors such as substituents or protonation/deprotonation by which the electron density or distribution is changed strongly influence the rates of heterolysis of the adducts. The observed structure-activity relations for heterolysis of the radical-molecule adducts and thus the electron transfer between the adduct components can be rationalized in terms of the classical solvolysis concepts.

INTRODUCTION: Defining the Concept "Electron Transfer by Heterolysis".

The pronounced tendency of radicals to engage in one-electron transfer reactions is well documented.[3] This reaction channel is a favored one because it provides the simplest way for radicals to lose their radical nature, i.e. to become species with an even number of electrons (closed-shell molecules). The direction of the electron flow between the radical $X^{\bullet}$ and the molecule Y depends on the oxidizing or reducing power of $X^{\bullet}$ and on the ability of Y to either donate or accept an electron: the final result of the interaction between $X^{\bullet}$ and Y will then be the either one-electron-reduced or -oxidized former radical (X^- or X^+) or the open-shell molecule ($Y^{\bullet -}$ or $Y^{\bullet +}$), cf. eq. 1:

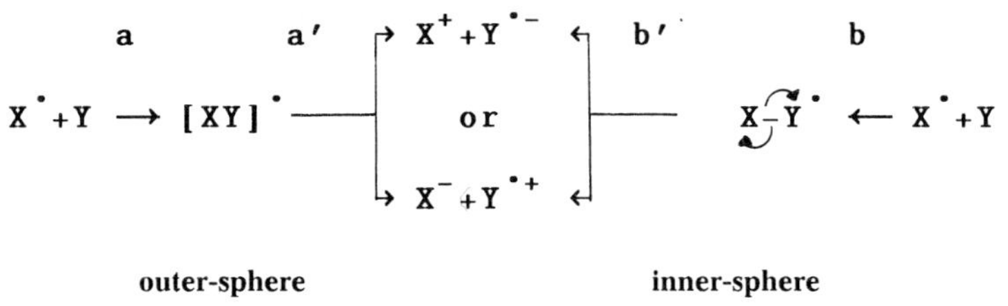

(1)

From a mechanistic point of view,[4] there are two extremes conceivable: if the interaction between $X^{\bullet}$ and Y is weak, $[XY]^{\bullet}$ symbolizes a transition state and the reaction (1a,a') is a case of outer-sphere electron transfer. If, however, the interaction is strong enough that it leads to, e.g., covalent-bond formation between X and Y (eq. 1b), the product of that interaction is an intermediate ("adduct") and the overall electron exchange between $X^{\bullet}$ and Y (via eq. 1b,b') is then an example of inner-sphere electron transfer.

Equation 1b,b' is the defining equation for the addition-elimination route for one-electron transfer between $X^{\bullet}$ and Y. It is important to note that although $X\text{-}Y^{\bullet}$ is a radical and the overall reaction results in the transfer of a single electron, in the actual electron transfer step an electron pair is shifted rather than a single electron.[5] This means that electron transfer is the consequence of a heterolysis reaction in which the electron pair joining X and $Y^{\bullet}$ ends up at either X or $Y^{\bullet}$, depending on which is the better electrophile. In this sense electron transfer is a sub-area of the general area of (heterolytic) solvolysis reactions, and in parts A and B (vide infra) examples will be presented that tend to support this view.

In addition to the addition route eq. 1b or 2b, the radicals $X\text{-}Y^{\bullet}$ are obtainable by, e.g., an abstraction path, eq. 2a:

$$
\begin{array}{ccccc}
 & \textbf{a} & & \textbf{b} & \\
X-Y-H & \xrightarrow{-H^{\bullet}} & X-Y^{\bullet} & \longleftarrow & X^{\bullet} + Y \\
 & & \downarrow c & & \\
 & & X^{+(-)} + Y^{\bullet -(+)} & &
\end{array}
\qquad (2)
$$

This enables one to use aliphatic systems as precursors to the radicals $X\text{-}Y^{\bullet}$ whose solvolytic (= redox) behavior can then be studied. Equations 2a,c describe what may be called "oxidative solvolysis". This reaction sequence, the first step of which is in many cases induced by the $\dot{O}H$ radical, is of great importance in radical (and radiation) chemistry. It extends from β-elimination reactions of monomeric radicals[6,7] to the mechanism of DNA strand breakage.[8] An example for eq. 2 in which it is shown that the radical $XY^{\bullet}$ can be produced by either step a or b is given in section B 3.

The analogy between electron-transfer via addition/elimination (eq. 2b,c) or abstraction/elimination (eq. 2a,c) and classical solvolysis involving closed-shell molecules (non-radicals) is seen by comparing scheme 1 with scheme 3, in which XY, the precursor of the ions $X^{+(-)}$ and $Y^{-(+)}$, is formally derived from the two radicals $X^{\bullet}$ and $Y^{\bullet}$. Analogous to scheme (1), on the way to the ionic products that result from the interaction between $X^{\bullet}$ and $Y^{\bullet}$ there are two possibilities: if XY denotes a transition state, the reaction (eq. 3a,a') is a case of outer-sphere electron transfer. If, however, a covalent bond is formed between X and Y, the path (eq. 3b,b') is an example of inner-sphere electron transfer. Obviously, part b' of the scheme describes the classical area of S_N1 solvolysis reactions (assuming either X or Y to be equal to C).[9,10] If a second reaction partner for C^+ (other than the solvent) is allowed for (the (partial) ions then represent transition states), then eq. 3b' also covers S_N2 reactions. If looked upon from the point of view of radical-radical reactivity, equations 3a and b show well-known reactions: radical disproportionation in eq. 3a,a' and combination in eq. 3b.

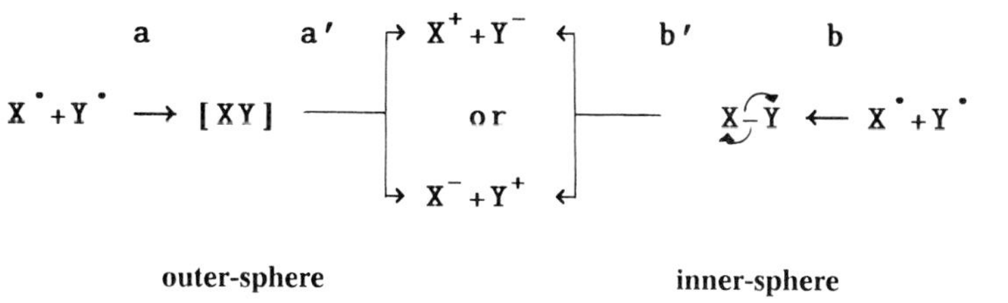

(3)

Concerning the general reaction scheme 1, in its discussion that follows attention will be restricted to two special areas: A), cases where $X^{\bullet}$ is a carbon-centered radical and Y is an oxygen atom joined by a double bond to some center Z, and B), cases where $X^{\bullet}$ is a hetero atom - in most cases: oxygen - centered radical and Y is a carbon, cf. eqs. 4 and 5.[11] One is then dealing with formation and heterolysis of a bond between a carbon- and a hetero-atom. Of the two, the hetero-atom is of course always more electron-affinic and therefore in the heterolysis the electron pair joining the two will go to the hetero-atom.

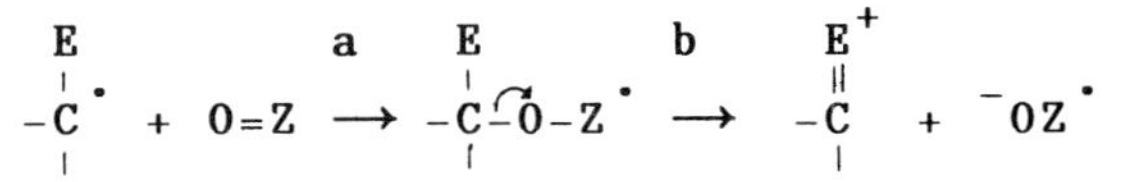

(4)

$Z = O, N(O)C(NO_2)_3, N(O)C_6H_4R$, quinone moiety

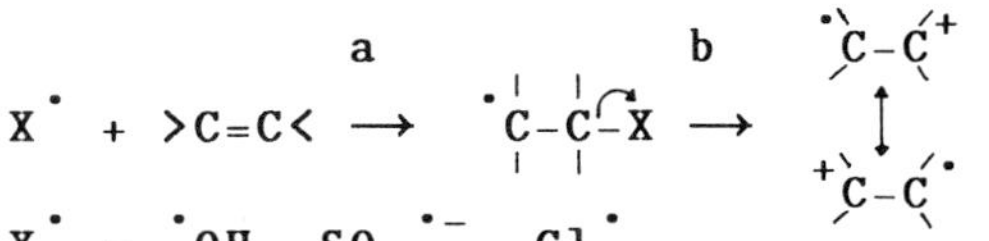

(5)

$X^{\bullet} = {}^{\bullet}OH,\ SO_4^{\bullet -},\ Cl^{\bullet}$

The transfer of the electron pair in the heterolysis reaction to the hetero atom (the nucleofugal group) leads to its reduction and to oxidation of the carbon (the electrofugal group). In eq. 4b the oxidized species is a cation[14] (or its solvolysis products), whereas in the case of eq. 5b it is a radical cation (or its solvolysis products). In the former case the reduced species (the nucleofugal leaving group) is a radical anion, and in the subsequent section **(A)** examples for this type of reaction will be presented. Examples for eq. 5b, where the nucleofugal leaving group is a closed-shell anion, will be given in section **B.**

A. OXIDATION OF CARBON-CENTERED RADICALS SUBSTITUTED BY A HETERO-ATOM E AT C_α (EQ. 4). 1. NITROBENZENES AS OXIDANTS.

These reactions belong to the most thoroughly studied ones in the field of radical-molecule reactions in aqueous solution. The interest in this area is to a large part due to its relevance to the understanding of the mechanism of action of nitroaromatics as sensitizers in the radiotherapy of cancer.[15]

The hetero atom E at C_α is required for two reasons: a) to make the addition (of C_α) to the nitro group possible (by providing the necessary "nucleophilicity"[16] to the radical,[17] and, b) to stabilize the (incipient) carbocation that results from the heterolysis. The features a) and b) are interrelated by the fact that not only the transition state for heterolysis but also that for addition has a large degree of electron transfer character.[18]

A.1.a. OH as the Hetero Group at C_α.

It has been recognized as early as 1968[19] that the interaction in aqueous solution of these radicals with nitroaromatics can lead to two types of (transient) product: α-alkoxynitroxyl radicals (produced by addition to the nitro group), and nitro radical anions, cf. eq. (6) and (7):

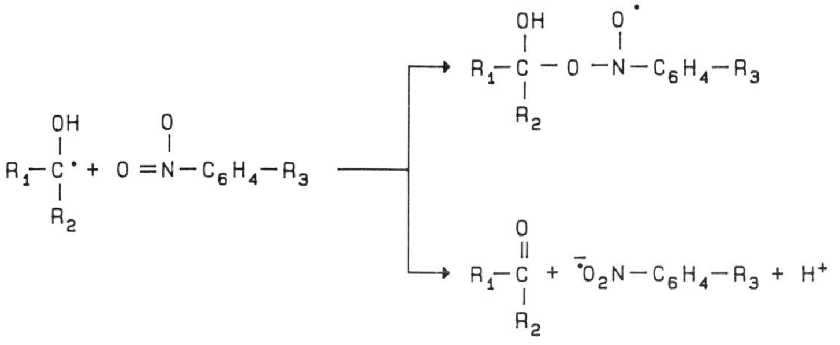

(6)

(7)

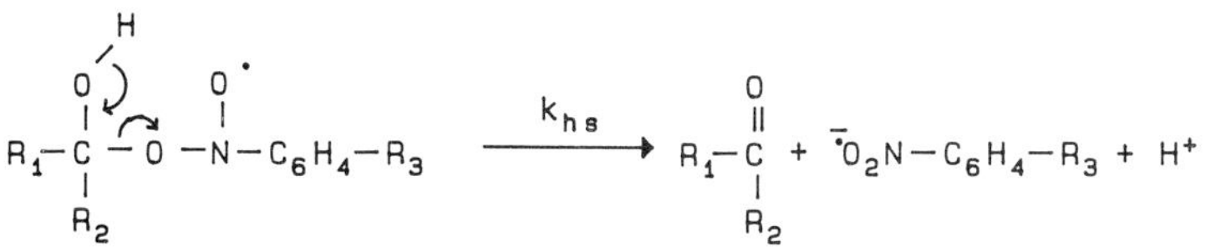

(8)

The tendency to react according to (6) or (7) depends on the stability of the (incipient) carbocation[19,20] and on the oxidizing power (redox potential) of the nitro compound.[15,21] It also depends on solvent, more polar solvents favoring the ionic path (7).[21]

It has recently been shown[21] that in aqueous solution the addition (via 6) and the electron transfer product (via 7) do not exclude each other and that (6) and (7) can occur simultaneously. This has been explained in terms of an electron-transfer/addition mechanism involving an ion pair type transition state with subsequent competition between combination of the ions (to give addition) and separation by solvent (leading to electron transfer).[18]

The extent to which the radicals react according to eq. 6 or 7 depends on the nature of R_1,R_2 and R_3. If $R_1 = R_2 = H$ and $R_3 = H$ to NO_2, the ratio (6) : (7) $\geq$ 20. The addition reactions observed with these systems are characterized by strongly negative activation entropies, which can be rationalized in terms of immobilization of water molecules by the positive charge at C_α in the transition state.[18] That the transition state for addition has pronounced electron-transfer character can also be concluded from the fact[18] that the rate constants for addition depend on the reduction potential of the nitrobenzene in a way describable by the Marcus relation for outer-sphere electron transfer.

The tetrahedral-type alkoxynitroxyl radicals formed in reaction (6) can undergo a C-O heterolysis, eq. 8. However, for $R_1 = R_2 = H$ at pH <6, k_{hs}, the rate constant for spontaneous heterolysis, is $\leq 10^2 s^{-1}$. However, at pH >7, a drastic increase in the heterolysis rate occurs, proportional to the concentration of $[OH^-]$. The increase is due to deprotonation of the hemiacetal.OH group by which the good electron-donor OH is converted to the excellent one O^-. The electron pair joining C_α and O is thereby "pushed out" leading to k (heterolysis) values of $\geq 5 \times 10^5 s^{-1}$,[21] cf. eq. 9:

$$\mathrm{H{-}\underset{H}{\overset{OH}{C}}{-}O{-}\overset{O^{\bullet}}{N}{-}..} \xrightarrow{+OH^-} \mathrm{H{-}\underset{H}{\overset{O^-}{C}}{-}O{-}\overset{O^{\bullet}}{N}{-}..} \xrightarrow[\geq 10^5 s^{-1}]{k_{hs}} \mathrm{H_2CO} + \mathrm{{}^{-}O{-}\overset{O^{\bullet}}{N}{-}..} \qquad (9)$$

If one hydrogen at C_α is replaced by the electron-donating methyl group, i.e. $R_1 = H$ and $R_2 = CH_3$, the ratio eq. (6) : (7) = 0.1 to ~3, depending on R_3. Due to the methyl group at C_α, the alkoxynitroxyl radicals are now able to undergo a spontaneous heterolysis (eq. 8) with rate constants ranging from 10^2 to $10^4 s^{-1}$, depending on R_3. The activation entropies for this process are very negative (-20 to -100 $Jmol^{-1}K^{-1}$), indicative of freezing of water molecules in the transition state. As judged by the solvent kinetic isotope effect (2.2)[21] and by comparison with the much more positive entropy changes observed on heterolysis of nitroxyls from α-alkoxyalkyl radicals[22] (see next section), the negative entropy changes are essentially due to deprotonation from the hemiacetal OH.

If $R_1 = R_2 = CH_3$, i.e. with two methyl groups at C_α, the rate of the spontaneous heterolysis (eq. 8) becomes $\geq 10^6 s^{-1}$ for all nitrobenzenes with R_3 more electron-withdrawing than CH_3O. This high heterolysis rate constant means that the reaction appears to proceed by electron transfer since an intermediate is not visible using detection techniques with μs time resolution. However, the addition/elimination sequence can be demonstrated to exist by using two approaches: a) reduce the reduction potential of the nitrobenzene (by introducing an electron-donating substituent R_3), or b) decrease the polarity of the solvent. An example for a) is the use of 4-nitroaniline which forms a nitroxyl radical with $(CH_3)_2\dot{C}OH$ which heterolyzes with $k_{hs} = 2.1 \times 10^3 s^{-1}$ in H_2O. An example for b) is the use of 95% i-propanol/5% water mixtures. In this solvent the rate of heterolysis of the nitroxyl from $(CH_3)_2\dot{C}OH$ and 4-nitrobenzonitrile is slowed down to $1.5 \times 10^4 s^{-1}$ from $\geq 10^6 s^{-1}$ in pure water.[21]

To summarize, by modifying in -$\overset{|}{\underset{|}{C}}$-O- either the leaving group abilities of the carbon moiety (the electrofuge) (e.g., by alkyl substitution at C_α or by ionization of OH) or those of the nitrobenzene (by substitution at the ring) it is possible to go all the way from pure addition to what appears to be pure electron transfer. The heterolysis rates increase with increasing electron "push" by the carbon and with increasing "pull"

(= reduction potential) of the nitrobenzene. Also the solvent has a strong influence on the rate constant for heterolysis of the C-O bond, as expected for S_N1 type reactions.

b. O-Alkyl as the Hetero Group at C_α.

On the basis of results obtained from (in-situ-radiolysis[22]) electron-spin-resonance[19,23] and pulse radiolysis with optical and conductance detection,[22] α-alkoxyalkyl radicals react in aqueous solution exclusively via addition to give alkoxynitroxyl radicals, cf. eq. 10. This is in contrast to the reactions of $CH_3\dot{C}HOH$ (see section a) and 5,6-dihydropyrimidine-6-methyl-6-yl radicals (see section c), where addition and redox products are formed.

$$R_1-\overset{\overset{\textstyle alkO}{|}}{\underset{\underset{\textstyle R_2}{|}}{C}}{}^{\bullet} + O=\overset{\overset{\textstyle O}{|}}{N}-C_6H_4-R_3 \longrightarrow R_1-\overset{\overset{\textstyle alkO}{|}}{\underset{\underset{\textstyle R_2}{|}}{C}}-O-\overset{\overset{\textstyle O^{\bullet}}{|}}{N}-C_6H_4-R_3 \qquad (10)$$

$R_1, R_2 = H, CH_3$, cyclo-alkyl; alk $= CH_3, C_2H_5, (CH_3)_2CH$;

$R_3 = H$ to NO_2

The rate constants for addition increase from e.g., 5.0×10^7 for $R_1 = R_2 = H$ to $1.8 \times 10^9\ M^{-1}\ s^{-1}$ for $R_1 = R_2 = CH_3$ and $R_3 = COCH_3$. The increase in this direction is due to the activation entropies becoming more positive as H is replaced as a substituent at C_α by CH_3. The rate-enhancing effect of the more positive activation entropies over-compensates the rate-decreasing effect of the higher activation enthalpies that result from the increasing number of methyl groups at C_α.[22]

On the basis of the very negative activation entropies, the transition states for the addition are highly ionic, i.e., there is a large degree of electron transfer in the transition state as with the hydroxyalkyl radicals (section a). In support of this is the fact that the rate constants for addition depend on the reduction potentials of the nitro-benzenes, varied by the substituent R_3, in a way describable by the Marcus equation for outer-sphere electron transfer.[22]

For those systems where $R_1 = R_2 = H$ or $R_1 = H$, $R_2 = CH_3$, i.e. where the number of alkyl groups at C_α is ≤ 1, and $R_3 = H$ to NO_2, the alkoxynitroxyl radicals formed according to eq. 7 under steady state ESR or pulse radiolysis conditions do not give rise to nitrobenzene radical anions. This means that the rate constants for heterolysis of the nitroxyls are $\leq 10^2 s^{-1}$. This is not only true in weakly acidic (pH 4) or neutral but also in strongly alkaline solution (pH 13-14). The latter shows that the nitroxyls are not susceptible to base catalyzed heterolysis. The rate constant for OH^- catalyzed decomposition can be estimated to be $\leq 10^2 M^{-1} s^{-1}$.[22] This low number for the hypothetical S_N2-type reaction is in line with the known general stability in basic solution of acetal-type compounds. A similar observation was made with acetalic nitroxyl radicals produced by addition of α-alkoxyalkyl radicals to tetranitromethane[24] (see section A 2.). However, these nitroxyl radicals are able to undergo a spontaneous heterolysis of the carbon-oxygen bond, a reaction that has all characteristics of S_N1.[24] The fact that the heterolysis is observable in the case of tetranitromethane as the electron (pair) acceptor and not with the nitrobenzenes is due to the much higher oxidizing or electron-deficient character of tetranitromethane as compared to the nitrobenzenes, including 1,4-dinitrobenzene. However, there is one case where heterolysis is observable (by time-resolved optical and conductance methods) also for a nitroxyl formed from a nitrobenzene and an ether radical mono-methylated at C_α, cf. eq. 11:[22]

$$CH_3\dot{C}HOC_2H_5 + O_2N-C_6H_4-N_2^+ \longrightarrow CH_3CH(OC_2H_5)-O-N(O^\bullet)-C_6H_4-N_2^+$$

$$\xrightarrow{+\ H_2O} CH_3CH(OC_2H_5)OH + O_2NC_6H_4^\bullet + N_2 + H^+ \qquad (11)$$

The reason for the rate increase of the heterolysis into a range ((4-5) x $10^5\ s^{-1}$) where it becomes detectible is the exceptionally large electron-accepting power of the N_2^+-substituent (Hammett $\sigma = 1.9$) which results in the electron pair being pulled strongly towards the nitrobenzene.

In order to enhance the C-O heterolysis it is also possible to increase the electron-donating properties of the electrofugal group (the carbon group) rather than enhance the electron-attracting power of the nucleofugal group (the oxygen group). This is most easily done by introducing another alkyl group at C_α. In S_N1 reactions of non-radicals an additional methyl group at C_α has been estimated to lead to a rate enhancement by the factor ~ 10^8.[25] In the nitrobenzene series, spontaneous heterolysis of the alkoxynitroxyl radicals formed by addition of α-alkoxyalkyl radicals is observed[22] for the acylic system $R_1 = R_2 = CH_3$, alk$=(CH_3)_2CH$ (eq. 10) and also for the cylic one:

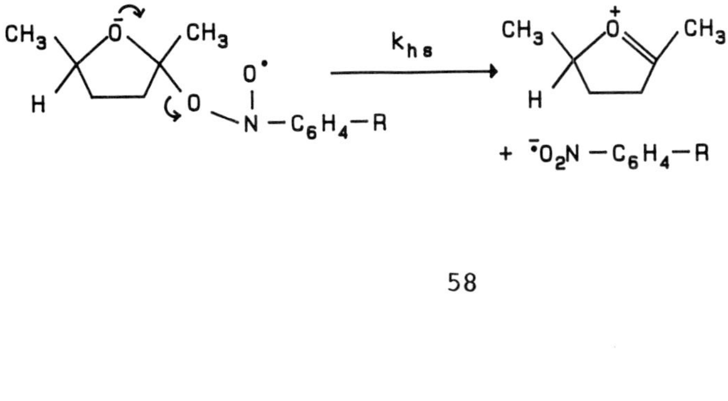

(12)

i.e. for systems with two alkyl groups and one alkoxyl group at C_α. The second alkyl group at C_α increases the electron density to an extent which makes the heterolysis rates at 20^0 C become of the order 10^5 s^{-1}. The rate constants increase with increasing electron-deficiency of the substituent R_3 on the nitrobenzene, i.e. with increasing electron pull by the electron acceptor. The heterolysis reaction is characterized by activation enthalpies of ~ 50 kJ/mol and by positive activation entropies, ~ 20 J/(mol K),[22] which is in contrast to the case of the heterolysis of the nitroxyls formed by addition of α-hydroxyethyl radical to the same nitrobenzenes, where the activation entropies are strongly negative (to ~ - 100 J/(mol K).[21] The difference between the two systems is probably due to the fact that in the case of the hydroxyethyl nitroxyls deprotonation from the hemiacetal OH group takes place in the transition state with hydration of the (incipient) proton thereby leading to freezing of water molecules which causes the loss of entropy, i.e.

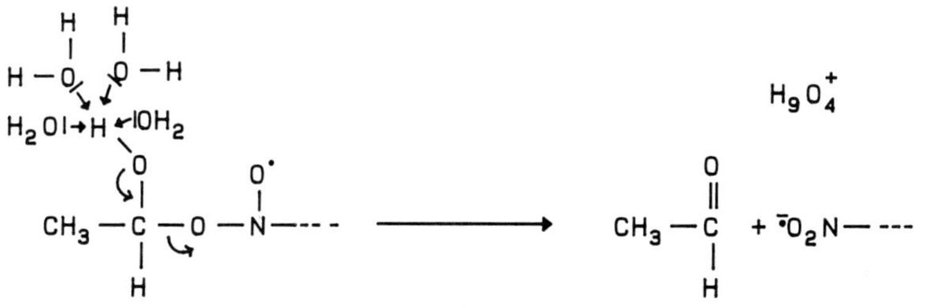

(13)

This means that C-O bond breaking is to a certain degree concerted with O-H bond breaking, or, expressed differently, solvent assisted O-H deprotonation provides the necessary electron density at C_α to make the C-O heterolysis possible. In agreement with this picture is the fact that replacement of the hemi-acetal H by an alkyl group (i.e. going to the C_α acetalic systems) makes the rate constant for heterolysis drop below 10^2 s^{-1}. Only after introducing an additional CH_3 group at C_α with the concomitant increase in electron density does the heterolysis become detectible again (k_{hs} ~ 10^5 s^{-1}).

c. -N(CO)- as the Hetero Group at C_α (5,6-Dihydropyrimidine-6-yl Radicals).

The 6-yl radicals produced by (the selective)[26,27] addition of $\dot{O}H$ to C(5) of the C(5)/(6) double bond of naturally occurring pyrimidine bases, nucleosides and nucleotides, or those formed by H-abstraction from C(6) of 5,6-dihydropyrimidines,[28] react with para-substituted nitrobenzenes by addition (k ~ 6 x 10^6 to 2 x 10^9 M^{-1} s^{-1}) to yield nitroxyl-type radicals which were identified by ESR and optical detection techniques,[29,30] cf. eq. 14:

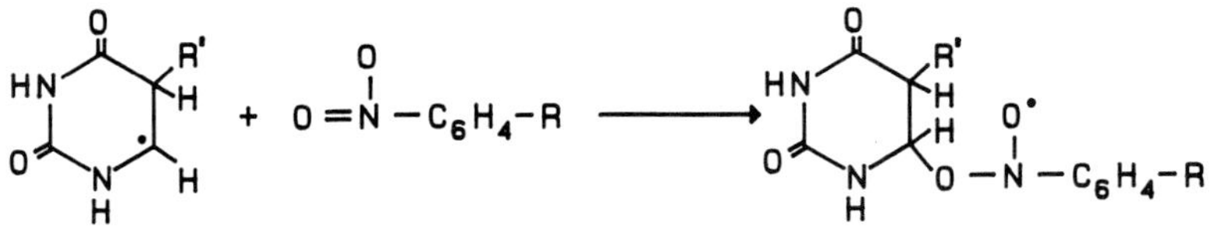 (14)

R'=H, OH

At pH < 7 the nitroxyl radicals do not undergo an observable heterolysis ($k_{hs} \lesssim 10^2\ s^{-1}$), but decay by bimolecular reactions. However, in basic solution takes place an OH^--catalyzed heterolysis to yield the radical anion of the nitrobenzene and an oxidized pyrimidine. In the case of the nitroxyls substituted at N(1) by H (i.e. those derived from the free bases), the OH^- catalysis involves deprotonation at N(1) which is adjacent to the reaction site (= C(6)), cf. eq. 15:[29]

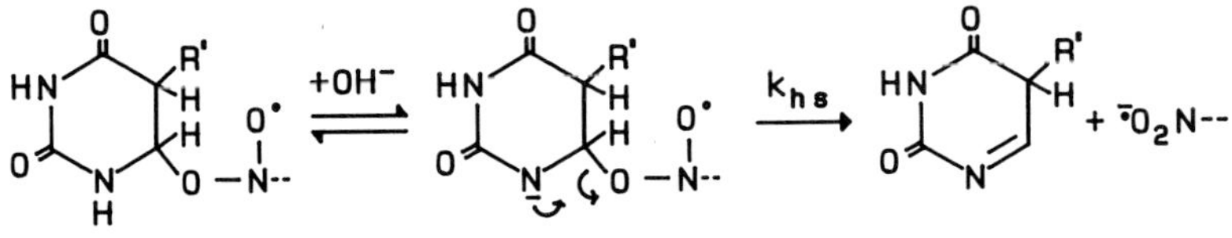 (15)

R'=H, OH

Deprotonation provides the necessary electron "push" to kick out the electron pair joining C(6) with the nitrobenzene oxygen. If, however, N(1) is alkylated (as with the nucleosides and nucleotides), OH^- catalysis is much less efficient since it now proceeds by deprotonation from N(3) (with the uracils) or from the amino group at C(4) (with the cytosines). In these cases the area of deprotonation is separated from the reaction site by a (hydroxy)methylene group, which means that the increase in electron density that results from deprotonation at N(3) is transferable to the reaction site only through the carbon skeleton (inductive effect), which, of course, is inefficient as compared to the electron-pair donation from N(1) (mesomeric effect).[29] Reaction 15 is a 1 : 1 model for the catalytic effect of OH^- on the heterolysis of peroxyl radicals from pyrimidine-6-yl radicals (see section A 4).

The ionized 6-yl radicals react with the nitrobenzenes also by addition, however with rate constants considerably higher ($k = 1 \times 10^8$ to $1 \times 10^9\ M^{-1}s^{-1}$) than those for the case of the neutral radicals. This indicates that also the transition state for the addition reaction is ionic.[29] The same conclusion can be reached from the increase of the rate constants for addition with increasing reduction potential of the nitrobenzenes and also from the very negative activation entropies for the addition reaction.[18]

The 5,6-dihydropyrimidine-6-yl radicals discussed above behave, in their reactions with nitrobenzenes, like the simpler radicals $\dot{C}H_2OH$ and $\dot{C}H$(alkyl)Oalkyl do, i.e. they react exclusively by addition to give nitroxyl radicals and uncatalyzed heterolysis is not observed ($k_{hs} \leq 10^2 s^{-1}$). If, however, a methyl group is introduced at C(6) (= C_α) of the pyrimidine-6-yl radical, the corresponding nitroxyl radicals heterolyze with rate constants at 20° of 10^3 to $5 \times 10^5 s^{-1}$ depending on the structure of the pyrimidine and of the nitrobenzene (eq. 16). This S_N1 type reaction is characterized by activation enthalpies of 30 - 40 $kJmol^{-1}$ and activation entropies of - 7 to - 89 $Jmol^{-1} K^{-1}$ (entropy control).[30] The rate-enhancing effect of the methyl group is, of course, due to stabilization of the (incipient) C(6) carbocation developing in the C-O heterolysis. The rate constants for the heterolysis reaction are a measure of the reducing power of 5,6-dihydro-6-methylpyrimidine-6-yl radicals. On this basis, the cytosine radicals are better reductants than the corresponding uracil radicals, and the radicals derived by hydrogen atom addition to pyrimidines are stronger reductants than those formed by OH radical addition.[30]

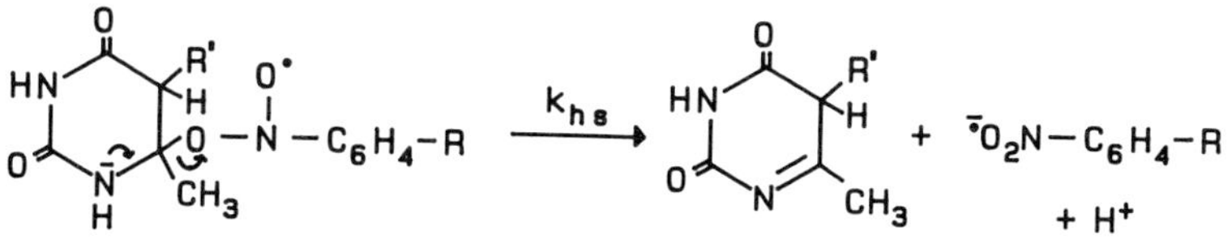

(16)

R' = H or OH

A 2. TETRANITROMETHANE AS AN OXIDANT.

It was mentioned above that acetalic nitroxyl radicals produced by addition of α-alkoxyalkyl radicals to tetranitromethane (TNM) undergo a spontaneous heterolysis with the carbon center being oxidized and TNM reduced (to nitroform anion (NF^-) and $\dot{N}O_2$). In order to see the addition-elimination sequence with acylic α-alkoxyalkyl radicals, there have to be two (electron-withdrawing) hydrogens at C_α. Even one alkyl group at C_α makes k_{hs} to be $\geq 10^6 s^{-1}$ and therefore too fast to measure. If, however, an alkyl group which is inductively deactivated is introduced at C_α, the k_{hs} values fall in the experimentally accessible range (10^2 - $10^6 s^{-1}$).[24] An example for this is the dioxolan-3-yl/TNM system, cf. eq. 17:

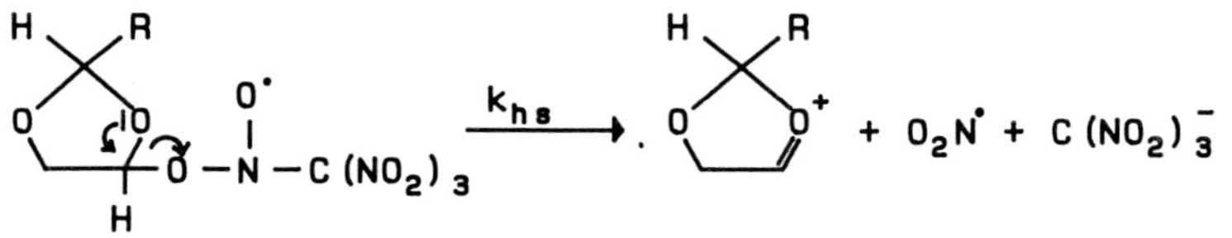 (17)

By varying R at C_γ (which is separated from the reaction site at C_α by an oxygen) it is possible to influence the electron density at C_α in a defined way to change the heterolysis rate constant. Corrected for the incomplete transmission of electronic effects through the C_γ methylene group and assuming that the transmission of the oxygen is 100%, the Taft ρ^* value for the effect of substitution on the heterolysis rate constant is equal to -3.95, comparable to that (ρ^* = -3.29) for the S_N1 hydrolysis of tertiary alkyl halides.[24]

That the heterolysis has pure S_N1 character was also deduced from the activation parameters of the heterolysis and from its solvent dependence.[24]

A 3. ANTHRAQUINONE-2,6-DISULFONATE AS AN OXIDANT.

This quinone reacts in aqueous solution with OH and H adducts of cytosines and uracils by an electron transfer/addition mechanism, similar to eq. 18.[31] Addition takes place at the quinone carbonyl oxygen to produce an anthroxyl radical. This then undergoes spontaneous C-O heterolysis:

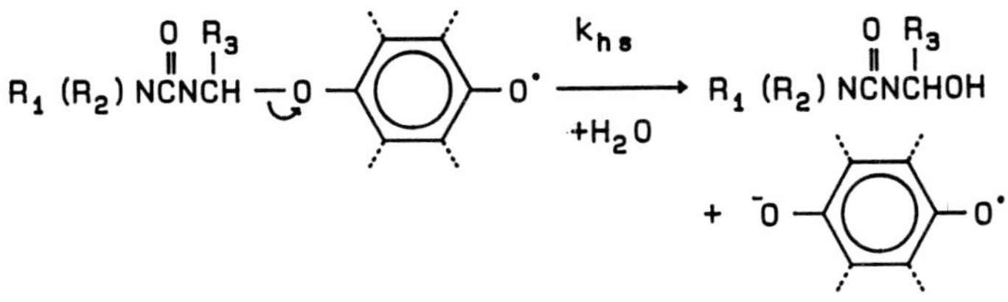 (18)

The activation parameters for the (bimolecular) addition and the (unimolecular) heterolysis steps have been determined[31] for the case of R_1, R_2, R_3 = H or CH_3 and the results are shown in Figure 1. It is obvious that the heterolysis reaction is entropy controlled which is the consequence of the highly ionic transition state which leads to freezing of water molecules with the concomitant loss of entropy.

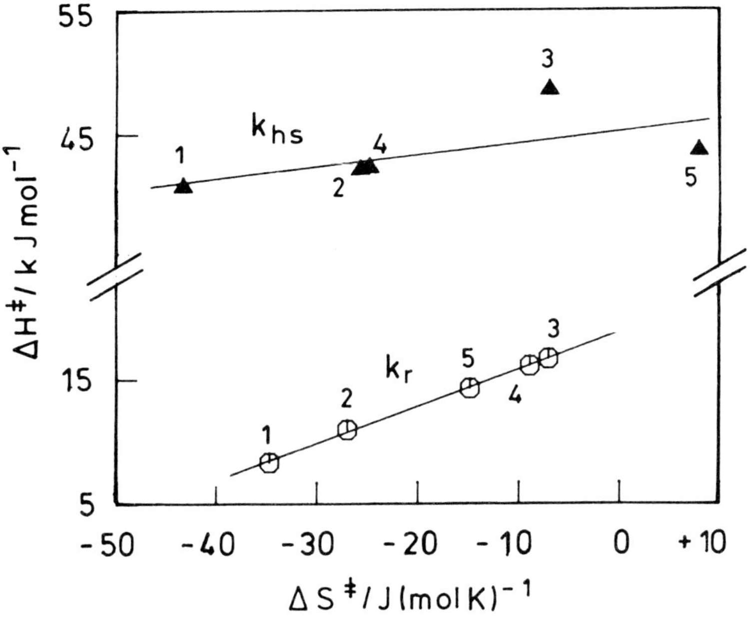

Figure 1. Isokinetic plot: The dependence on R of the activation parameters for formation in H_2O, k_r, and those for heterolysis (acc. to eq. 18) of the anthroxyl radical, k_{hs}.

Key: 1: $R_1 = R_2 = R_3 = H$. 2 and 3: $R_1 = R_2 = H$, $R_3 = CH_3$.
3: solvent D_2O. 4: $R_1 = R_3 = CH_3$, $R_2 = H$. 5: $R_1 = R_2 = R_3 = CH_3$.

A 4. O_2 AS AN OXIDANT.

The way O_2 reacts with many radicals from biologically important precursors is quite similar to that described in sections A 1-3 for nitro compounds and the quinone. It is for this reason that quinones and nitro compounds can to a certain degree replace O_2 as a radiation sensitizer. O_2 reacts with organic radicals by addition,[32] and this in spite of the fact that electron transfer is often thermodynamically highly favorable. E.g., in the case of the reaction of O_2 (E^1_7 = -0.155 V/NHE) with $(CH_3)_2\dot{C}OH$ (E^2_7 = -2.2 V/NHE),[33] the difference in the reduction potentials of the reactants (~ 2 V) corresponds to a driving force of ~ 46 kcalmol^{-1} for the hypothetical outer-sphere electron transfer according to eq. 19:

$$(CH_3)_2\dot{C}OH + O_2 \not\rightarrow (CH_3)_2CO + O_2^{\bullet -} + H^+ \quad (19)$$

However, in spite of this enormous exothermicity, the reaction does not proceed by electron transfer but by addition,[34] eq. 20a,

$$(CH_3)_2\dot{C}OH + O_2 \xrightarrow{a} (CH_3)_2C(OH)\text{-}O_2^{\bullet} \xrightarrow{b} (CH_3)_2CO + O_2^{\bullet -} + H^+ \quad (20)$$

and the adduct (the hydroxyperoxyl radical) then undergoes spontaneous C-O heterolysis (k_{hs} = 670 s^{-1}), cf. eq. 20b, to give the oxidized organic, i.e. acetone, and the reduced O_2, i.e. $O_2^{\bullet -}$.[34] The hydroxyperoxyl radical is of the hemi-acetal type, and base catalysis of its heterolytic decomposition is therefore possible. It involves deprotonation of the OH group to give the substituent $-O^-$ whose greatly increased electron density (as compared to -OH) increases the heterolysis rate to > 10^5 s^{-1}.[34,35]

The heterolysis rate is also increased by introducing a second OH[36] or Oalkyl group[37] at C_α, e.g.

$$HC(OH)_2\text{-}O_2^{\bullet} \longrightarrow HC(O)OH + O_2^{\bullet -} + H^+;\ k_{hs} \geq 10^6\ s^{-1}. \quad (21)$$

The reason for the rate enhancement compared to eq. 20 is the added stabilization of the (incipient) carbocation by the two OH groups at C_α.

At this stage it may be interesting to compare the one-electron oxidizing efficiency of O_2 with that of, e.g., nitrobenzene, taking $(CH_3)_2\dot{C}OH$ as the common reductant. In aqueous solution the rate constant for heterolysis of the adduct of $(CH_3)_2\dot{C}OH$ to the weak oxidant nitrobenzene (E_7^1 = -0.486 V/NHE)[33] is ≥ 10^6 s^{-1},[21] whereas that for heterolysis (see eq. 20b) of the adduct of $(CH_3)_2\dot{C}OH$ to the strong oxidant O_2 (E_7^1 = -0.155 V/NHE)[33] is only 670 s^{-1}.[34] The weak oxidant nitrobenzene thus leads to product (acetone) considerably faster than does the strong oxidant O_2. This puzzling result can however be understood if the leaving group abilities of the reduced oxidants are considered. On the basis of the pK_a values of the conjugate acids, the nitrobenzene radical anion (pK = 3.2)[38] is a better leaving group than $O_2^{\bullet -}$ (pK = 4.8)[39].

A further example for oxidation by O_2 are 5,6-dihydropyrimidine-6-yl radicals. These react with O_2 to give the 6-peroxyl radicals shown.[26-28,40]

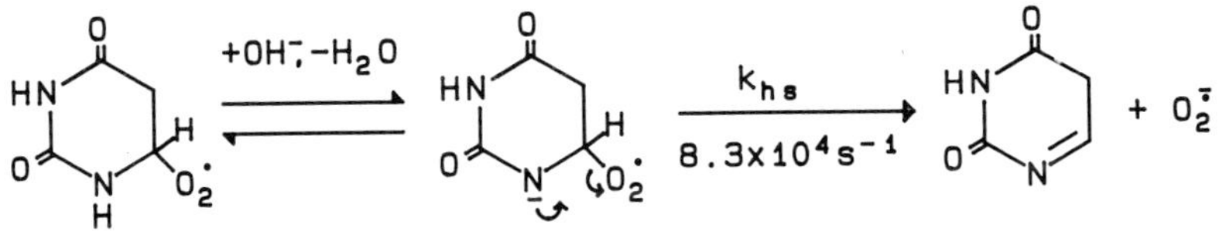

(22)

With these radicals spontaneous C(6)-O heterolysis is slow ($< 10^3\ s^{-1}$). However, if the electron density of the system is increased by OH^- induced deprotonation of N(1)-H, $O_2^{\bullet -}$ elimination is observed.[26-28] With the peroxyl radical from 5,6-dihydrouracil-6-yl, the heterolysis rate constant is $8.3 \times 10^4\ s^{-1}$, the reaction leading to the isopyrimidine derivative shown.[40] The reaction is perfectly analogous to the eliminations of the radical anions of nitrobenzenes (eq. 15) or anthraquinone-2,6-disulfonate (eq. 18).

B. OXIDATION OF OLEFINIC AND AROMATIC COMPOUNDS (EQ. 5).

1. $\dot{O}H$ AS THE OXIDANT.[41]

The $\dot{O}H$ radical is a very powerful one-electron oxidant. Its reduction potential at pH 0 is 2.7 V/NHE,[42] and at pH 7 it is still 2.3 V. That this number indicates strong oxidizing power is evident on comparing it with those of some well known oxidants such as $IrCl_6^{2-}$ (0.87 V) or Tl^{2+} (2.2 V). In spite of this $\dot{O}H$ does usually not react by electron transfer but by addition, not only with organic substrates (containing double bonds), but also with anions[12] and even metal ions.[13] This tendency to add rather than to oxidize is probably caused by the transition state for addition being stabilized by contributions from bond making, whereas electron transfer requires pronounced bond and solvent reorganization with a correspondly large free energy change to reach the transition state.

$\dot{O}H$ addition leads to the "OH adduct" $HO\text{-}Y^{\bullet}$ (eq. 23). **In order for this adduct to yield electron transfer products, heterolysis of the bond joining HO and Y has to occur.** However, due to the fact that OH^- is a very bad leaving group (as evidenced by the high pK_a (15.7) of its conjugate acid, H_2O), the rate of the spontaneous heterolysis, k_{hs} (eq. 23a), is very low (very often, $\ll 10^2\ s^{-1}$). As a consequence, the final (non-radical) products from $\dot{O}H$ reactions with Y are typically derived from dimerization or disproportionation of $HOY^{\bullet}$.

One-electron oxidation of Y by $\dot{O}H$ is, however, possible by changing the leaving-group abilities of the adduct components, HO- and $Y^{\bullet}$. As shown in eq. 23b, protonation of HO- to give $H_2\overset{+}{O}$- converts the bad leaving group OH^- into the excellent one H_2O (pK_a (H_3O^+) = -1.7). If it is assumed that the Brönsted catalysis law is applicable to this case and the Brönsted coefficient, α, is equal to 0.5, a rate enhancement of $10^{8.7}$ induced by protonation of the leaving group -OH is calculated from the difference in the pK_a values of H_2O and H_3O^+. A somewhat similar number (10^7) is obtained[3d] by considering the difference in reduction potential of $\dot{O}H$ at pH 7 and at pH 0.

$$HO^{\bullet} + Y \longrightarrow HO{-}Y^{\bullet} \xrightarrow{k_{hs}} HO^- + Y^{\bullet +} \qquad a \qquad (23)$$

$$HO{-}Y^{\bullet} \underset{-H^+}{\overset{+H^+}{\rightleftharpoons}} H_2\overset{+}{O}{-}Y^{\bullet} \xrightarrow{k_h} H_2O + Y^{\bullet +} \qquad b$$

Protonation is a very effective method to improve the nucleofugacity of the leaving group -OH, which results in an increase in the rate of heterolysis of the HO-Y$^•$ bond. The reciprocal way, which can be even more efficient, is to improve the electrofugacity of Y$^•$. A way to achieve this is to introduce electron-donating substituents into Y$^•$. The most elegant (and an important) method is to increase the electron density by ionization of a substituent on Y$^•$ which is a Brönsted acid. An example for this, which serves also to summarize the mechanisms of H^+ and OH^- supported dehydration, is shown in eq. 24.

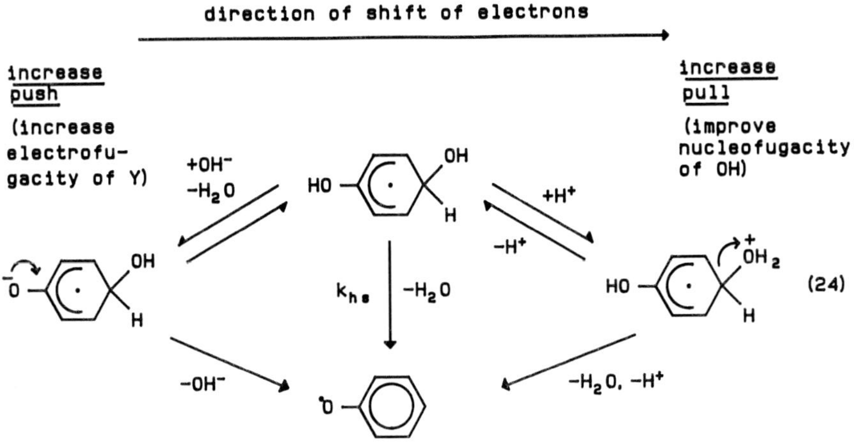

(24)

Another example relates to OH^- aided one-electron oxidation of cytosine. With cytosine, the $\dot{O}H$ reaction proceeds by addition to C(5), a process that has a selectivity of 90%.[27] The 5-hydroxy-6-yl radical is an excellent reductant, and the same is true for the ionized 6-yl radical formed by deprotonation from N(1). This radical anion now contains sufficient electron density to eliminate the OH group at C(5) as OH^-. The result is the cytosine-1-yl radical which is oxidizing, probably due to appreciable spin density at the hetero atoms N(1) and O^2.[27]

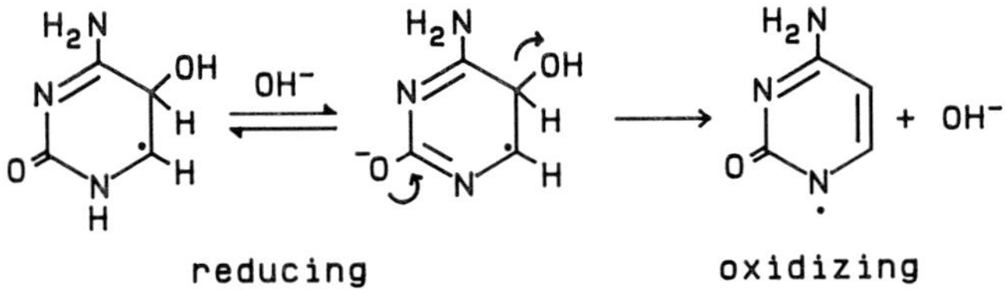

(25)

Reactions 24 and 25 are examples for the general phenomenon of "redox inversion"[41] by dehydration of OH adducts.

The reaction between $\dot{O}H$ and phenol lends itself to an analysis of its thermochemistry. On the basis of $E^2_7(\dot{O}H)$ = 2.3 V/NHE and an estimated[33] $E^2_7(C_6H_5O^\bullet)$ of 1.3 V/NHE the formation of $C_6H_5O^\bullet$ and H_2O via an electron-transfer mechanism is exothermic by 1 V=23 kcal/mol. However, the reaction proceeds by addition, as outlined in eq. 24. The propensity of $\dot{O}H$ to add rather than to oxidize is probably caused by the transition state for addition being stabilized by contributions from bond making, in contrast to electron transfer which requires pronounced bond and solvent reorganization which results in a large (entropy caused) free energy change.

2. $SO_4^{\bullet -}$ AS AN OXIDANT.

This radical, which, like $\dot{O}H$, is a strong one-electron oxidant (E ~ 2.5 - 3.1 V/NHE)[4c], has been proposed to react by outer-sphere electron transfer with aromatics.[43] However, on the basis of a Marcus treatment considering the redox potentials of the species involved, some scepticism concerning this mechanism is appropriate.[4c] With certain simple alkenes $SO_4^{\bullet -}$ reacts by addition,[44] whereas with alkenes carrying (electron-donating) alkyl groups, one-electron-transfer products have been observed.[44b] Electron transfer products are also seen on reaction with electron-rich benzenes.[45] The question thus arises whether there is a duality of reaction mechanism or whether the electron-transfer products are the result of an addition - elimination sequence.[6a,44b] Using the method of time-resolved conductance and appropriately substituted alkenes it is possible to show that their oxidation proceeds by addition followed by heterolysis of the so-formed SO_4^- adduct.[2] For instance, in the case of cyclohexene in aqueous solution, the SO_4^- adduct formed in the addition step hydrolyzes at 20 °C with a rate constant k_{hs} = 3.0 x 10^4 s^{-1}, an activation enthalpy of 17 $kJmol^{-1}$ and an activation entropy of -103 $J(mol\ K)^{-1}$.[46] This negative value for a decomposition reaction (with an intrinsic entropy gain) can be explained by assuming that water molecules are immobilized in the ionic heterolysis transition state. The reaction is of the S_N1 type, a conclusion that is supported by a), the lack of effect of the strong nucleophile OH^- on the heterolysis rate and, b), the strong effect of methyl substitution at C_α (the electrofuge): in the case of 1-methylcyclohexene-1 the $SO_4^{\bullet -}$ adduct heterolyzes with $k_{hs} \geq 5 \times 10^6\ s^{-1}$.[46]

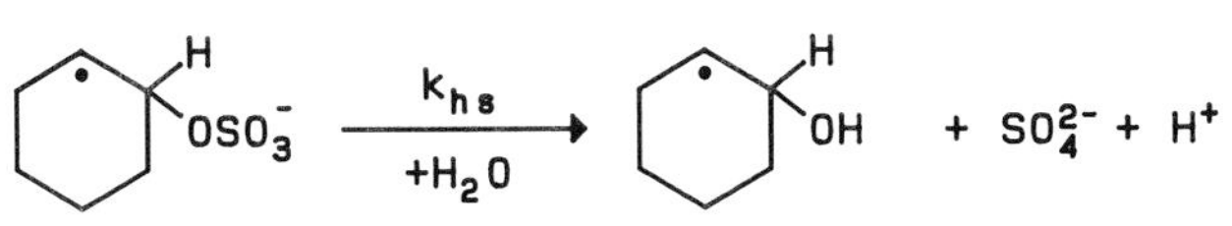

(26)

The addition/elimination type of reaction that $SO_4^{\bullet -}$ undergoes with alkenes may be compared with the pure addition behavior of $\dot{O}H$: The reduction potentials of the two radicals being similar means that their oxidizing power is comparable. **The difference between the two thus lies in the <u>leaving group abilities</u> of their redox partners, i.e. SO_4^{2-} and OH^-, respectively.** On the assumption that the Brönsted catalysis law is valid in this case and that Brönsted $\alpha = 0.5$, the difference between the pK_a values of the conjugate acids, HSO_4^- ($pK_a = 1.9$) and H_2O ($pK_a = 15.7$) translates into a difference in the rates of heterolysis of SO_4^- and OH^- adducts corresponding to a factor of $10^{6.9}$. Such a large factor, of course, makes $SO_4^{\bullet -}$ appear to be the better oxidant relative to $\dot{O}H$.

In this connection it is interesting to compare the leaving group abilities of SO_4^{2-} and H_2O. On the basis of the pK_a values ($pK_a(H_3O^+) = -1.7$) and again assuming that Brönsted $\alpha = 0.5$, H_2O is a better leaving group than SO_4^{2-} by the factor $10^{1.8}$. This explains why in strongly acidic solutions $\dot{O}H$ behaves as an oxidant[3d,6,44b,45] of a strength roughly comparable to that of $SO_4^{\bullet -}$.

Concerning $SO_4^{\bullet -}$ reactions with aromatics, adducts have so far not been found. For instance, if an SO_4^- adduct is formed with benzene, it hydrolyzes to give the hydroxycyclohexadienyl radical with a rate constant of $> 10^7$ s^{-1}.[43] Even if the strongly electron-withdrawing CN group is introduced into the benzene moiety to reduce its electrofugal leaving group ability (i.e. the benzonitrile/$SO_4^{\bullet -}$ system), the rate constant for heterolysis is still $> 5 \times 10^6$ s^{-1}.[47]

3. $Cl_2^{\bullet -}$ AS AN OXIDANT.

This radical has been shown[48] to react by oxidation or by addition, depending on the nature of the substrate. The reduction potential in aqueous solution, $E^2(Cl_2^{\bullet -} + e^- \rightarrow 2Cl^-)$ is equal to 2.1 V/NHE, i.e. less than that of $\dot{O}H$ or $SO_4^{\bullet -}$. Since the reduction potential of the related system, $E^2(Cl^\bullet + e^- \rightarrow Cl^-) = 2.4$ V/NHE is still less than those of $\dot{O}H$ and $SO_4^{\bullet -}$, it is thermodynamically possible to oxidize Cl^- with these radicals to give $Cl^\bullet$, and subsequently, by reaction of $Cl^\bullet$ with Cl^-, $Cl_2^{\bullet -}$ is obtained.[44a] It may be interesting at this point to recall that the oxidation by $\dot{O}H$ of Cl^- (and other halides and pseudohalides) proceeds by an addition/elimination sequence[12] with H^+ serving to convert the bad leaving group OH^- into the good one H_2O, as outlined for the general case in scheme 23.

The rate constant for reaction of $SO_4^{\bullet -}$ with Cl^- is 3.1×10^8 $M^{-1}s^{-1}$.[44a] By introducing an excess of Cl^- over that of an organic substrate in a system in which $SO_4^{\bullet -}$ radicals are produced it is thus possible to produce $Cl_2^{\bullet -}$ ($k(Cl^\bullet + Cl^-) = 2.1 \times 10^{10}$ $M^{-1}s^{-1}$)[49] and make it react with the organic.[48]

If an experiment of this type is performed with an aqueous solution saturated with isobutene and containing 20 mM Cl^-, a unimolecular rise of conductance of the solution occurs after production (< 1 μs) of $SO_4^{\bullet -}$ radicals. At 20 ^{o} the rate constant of this conductance change, which is independent of pH between 4 and 11, is 3.1 x 10^4 s^{-1}.[46] These results are explained by reactions 27-30, eq.'s 29 and 30 constituting the actual addition/elimination sequence:

$$SO_4^{\bullet -} + Cl^- \longrightarrow SO_4^{2-} + Cl^\bullet \quad (27)$$

$$Cl^\bullet + Cl^- \longrightarrow Cl_2^{\bullet -} \quad (28)$$

$$Cl_2^{\bullet -} + (CH_3)_2C{=}CH_2 \longrightarrow (CH_3)_2\dot{C}CH_2Cl + Cl^- \quad (29)$$

$$(CH_3)_2\dot{C}CH_2Cl + H_2O \xrightarrow{k_{hs}} (CH_3)_2\dot{C}CH_2OH + Cl^- + H^+ \quad (30)$$

The identification of reaction 30 as that causing the first-order conductance increase is supported by the fact that the same first-order conductance change is observed on producing $(CH_3)_2\dot{C}CH_2Cl$ by H-abstraction by $\dot{O}H$ from iso-butylchloride, eq. 31:

$$(CH_3)_2CHCH_2Cl + \dot{O}H \rightarrow (CH_3)_2\dot{C}CH_2Cl + H_2O \quad (31)$$

The oxidative solvolysis steps, eq. 31/30, have previously been demonstrated to occur, and the room temperature rate constant given (k_{hs} = 3.5 x 10^4 s^{-1})[7] is in good agreement with that measured on producing the β-chloroalkyl radical by the addition route, eq. 29.

A further support for the identification of the species responsible for the unimolecular conductance increase in terms of a chlorine-containing radical is the fact that in a blank experiment, i.e. one in which chloride is left out, a slow conductance increase is not observed and the overall conductance yield is only half of that in the presence of chloride. Since in isobutene-saturated aqueous solution the lifetime of $SO_4^{\bullet -}$ is only 90 ns due to its rapid reaction with the alkene (as determined by optical experiments at 450 nm),[46] the non-observability of a unimolecular conductance increase means that the rate constant for heterolysis of the $SO_4^{\bullet -}$ adduct to isobutene is $\leq 10^2$ s^{-1}; cf. eq. 32:

$$(CH_3)_2\dot{C}CH_2OSO_3^- + H_2O \rightarrow (CH_3)_2\dot{C}CH_2OH + H^+ + SO_4^{2-};\ k_{hs} \leq 10^2\ s^{-1} \quad (32)$$

It is thus seen that in their reactions with isobutene the weaker oxidant $Cl_2^{\bullet -}$ gives rise to product $((CH_3)_2\dot{C}CH_2OH)$ in a considerably faster reaction than does the stronger oxidant $SO_4^{\bullet -}$. **In this sense the activity of the oxidant is not determined by its oxidizing power but by the leaving group quality of the conjugate redox partner.**

Obviously, Cl^- is a much better leaving group (pK_a(HCl) = -7) than SO_4^{2-} (pK_a(HSO_4^-) = 1.9). That Cl^- is an excellent leaving group also in solvolyses of radicals is well documented.[7]

SUMMARY AND CONCLUSIONS

1. Examples have been given for one-electron redox reactions between organic molecules and reducing or oxidizing organic or inorganic radicals, reactions that proceed by addition followed by elimination whereby the process separating the components of the adducts is a heterolysis, in most cases one involving a carbon and a hetero atom. Addition takes place even in cases where electron transfer is strongly exothermic. The reason for this preference for addition is probably that the transition state for addition is lower in energy than that for electron transfer because addition profits from bond making whereas electron transfer requires entropically expensive bond and solvent reorganization.[4]

2. The efficiency of an oxidant or a reductant does not necessarily depend solely on its redox potential. It is a characteristic of any inner-sphere or "bonded"[50] electron transfer mechanism that the leaving group properties of the conjugate redox partner, i.e. the reduced oxidant or the oxidized reductant, respectively, are of great importance in determining the overall efficiency of the oxidant or the reductant. The leaving group properties are not necessarily related in a simple way to the reduction potentials. This is immediately clear if equations 33 and 34 are compared. Equation 33 defines the leaving group properties: If Y=H, the reaction defines the Brönsted acidity (of Y-H) or basicity (of X^-) (eq. 33a). If Y=$-\overset{|}{\underset{|}{C}}$, the electro or nucleofugacities and the electro or nucleophilicities are defined (eq. 33b). Obviously, a) and b) are related since they differ only by the electrofuge (H^+ vs $-\overset{|}{\underset{|}{C}}{}^+$). The reaction (=heterolysis[5]) consists in a shift of an electron pair between X and Y with more (S_N2) or less (S_N1) participation of a second reaction partner (other than the solvent). The common denominator is the propensity of X^+ to accept an electron **pair.**

$$\mathbf{Y - X \rightleftarrows Y^+ + :X^-} \qquad \mathbf{(33)}$$

a) Y = H
b) Y = $-\overset{|}{\underset{|}{C}}$

In contrast to this is the defining reaction for the reduction potential, eq. 34.

$$\mathbf{e^{\bullet -} + {}^{\bullet}X \rightleftarrows :X^-} \qquad \mathbf{(34)}$$

This is a reaction between two open-shell systems. The reaction involves **not the shift** of an electron pair between two centers **but the formation of an electron pair,** i.e. the production of a closed-shell system from two radical precursors. In view of these differences in the nature of reactions 33 and 34 a simple relation between the electron **pair** accepting ability of X^+ (eq. 33) and the **single** electron accepting power of $X^{\bullet}$ (eq. 34) is not to be expected. Very likely, the dependences on the structure of X of the reorganization free energies for eq. 33 and 34 are different. As pointed out in sections **A** and **B,** the leaving group properties can often be changed drastically by substitution or by protonation or deprotonation (as can the redox potentials).

REFERENCES AND NOTES

(1) For a similar discussion of inner-sphere electron transfer mechanisms see ref. 2.

(2) Steenken, S. in: Radiation Research, Proceedings 8th Internat. Congr. Radiation Research, Edinburgh 1987; Vol. 2, Fielden, E.M.; Fowler, J.F.; Hendry, J.H.; Scott, D. (eds), Taylor and Francis, London 1987.

(3) For reviews see, e.g., a) Neta, P. Adv. Phys. Org. Chem. 1976, **12,** 2. b) Henglein, A. Electroanal. Chem. 1976, **9,** 163. c) Swallow, A. J. Progr. React. Kinet. 1978, **9,** 195. d) Steenken, S. J. Chem. Soc., Faraday Trans. 1, 1987, **83,** 113.

(4) For recent reviews on electron transfer mechanism see, e.g., a) Todres, Z.V. Russ. Chem. Rev. 1978, **47,** 148. b) Albery, W. J. Ann. Rev. Phys. Chem. 1980, **31,** 227. c) Eberson, L. Adv. Phys. Org. Chem. 1982, **18,** 79. d) Pross, A. Acc. Chem. Res. 1985, **18,** 212. e) Shaik, S. S. Prog. Phys. Org. Chem. 1985, **15,** 197. f) Eberson, L.; Radner, F. Acc. Chem. Res. 1987, **20,** 53. g) Chanon, M. Acc. Chem. Res. 1987, **20,** 214.

(5) For a discussion of single electron "shifts" in two-electron transfer processes see ref.'s 4d and e.

(6) a) Norman, R.O.C. in: Essays in Free-Radical Chemistry, J. Chem. Soc., Special Publication No. 24, London, 1970, p. 117. b) Gilbert, B.C. Electron Spin Resonance 1973, **1,** 206. c) Steenken, S.; Davies, M.J.; Gilbert, B.C. J. Chem. Soc. Perkin Trans 2, 1986, 1003 and references therein; see also ref. 44b).

(7) Koltzenburg, G.; Behrens, G.; Schulte-Frohlinde, D. J. Am. Chem. Soc. 1982, **104,** 7311; ibid. 1983, **105,** 5168.

(8) Schulte-Frohlinde, D. in: "Radioprotectors and Anticarcinogens", Nygaard, O.F.; Simic, M.G. (eds), Academic, New York 1983, p. 53.

(9) Solvolysis reactions have in fact been interpreted in terms of electron transfer, cf. a) ref. 4b-e. b) Murdoch, J.R.; Magnoli, D.E. J. Am. Chem. Soc. 1982, **104,** 3792. c) Chanon, M. Bull. Soc. Chim. Fr. 1982, 216. d) Lewis, E.S. J. Phys. Chem. 1986, **90,** 3756.

(10) An excellent recent review of solvolysis reactions is to be found in Vogel, P. "Carbocation Chemistry", Studies in Organic Chemistry, **21,** Elsevier, Amsterdam 1985.
(11) The validity of reaction scheme 1 is not limited to heterolyses of covalent bonds to carbon. It covers also cases where both X and Y are hetero atoms (e.g. oxidation of halides and pseudohalides by $^{\bullet}OH$, cf. ref. 12) or where X is a hetero atom and Y is a metal (cf. ref. 13).
(12) Fornier de Violet, P. Rev. Chem. Intermediates 1981, **4,** 121.
(13) Asmus, K.-D.; Bonifacic, M.; Toffel, P.; O'Neill, P.; Schulte-Frohlinde, D.; Steenken, S. J. Chem. Soc. Faraday Trans. 1 1978, **74,** 1820.
(14) The species is oxidized since it contains one electron less than its precursor, the radical.
(15) For reviews see, e.g., Wardman, P.; Clarke, E.D. in: "New Chemo and Radiosensitizing Drugs", Breccia, A.; Fowler, J. F. (eds), Edizione Scientifiche <Lo Scarabeo>, 1985, p. 21. Wardman, P. in: "Radiation Chemistry: Principles and Applications", Farhataziz; Rodgers, M.A.J. (eds), Verlag Chemie, Weinheim 1987, p. 565.
(16) Minisci, F.; Citterio, A. Adv. Free Radical Chem. 1980, **6,** 65.
(17) Non-"nucleophilic" radicals, e.g. alkyl radicals not substituted by a hetero atom at C_{α}, do not seem to react by addition to the nitro group ($k \leq 10^7\ M^{-1}\ s^{-1}$). Whether they add to the benzene ring seems to be essentially unknown.
(18) Jagannadham, V.; Steenken, S. J. Am. Chem. Soc. in press.
(19) McMillan, M.; Norman, R.O.C. J. Chem. Soc. B 1968, 590.
(20) Adams, G.E.; Willson, R.L. J. Chem. Soc. Faraday Trans. 2 1973, **69,** 719.
(21) Jagannadham, V.; Steenken, S. J. Am. Chem. Soc. 1984, **106,** 6542.
(22) Steenken, S. unpublished material.
(23) Janzen, E.G.; Gerlock, J. L. J. Am. Chem. Soc. 1969, **91,** 3108; Sleight, R.B.; Sutcliffe, L.H. Trans. Faraday Soc. 1971, **67,** 2195; Wong, S.K.; Wan, J.K. Can. J. Chem. 1973, **51,** 753.
(24) Eibenberger, J.; Schulte-Frohlinde, D.; Steenken, S. J. Phys. Chem. 80, **84,** 704.
(25) see Lowry, T.H.; Richardson, K.S. "Mechanism and Theory in Organic Chemistry", 2nd edition, Harper and Row, New York 1981, p. 351 ff.
(26) Fujita, S.; Steenken, S. J. Am. Chem. Soc. 1981, **103,** 2540.
(27) Hazra, D.K.; Steenken, S. J. Am. Chem. Soc. 1983, **105,** 4380.
(28) Schuchmann, M.N.; Steenken, S.; Wroblewski, J.; von Sonntag, C. Int. J. Radiat. Biol. 1984, **46,** 225.
(29) Steenken, S.; Jagannadham, V. J. Am. Chem. Soc. 1985, **107,** 6818.
(30) Jagannadham, V.; Steenken, S. J. Phys. Chem. 1987, in press.
(31) Steenken, S.; Renganathan, R. manuscript in preparation.
(32) Adams, G.E.; Willson, R.L. Trans. Faraday Soc. 1969, **65,** 2981; Willson, R.L. Int. J. Radiat. Biol. 1970, **17,** 349.

(33) For a collection of redox potentials in aqueous solution see Steenken, S. in Landolt-Börnstein 1985, **13e,** 147.
(34) Bothe, E.; Behrens, G.; Schulte-Frohlinde, D. Z. Naturforsch. 1977, **32b,** 886.
(35) Rabani, J.; Klug-Roth, D.; Henglein, A. J. Phys. Chem. 1974, **78,** 2089.
(36) Bothe, E.; Schulte-Frohlinde, D., Z. Naturforsch. 1980, **35b,** 1035.
(37) Bothe, E.; Schuchmann, M.N.; Schulte-Frohlinde, D.; von Sonntag, C. Photochem. Photobiol. 1978, **28,** 639.
(38) Asmus. K.-D.; Wigger, A.; Henglein, A. Ber. Bunsenges. Phys. Chem. 1966, **70,** 862.
(39) Bielski, B.H.J.; Cabelli, D.E.; Arudi, R.L.; Ross, A. J. Phys. Chem. Ref. Data 1985, **14,** 1041.
(40) Al-Sheikley, M.I.; Hissung, A.; Schuchmann, H.-P.; Schuchmann, M.N.; von Sonntag, C.; Garner, A.; Scholes, G. J. Chem. Soc. Perkin Trans. 2 1984, 601.
(41) For a review see ref. 3d.
(42) Schwarz, H.A.; Dodson, R.W. J. Phys. Chem. 1984, **88,** 3643; Kläning, U.K.; Sehested, K.; Holcman, J. J. Phys. Chem. 1985, **89,** 760.
(43) Neta, P.; Madhavan, V.; Zemel, H.; Fessenden, R.W. J. Am. Chem. Soc. 1977, **99,** 163.
(44) a) Chawla, O.P.; Fessenden, R.W. J. Phys. Chem. 1975, **79,** 2693. b) Davies, M.J.; Gilbert, B.C. J. Chem. Soc. Perkin Trans. 2 1984, 1809.
(45) O'Neill, P.; Steenken, S.; Schulte-Frohlinde, D. J. Phys. Chem. 1975, **79,** 2773. Walling, C. Acc. Chem. Res. 1975, **8,** 125. O'Neill, P.; Steenken, S.; Schulte-Frohlinde, D. J. Phys. Chem. 1977, **81,** 31. Sehested, K.; Holcman, J.; Hart, E.J. J. Phys. Chem. 1977, **81,** 1363. Sehested, K.; Holcman, J. Nukleonika 1979, **24,** 941.
(46) Steenken, S.; Koltzenburg, G. manuscript in preparation.
(47) This number is based on the observation that in the reaction with $SO_4^{\bullet -}$ k(observed) for formation of optical density at 340 nm (where the $\dot{O}H$ adduct of benzonitrile absorbs) is proportional to [benzonitrile] up to the saturation limit (~ 20 mM, k(observed) = $5 \times 10^6\ s^{-1}$), as shown by 248 nm laser experiments.
(48) Hasegawa, K.; Neta, P. J. Phys. Chem. 1978, **82,** 854.
(49) Jayson, G.G.; Parsons, B.J.; Swallow, A.J. J. Chem. Soc. Faraday Trans. 1 1973, **9,** 1597.
(50) Littler, J.S. in: Essays in Free Radical Chemistry, J. Chem. Soc., Special Publication No. 24, London 1970, p. 383.

FREE RADICALS: CHEMISTRY, PATHOLOGY AND MEDICINE
edited by Catherine Rice-Evans & Thomas Dormandy, © 1988 Richelieu Press, London

Transformation of Lipid Hydroperoxides

H. W. Gardner
Northern Regional Research Center, Peoria, Illinois 61604, U.S.A.

Hydroperoxides are readily formed from unsaturated lipids by either autoxidation or catalytic oxidation with the enzyme, lipoxygenase. Because stereochemically pure lipid hydroperoxides can be obtained using lipoxygenase, these enzyme products are preferred for studying mechanistic aspects of lipid hydroperoxide reactions. As a result of the conjugated hydroperoxydiene structure, which often includes additional methylene interrupted olefinic groups, lipid hydroperoxides undergo an exceptionally diverse number of reactions. Rearrangements catalyzed either homolytically or heterolytically are particularly noteworthy. The more commonly observed reactions are described. Although monomeric products of lipid hydroperoxides are now well known, dimeric and polymeric products have received relatively little attention. Many of the purely chemical transformations of lipid hydroperoxides have similar counterparts in the plant and animal kingdoms via enzyme catalysis. These comparisons are described briefly with emphasis on plant enzymes.

Formation of hydroperoxides.

Due to cyclic regeneration of R• radicals through H-abstraction by peroxy radicals, the familiar process of autoxidation results in an ever accelerating oxidation of polyunsaturated lipids, provided that the process is not arrested by radical termination. The hydroperoxidation of a simple diunsaturated fatty acid, such as linoleic acid, affords a mixture of eight different isomers; that is, all permutations of two positional isomers (9- and 13-hydroperoxides), two geometric isomers of the double bonds (Z,E and E,E) and two stereoisomers (R and S hydroperoxides). Autoxidation mixtures are made even more complex by decomposition of hydroperoxides into peroxidic and non-peroxidic oxygenated fatty acids, dimers, oligomers and volatile scission products.

In order to decipher the complexities of lipid hydroperoxide transformation, many laboratories, including our own, have utilized isomerically pure hydroperoxides obtained from the action of lipoxygenases for mechanistic studies. Regardless of the carbon oxygenated, most, but not all, lipoxygenase isoenzymes form the S hydroperoxide enantiomer. For example, at high pH soybean lipoxygenase oxidized linoleic acid to 94% (13S)-(9Z,11E)-13-hydroperoxy-9,11-octadecadienoic acid (1). Lipoxygenase isoenzymes from maize (2) and tomato (3) afforded high percentages of the (9*S*)-(10*E*,12*Z*)-9-hydroperoxy-10,12-octadecadienoic acid at neutral and/or acidic pH's. These pure stereoisomers, as well as similar 9*S* and 13*S*-hydroperoxides of linolenic acid, have been utilized to investigate the homolytic and heterolytic transformations described below in this review of selected investigations.

Peroxy radical reactions.

Because of the relatively high dissociation energy of the O-H bond, peroxy radicals are not likely to generate spontaneously. Peroxy radicals arise by two important pathways, (a) H-abstraction from hydroperoxides, and (b) combination of an alkyl radical with the diradical, O_2:

(a) $ROOH \xrightarrow{X\bullet \;\; XH} ROO\bullet$

$X\bullet$ = A generic radical

(b) $R\bullet + O_2 \longrightarrow ROO\bullet$

Unlike oxy radicals, peroxy radicals can be transformed back to hydroperoxides by a reversible process:

$$ROO\bullet \underset{XH \;\; X\bullet}{\overset{XH \;\; X\bullet}{\rightleftharpoons}} ROOH$$

In addition, oxidized metal ions are known to form peroxy radicals from hydroperoxides:

$$ROOH \xrightarrow{M^{n+1} \;\; M^{n}} ROO\bullet + H^+$$

Some metal ions, like ferric, are relatively inefficient oxidizers of the hydroperoxide group. Metal redox reactions can be affected by numerous parameters, such as the type of ligands associated with the metal, the pH, the type of transition metal ions, and solvent.

Once formed, lipid peroxy radicals react in diverse competitive pathways, and the predominance of one or more of these pathways can often be controlled by the conditions of reaction. For example, aerobic conditions promote secondary hydroperoxidation through O_2 scavenging, but exclusion of O_2 causes other radical combination reactions to prevail. And, dilution by solvent or inclusion of antioxidants usually greatly inhibits polymerization reactions.

β-Scission or elimination is a well-known reaction of radical chemistry. With this retrospective view, it is not surprising that the peroxy radical should eliminate O_2. When Chan and coworkers (4) reported that pure methyl (9Z,11E)-13-hydroperoxy-9,11-octadecadienoate, as well as the corresponding pure 9-hydroperoxide, was rearranged into mixtures of (Z,E) and (E,E) methyl 9-hydroperoxy-10,12-octadecadienoates and methyl 13-hydroperoxy-9,11-octadecadienoates, their finding greatly increased our knowledge of the peroxidation process (Fig. 1). Porter and coworkers (5, 6) have further defined the kinetics of

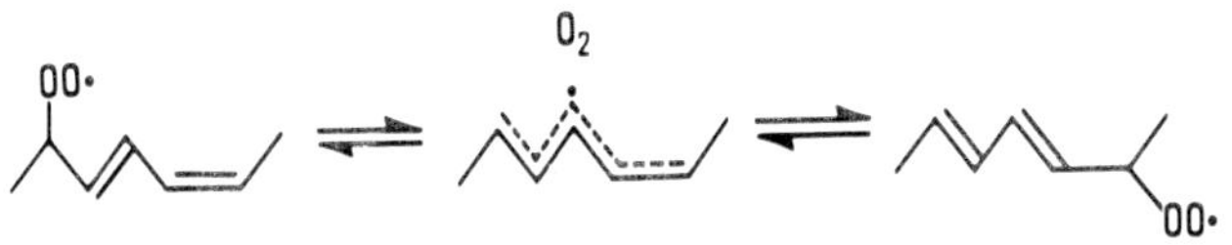

Fig. 1. β-Scission of the peroxy radical derived from linoleic acid hydroperoxide followed by reversible O_2 uptake. Structures are abbreviated to show only unsaturation.

both O^2 β-scission and the attendant process of diene isomerization. The latter workers (5, 6) recognized that Z,E to E,E isomerization during autoxidation is under kinetic control, and the hydroperoxides themselves do not significantly isomerize after they are formed; that is, the initially formed peroxy radical is the species which is most subject to β-scission and double bond isomerization. Since the important competitive reaction of peroxy radical rearrangement (see below) is not possible for conjugated diene hydroperoxides of linoleic acid, these hydroperoxides seem to have a greater propensity for the β-scission reaction. Recently, the β-scission reaction was utilized to measure the formation of peroxy radicals from the 13-hydroperoxide of linoleic acid exposed to manganese ions (7). That is, analyses of the appearance of the 9-hydroperoxide isomer with time constituted a cumulative assessment of peroxy radicals,a measurement not easily obtained by other methods.

Rearrangement is an important reaction of peroxy radicals derived from the "inner" hydroperoxides of polyunsaturated fatty acids. "Inner" hydroperoxides are those fatty acids possessing a double bond β,γ to the hydroperoxide group, such as the 12- and 13-hydroperoxide of α-linolenic acid. As shown in Fig. 2,

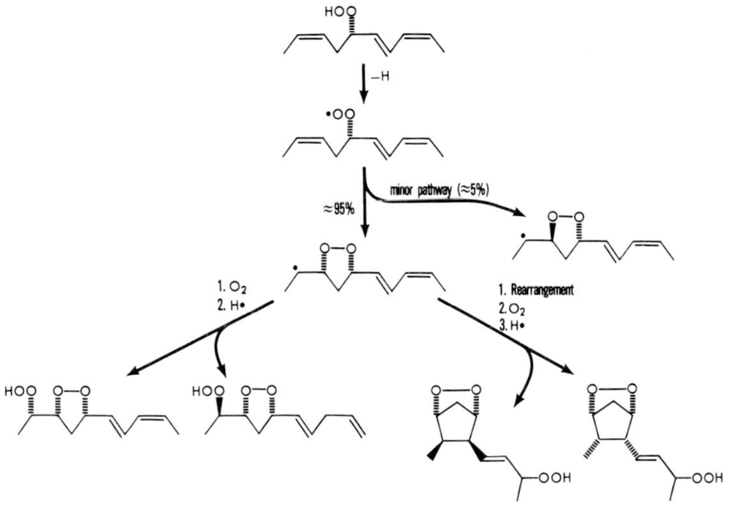

Fig. 2. Peroxy radical rearrangement of the 13S-hydroperoxide of linolenic acid (8, 9). Reprinted from Gardner (40).

peroxy radicals derived from these fatty acids can add intramolecularly to the β,γ olefin leading to hydroperoxy-cyclic peroxides and prostaglandin-like bicyclo endoperoxides. By using pure isomers from lipoxygenase oxygenation, the

13S-hydroperoxide of α-linolenic acid and 9S-hydroperoxide of γ-linolenic acid, O'Connor et al. (8, 9) defined the stereochemical course of the rearrangement. As seen in Fig. 2, there was a bias toward formation of the cis configuration of the cyclic peroxide, as well as the alkyl chains of the bicyclo endoperoxides. Since the natural prostaglandins have their alkyl chains trans, it is interesting that the trans isomers in the autocatalytic mixture were found in only minor amounts (9).

It is likely that reactions of peroxy radicals figure prominently in low temperature polymerization reactions of polyunsaturated lipids, particularly since the dimers produced during polymerization contain a high content of intermolecular peroxide bonds (10, 11). On the other hand, high temperatures ($>60^{\circ}C$) afford ether and carbon-carbon cross-links indicating a mechanistic change probably caused by heat induced peroxide bond rupture. At this time it is largely conjecture how peroxide cross-links form during low temperature polymerization. Two potential candidates are peroxy radical intermolecular addition to a double bond and peroxy radical combination. The former possibility would permit polymerization, but the latter would afford only dimers. A dimer isolated from methyl linoleate hydroperoxides incubated at $30^{\circ}C$ and exposed to O_2 appeared to have been formed by the addition mechanism (12):

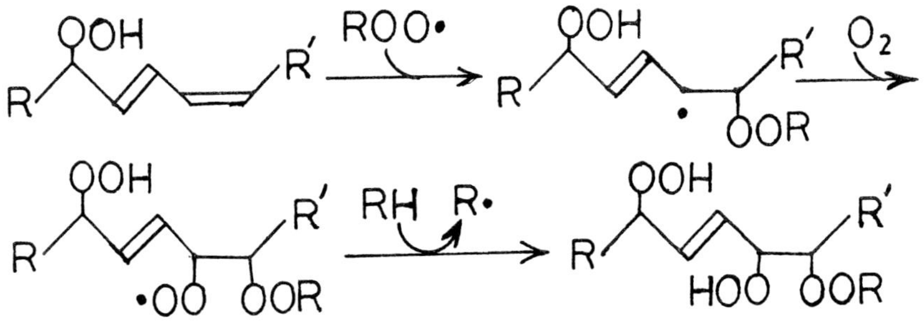

A peroxy radical combination mechanism was indicated in dimer formation from the 13-hydroperoxide of methyl linoleate incubated under anaerobic conditions with a peroxy radical initiator (13, 14):

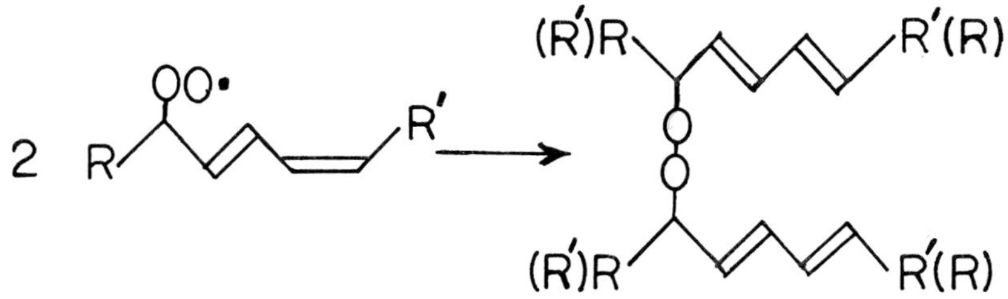

Schieberle et al. (13) proposed a peroxy radical combination followed by O_2 elimination and oxy radical recombination:

$$2\ ROO\cdot \longrightarrow \underset{\text{}}{\overset{\text{Solvent cage}}{\boxed{RO\cdot\ O_2\cdot OR}}} \longrightarrow ROOR + O_2$$

Although the actual mechanism has not been demonstrated at this time, a more attractive proposal in my view is β-scission followed by combination:

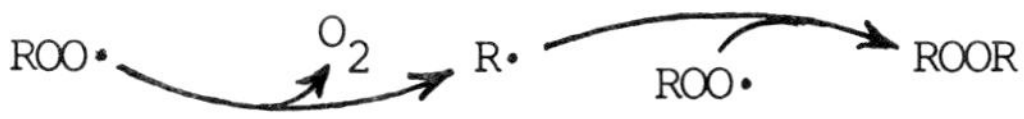

The two possibilities could be sorted out using a 1/1 mixture of $^{18}O_2$- and $^{16}O_2$-labeled linoleate hydroperoxides and analyzing for isotope scrambling in the products. The former mechanism predicts O_2 scrambling, but the latter does not. Less than 1.5% isotope mixing found in $^{16}O_2/^{18}O_2$ mixtures used to autoxidize linoleic acid by the same workers (15) would appear to argue against their oxy radical combination proposal. Neff and Frankel (11) also recently reported peroxide-linked dimers derived from linolenate hydroperoxides which were similar in structure to the linoleate dimers described above (13, 14). In any case, more definitive research is needed to understand the mechanisms at work during polymerization of polyunsaturated lipids.

Simple epoxy fatty acids usually comprise a small portion of lipid peroxidation mixtures. The epoxide group reflects the position and geometry of the double bond contained in the parent fatty acid. It is generally thought that peroxy radicals directly add to the double bond homolytically. In addition, heterolytic mechanisms have been proposed for catalytic epoxidations of olefins involving hydroperoxides and certain transition metal ions, such as Mo, Cr and V (16).

Oxy radical reactions.

Homolysis of the hydroperoxide group into the alkyl oxy radical requires energy input, such as UV irradiation (17), heat, or metal ion reduction. Ferrous ion, which often is a potent free radical catalyst in both biological systems and oil autoxidation, is particularly efficient in reducing hydroperoxide into the oxy radical. Our laboratory (18) has utilized the redox ability of trace amounts of ferrous/ferric ion fueled by cysteine and linoleic acid hydroperoxide to afford an efficient method of generating lipid oxy radicals. Oxy radical reactions, described below, sometimes follow pathways that could not have been readily predicted by conventional theory.

Certain oxy radicals derived from organic hydroperoxides, such as τ-butylhydroperoxide, are effective in the process of H-abstraction:

However, H-abstraction appears to be of little importance to oxy radicals derived from lipid hydroperoxides. Instead, the double bonds vicinal to the hydroperoxide promote oxy radical rearrangement into an epoxyallylic radical (Fig. 3). The epoxyallylic radical is subsequently susceptible to O_2 scavenging leading eventually to epoxyhydroxyoctadecenoic and epoxyoxooctadecenoic

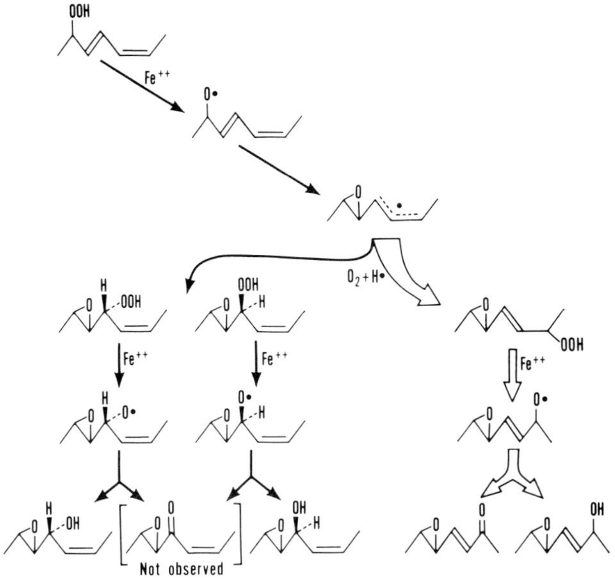

Fig. 3. Oxy radical rearrangement of 13S-hydroperoxide of linoleic acid (structures are abbreviated). Main pathway is combination of O_2 with carbon-9 as indicated by size of arrows. The fatty acid enclosed in brackets is observed only with 9E unsaturation instead of 9Z as shown. Reprinted from Gardner and Kleiman (19).

acids (19). When the optically active 13S-hydroperoxide oflinoleic acid was used, the products were also optically active showing that it was an authentic rearrangement that did not break the C-O bond at carbon-13 (19). As shown in Fig. 3, O_2 scavenging of the epoxyallylic radical promotes the formation of secondary hydroperoxides (19, 20, 21), but with the exclusion of

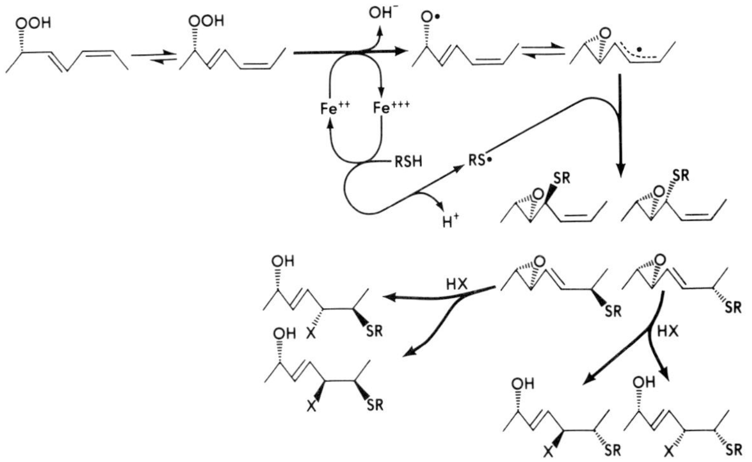

Fig. 4. Proposed mechanism of formation of N-acetylcysteine (RSH) combination products with the 13S-hydroperoxide of linoleate catalyzed by a ferrous/ferric redox cycle (structures are abbreviated). Reprinted from Gardner et al. (22).

O_2 the epoxyallylic radical has an opportunity to combine with another available radical. Thus, Gardner et al. (22)

demonstrated that excluding O_2 from a N-acetylcysteine-hydroperoxide ferrous/ferric redox system resulted in combination of the thiyl radical with the epoxyallylic radical (Fig. 4). Similarly, replacement of N-acetylcysteine with the α-tocopherol model compound, 2,2,5,7,8-pentamethyl-6-hydroxychroman, resulted in combination of the oxy radical of the α-tocopherol model with the epoxyallylic radical to afford mainly isomers (23, 24) identical to both the thiyl radical combination products and O_2 scavenging products. And, anaerobic UV irradiation of the 13S-hydroperoxide of methyl linoleate in methanol also resulted in a similar combination of the epoxyallylic radical with $\cdot CH_2OH$ (17); $\cdot CH_2OH$ presumably arose by H-abstraction from methanol.

β-Scission or elimination by the oxy radical accounts for the ubiquitous appearance in autoxidation of volatiles from chain scission. Fig. 5 shows the expected scission of an oxy radical derived from the 13-hydroperoxide of methyl linoleate. In theory, only the carbon-13, -14 bond should cleave, and indeed, pentane and methyl 13-oxotridecadienoate isomers are readily observed after thermal treatment of the hydroperoxide (Fig. 6). Cleavage between carbon-13 and olefinic carbon-12 would be energetically unfavorable causing some workers to question the vinyl radical shown in Fig. 5. The appearance of hexanal as a major volatile (Fig. 6) demonstrated that the carbon-12, -13 bond does cleave; however, a more plausible explanation is a heterolytic cleavage originating from a process like the acid-catalyzed reaction described below. Hypothetically, thermal conditions would accelerate both heterolytic and homolytic reactions.

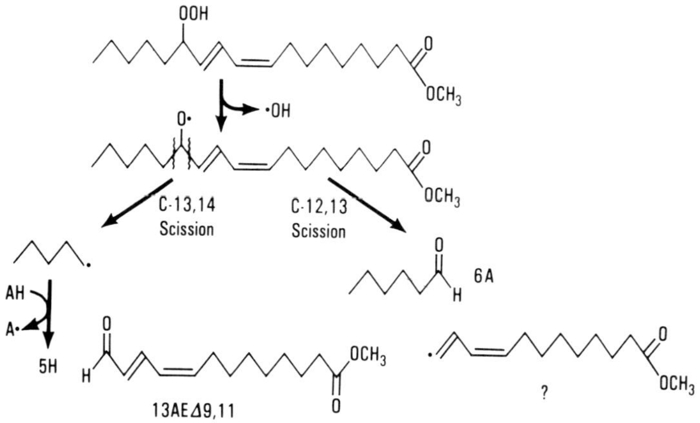

Fig. 5. β-Scission of the oxy radical derived from the 13-hydroperoxide of methyl linoleate. 5H = pentane; the question mark denotes a vinyl radical, which should be energetically disfavored. Reprinted from Gardner and Plattner (34).

Thermal treatment also causes formation of ether-linked dimers and polymers (11), possibly from oxy radical reactions. Information about the structure of these compounds is sparse.

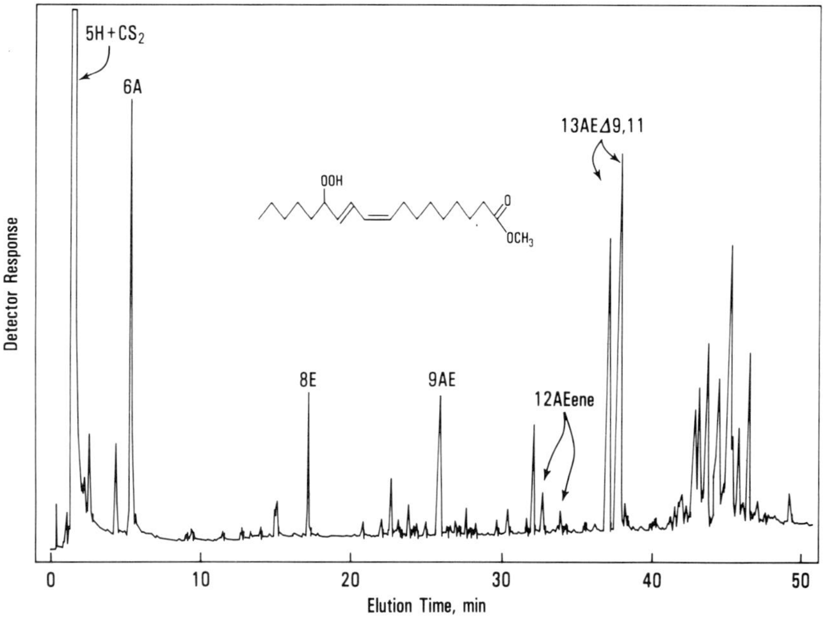

Fig. 6. Capillary GC of heat-induced volatiles from scission of the 13-hydroperoxide of methyl linoleate. Abbreviations: 5H, pentane; 6A, hexanal; 8E, methyl octanoate; 9AE, methyl 9-oxononanoate; 12AEene, methyl 12-oxododecenoate isomers; 13AEΔ9,11, methyl 13-oxo-9, 11-tridecadienoate isomers. Reprinted from Gardner and Plattner (34).

Heterolytic reactions.

With so much attention focussed on homolytic reactions of hydroperoxides, there often is a general disregard of the potential for heterolytic reactions. Current work indicates that several plant enzymes may catalyze reactions of lipid hydroperoxides through heterolytic mechanisms.

A heterolytic mechanism was probable in the formation of 12,

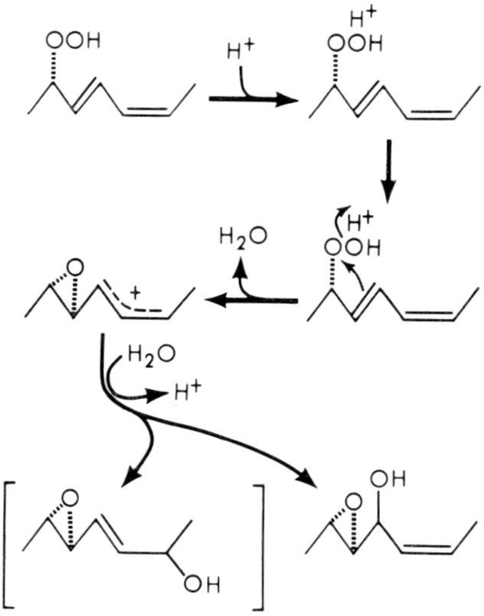

Fig. 7. Heterolytic formation of epoxyhydroxy fatty acids from 13S-hydroperoxide of linoleic acid in aqueous acid solutions (structures are abbreviated). The bracket indicates that this epoxide is very labile to solvolysis. Reprinted from Gardner (41).

13-epoxy-11-hydroxy-9-octadecenoic acid from the 13-hydroperoxide of linoleic acid by heating in ethanol-H_2O under anaerobic conditions (25). The 11-hydroxyl group, rather than originating from a radical source, was derived from H_2O solvent. Recently, our laboratory (26) found that acids in protic solvents will catalyze a similar transformation (Fig. 7). It is possible that

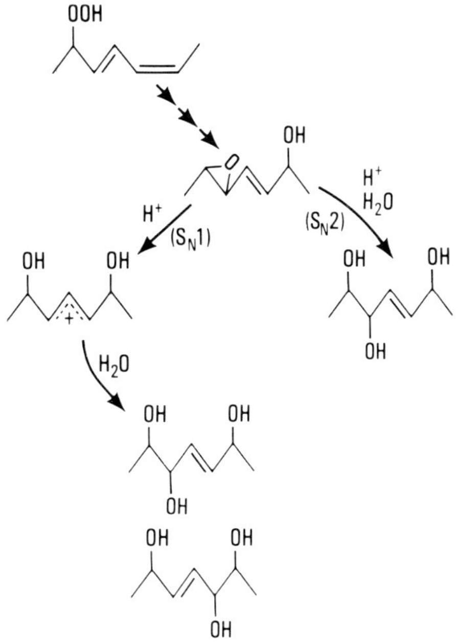

Fig. 8. Formation of trihydroxyene fatty acid isomers from the 13-hydroperoxide of linoleic acid through an epoxyhydroxyene intermediate (structures are abbreviated). Reprinted from Gardner (27).

the heat-induced reaction (25) was catalyzed by the heat-accelerated action of incipient acids in solution.

Because a double bond is allylic to the epoxide, the epoxyhydroxyoctadecenoic acid, bracketed in Fig. 7, is particularly sensitive to solvolysis even in mild acids. As seen by Fig. 8, solvolysis results in a mixture of isomeric trihydroxyoctadecenoic acids (26). Similar epoxyhydroxyoctadecenoic and trihydroxyoctadecenoic acids are formed from linoleic acid hydroperoxides by enzymes from cereal grains (27) and potatoes (28), but is it not known whether these enzymic products originate by either a homolytic or heterolytic rearrangement.

The epoxyallylic cation proposed as an intermediate in the acid-catalyzed rearrangement (Fig. 7) has been suggested as an intermediate in the formation of ketols by hydroperoxide isomerase (29). As proposed, the epoxyallylic cation shifts a hydride from the epoxide to adjacent carbons affording the α- and γ-ketols. According to recent work (30, 31), the hydride shift does not occur. Hamberg (31) found that the immediate precursor of the ketols is an unstable allene epoxide, 12,13-epoxy-9, 11-octadecadienoic acid, which accounts for the absence of a hydride shift:

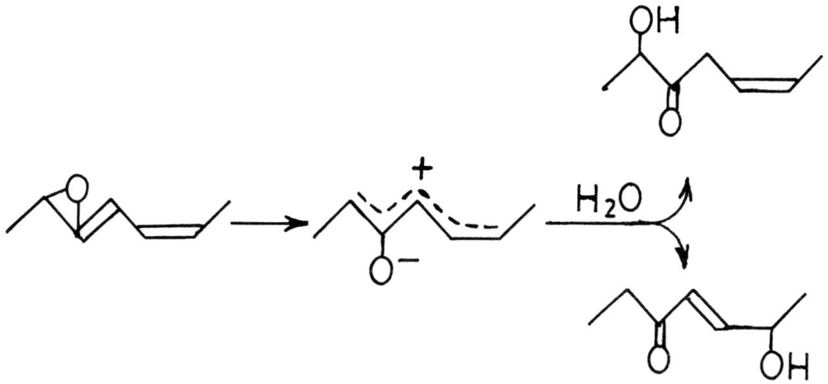

Nevertheless, the epoxyallylic cation can not be discounted as a viable mechanistic precursor to the allene epoxide:

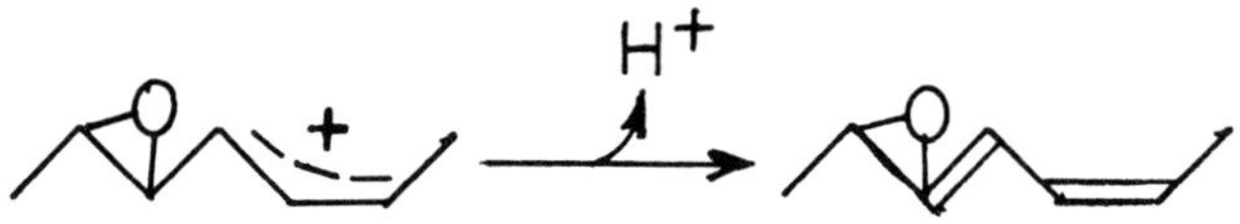

It is also of interest that hydroperoxide cyclase has mechanistic similarities to hydroperoxide isomerase (32, 33); it is possible that the two enzymes are identical. It has not yet been directly demonstrated that the allene epoxide is the immediate precursor to the cyclic fatty acid, 12-oxophytodienoic acid.

It has been known for over a half century that acids cause rearrangement and carbon-carbon cleavage of organic hydroperoxides, such as the acid cleavage of cumene hydroperoxide to acetone and phenol. In theory then, the 13-hydroperoxide of linoleic acid should cleave to hexanal and (9Z)-12-oxo-9-dodecenoic acid (Fig. 9), but in acid the double bond of the latter compound would isomerize from 9Z to 10E. Exposure to protic acid in protic solvent caused only a minor amount of carbon-carbon cleavage, but instead, resulted mainly in the rearrangement shown in Fig. 7. A Lewis acid, BF_3, in ether (34) was shown to be more efficient in cleaving methyl linoleate hydroperoxides (Fig. 10). The products obtained by BF_3-ether treatment are reminiscent of those obtained by the action of the plant enzyme, hydroperoxide lyase. As a result of $^{18}O_2$-labeling studies, Hatanaka et al. (35) concluded that hydroperoxide lyase does appear to utilize the heterolytic cleavage mechanism. Formation of vinyl ether fatty acids from hydroperoxides by a potato enzyme may also be explained by this heterolytic process (Fig. 9). Recent work by Crombie et al. (36) with the vinyl ether-forming enzyme gave credence to this possibility.

A more detailed accounting of the subject matter contained herein can be found in other reviews (37-39).

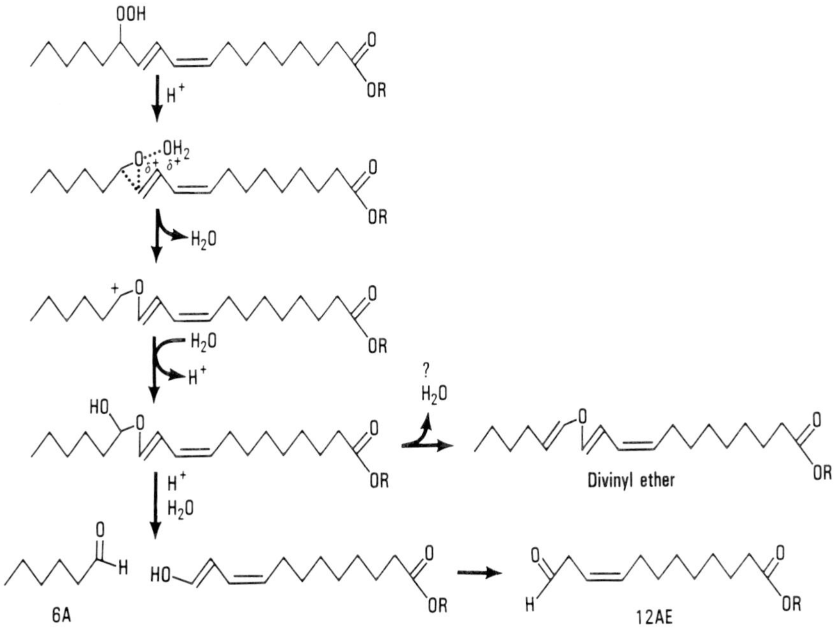

Fig. 9. Proposed mechanism of heterolytic cleavage of the 13-hydroperoxide of linoleate and a postulated route to divinyl ethers. Although a proton is shown as the catalyst, the Lewis acid, BF_3, in aprotic solvent was more effective. Reprinted from Gardner (41).

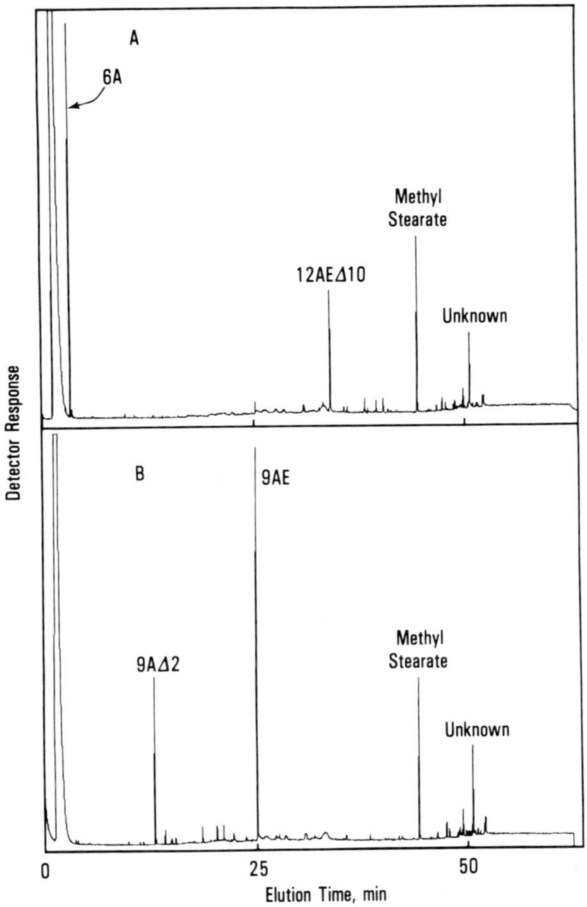

Fig. 10. Capillary GC-MS of products from BF_3-ether treated 13-hydroperoxide of methyl linoleate (A) and 9-hydroperoxide of methyl linoleate (B) (methyl stearate, an internal standard). Abbreviations: 6A, hexanal; 12AEΔ10, methyl (10E)-12-oxo-10-dodecenoate; 9AΔ2, 2-nonenal; 9AE, methyl 9-oxononanoate. Reprinted from Gardner and Plattner (34).

References

1. Hamberg, M. Anal. Biochem. 43 (1971) 515.

2. Gardner, H. W., and Weisleder, D. Lipids 5 (1970) 678.

3. Matthew, J. A., Chan, H. W.-S., and Galliard, T. Lipids 12 (1977) 324.

4. Chan, H. W.-S., Levett, G., and Matthew, J. A. J. Chem. Soc. Chem. Commun. (1978) 756.

5. Porter, N. A., Weber, B. A., Weenen, H., and Khan, J. A. J. Am. Chem. Soc. 102 (1980) 5597.

6. Porter, N. A., and Wujek, D. G. J. Am. Chem. Soc. 106 (1984) 2626.

7. Gardner, H. W., and Newton, J. W. Phytochemistry 26 (1987) 621.

8. O'Connor, D. E., Mihelich, E. D., and Coleman, M. C. J. Am. Chem. Soc. 103 (1981) 223.

9. O'Conner, D. E., Mihelich, E. D., and Coleman, M. C. J. Am. Chem. Soc. 106 (1984) 3577.

10. Miyashita, K., Fujimoto, K., and Kaneda, T. Agric. Biol. Chem. 46 (1982) 751.

11. Neff, W. E., and Frankel, E. N. Am. Chem. Soc. National Meeting, Denver, CO (1987) Abstract No. 82.

12. Miyashita, K., Hara, N., Fujimoto, K., and Kaneda, T. Lipids 20 (1985) 578.

13. Schieberle, P., Tsoukalas, B., and Grosch, W. Z. Lebensm. Unters. Forsch. 168 (1979) 448.

14. Schieberle, P., and Grosch, W. Z. Lebensm. Unters. Forsch. 173 (1981) 192.

15. Schieberle, P., Grosch, W., Kexel, H., and Schmidt, H.-L. Biochem. Biophys. Acta 666 (1981) 322.

16. Chong, A. O., and Sharpless, K. B. J. Org. Chem. 42 (1977) 1587.

17. Schieberle, P., Trebert, Y., Firl, J., and Grosch. W. Chem. Phys. Lipids 37 (1985) 99.

18. Gardner, H. W., Kleiman, R., and Weisleder, D. Lipids 9 (1974) 696.

19. Gardner, H. W., and Kleiman, R. Biochem. Biophys. Acta 665 (1981) 113.

20. Gardner, H. W., and Jursinic, P. A. Biochem. Biophys. Acta 665 (1981) 100.

21. Schieberle, P., Trebert, Y., Firl, J., and Grosch, W. Chem Phys. Lipids 41 (1986) 101.

22. Gardner, H. W., Plattner, R. D., and Weisleder, D. Biochem. Biophys. Acta 834 (1985) 65.

23. Gardner, H. W., unpublished; see Gardner, H. W. Xenobiotics in Foods and Feeds, Am. Chem. Soc. (1983) p.78.

24. Kaneko, T., and Matsuo, M. Chem Pharm. Bull. 33 (1985) 1899.

25. Hamberg, M., and Gotthammar, B. Lipids 8 (1973) 737.

26. Gardner, H. W., Nelson, E. C., Tjarks, L. W., and England, R. E. Chem. Phys. Lipids 35 (1984) 87.

27. Gardner, H. W. Advances in Cereal Science and Technology, Am. Assoc. Cereal Chem. Vol. 9 (in press)

28. Galliard, T., Phillips, D. R., and Matthew, J. A. Biochem. Biophys. Acta 409 (1975) 157.

29. Gardner, H. W. Lipids 14 (1979) 208.

30. Crombie, L., and Morgan, D. O. J. Chem. Soc. Chem. Commun. (1987) 503.

31. Hamberg, M. J. Am. Oil Chem. Soc. 64 (1987) 652; Abstract No. 141A.

32. Vick, B. A., Feng P., and Zimmerman, D. C. Lipids 15 (1980) 468.

33. Corey, E. J., d'Alarcao, M., Matsuda, S. P. T., and Lansbury, P. T. Jr. J. Am. Chem. Soc. 109 (1987) 289.

34. Gardner, H. W., and Plattner, R. D. Lipids 19 (1984) 294.

35. Hatanaka, A., Kajiwara, T., Sekiya, J., and Toyota, H. Z. Naturforsch. 41c (1986) 359.

36. Crombie, L., Morgan, D. O., and Smith, E. H. J. Chem. Soc. Chem. Commun. (1987) 502.

37. Chan, H. W.-S. Autoxidation of Unsaturated Lipids, Academic Press (1987).

38. Simic, M. G., and Karel, M. Autoxidation in Food and Biological Systems, Plenum Publishing (1980).

39. Min, D. B., and Smouse, T. H. Flavor Chemistry of Fats and Oils, Am. Oil Chem Soc. (1985).

40. Gardner, H. W. Xenobiotics in Foods and Feeds, Am. Chem. Soc. (1983) 63-84.

41. Gardner, H. W. Chemical Changes in Food During Processing, AVI Publishing Co. (1985) 177-203.

ANTIOXIDANT ACTION IN CHEMICAL SYSTEMS, BIOLOGICAL PROCESSES AND PATHOLOGICAL STATES

FREE RADICALS: CHEMISTRY, PATHOLOGY AND MEDICINE
edited by Catherine Rice-Evans & Thomas Dormandy, © 1988 Richelieu Press, London

Antioxidants: Can Man Improve on Nature ?

Gerald Scott

Aston University, Birmingham, B4 7ET.

Abstract.

The discovery and investigation of the antioxidant activity of the tocopherols in unsaturated fatty esters preceded the recognition that the same antioxidants are important in vivo by some twenty years. In the meantime chemists have elucidated the mechanisms by which antioxidants act and it is now possible to suggest ways in which this knowledge may be applied to modify and improve the performance of analogous antioxidants in vivo. The peroxidases and catalases represent an exception to this generalisation. Here man has yet to learn how to control the single electron reductions of hydroperoxides without the generation of damaging free radicals.

Antioxidants - Discovery and Significance.

The word "antioxidant" is a comprehensive term used to describe all types of chemical agent which inhibit the free radical chain reaction of oxygen with organic substrates [1]. The phenomenon of rubber "ageing" and the effects of very small amounts of added chemicals in inhibiting this process was observed at an early stage in the development of the rubber industry, and was recognised to be associated with the absorption of oxygen by Ostwald in the early 1900's [2]. However, the understanding of the mechanism by which antioxidants inhibited oxidation remained obscure until the free radical chain reaction of autoxidation was proposed by Bäckström for benzaldehyde in the mid 1920's [3]. A key feature of this process was the formation of an intermediate peroxide [4,5] since this provided an explanation for the autoaccelerating oxidation rate curve which had first been observed by Genthe in 1906 in a study of the autoxidation of linseed oil [6]. The rate of autoxidation was found to be directly related to the concentration of this species, and once it was present in the autoxidising system, no other initiator was required.

Initiation

$$ROOH \xrightarrow[(a)]{\Delta, h\nu} RO^{\bullet} + {}^{\bullet}OH \qquad (1)$$

$$RH + \begin{Bmatrix} \cdot OH \\ \cdot OR \end{Bmatrix} \xrightarrow[(b)]{} R\cdot + \begin{Bmatrix} H_2O \\ ROH \end{Bmatrix}$$

Propagation

$$R^{\bullet} + O_2 \longrightarrow ROO^{\bullet} \qquad (2)$$

$$ROO^{\bullet} + RH \longrightarrow ROOH + R^{\bullet} \qquad (3)$$

The recognition of the importance of peroxides led to one of the earliest theories of antioxidant action. Moureau and Dufraisse [7] postulated the formation of a primary peroxide which could be intercepted by an "antioxygen", reaction 4.

$$RH + O_2 \longrightarrow [RHO_2] \xrightarrow{A} RHO + AO \qquad (4)$$

where A is an antioxidant.

The Moureau and Dufraisse mechanism was proposed before the kinetic chain theory of autoxidation had been developed and they were not in a position to define the reactive peroxidic intermediates. The significance of their proposal was therefore subsequently observed by the mechanistically more intelligible proposal due to Lowry and his co-workers [8] and later elaborated by Bolland [9] that the alkylperoxyl radical and not a stable peroxide is reduced by hydrogen transfer from phenols and amines (AH), see Table 1.

TABLE 1. SOME CHAIN-BREAKING DONOR (CB-D) ANTIOXIDANTS

CHEMICAL STRUCTURE		CODE	TYPICAL SUBSTRATES
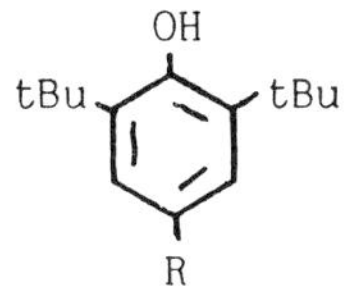	(a) R = CH_3	BHT	FOODSTUFFS, PLASTICS, RAW RUBBERS AND RUBBER-MODIFIED PLASTICS
	(b) R = $C_{12}H_{25}$	1076	
	(c) R = OCH_3	DBHA	
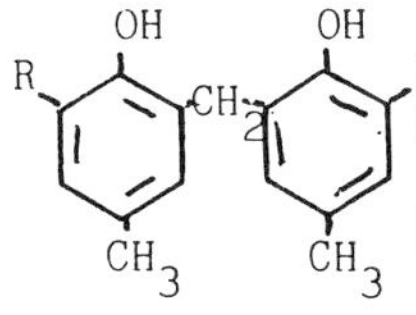	(a) R = tBu	2246	VULCANISED RUBBERS AND RUBBER-MODIFIED PLASTICS
	(b) R = methyl cyclohexyl	WSP	
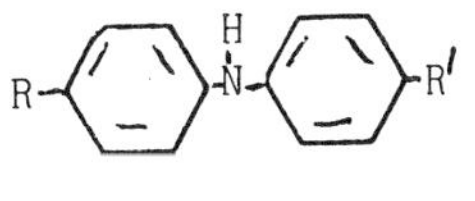	(a) R = H, R′ = NHiPr	IPPD	TYRE RUBBERS AND LUBRICATING OILS
	(b) R = R′ = isooctyl	OD	
	(c) R = H, R′ = NHPh	DPPD	
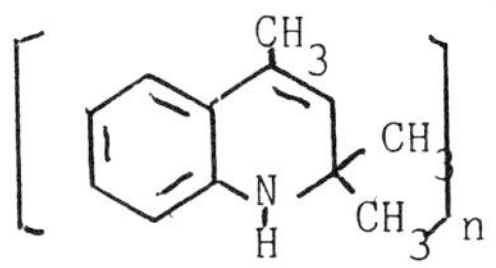		TMDHQ	RUBBERS FOR SEALS, HOSES AND GASKETS

This process (reaction 5) is in competition with reaction 3.

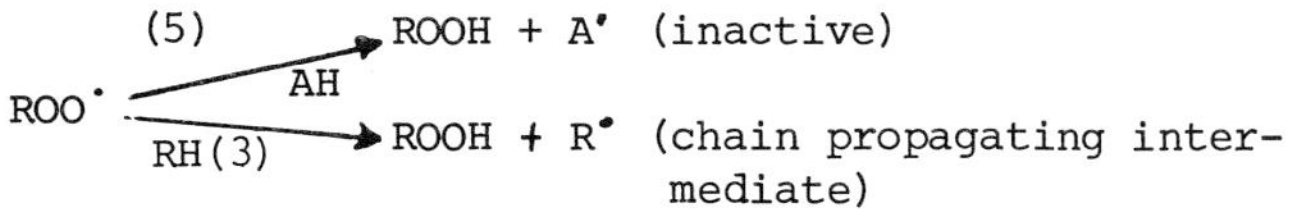

Lowry showed [8] that there was a relationship between the antioxidant efficiency of phenols in hydrocarbon substrates and their critical oxidation reduction potentials. More recently it has been shown that some antioxidants do act by destroying hydroperoxides by a non-radical mechanism [10,11] and today both the peroxidolytic (PD) and chain-breaking hydrogen (electron) donor (CB-D) mechanisms of antioxidant action are equally important. In addition it has recently been shown [12,13] that under certain conditions (notably in low oxygen concentrations), and when the substrate is highly oxidisable, oxidising agents can also act as antioxidants by trapping alkyl radicals or oxidising them to olefins. This process (reactions (6), (7)) is also in competition with the other chain propagating step (reaction 2) in the chain reaction.

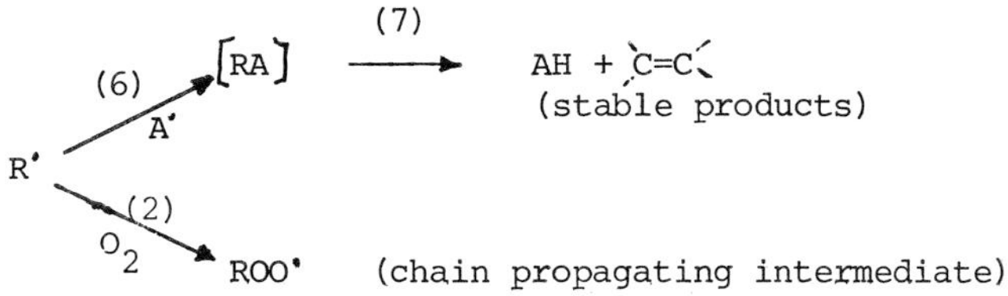

Since reactions (2) and (3) are cyclical in which R• and ROO• are continuously regenerated, it follows that reactions (5) and

(6) can and do occur reciprocally, leading to a catalytic antioxidant mechanism.

SCHEME 1. MECHANISM OF CATALYTIC CHAIN-BREAKING ANTIOXIDANTS

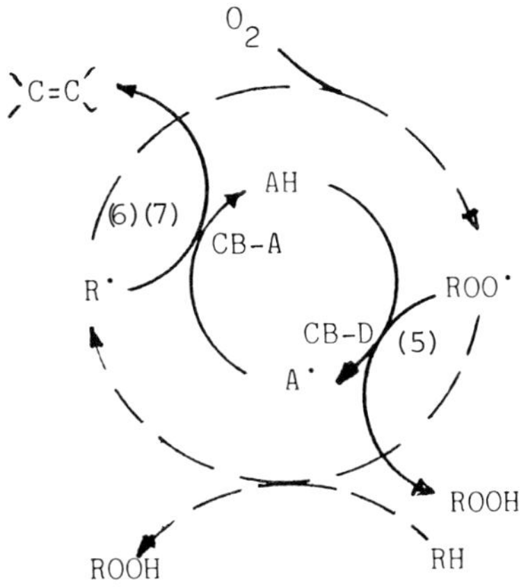

A· = PHENOXYL,NITROXYL,QUINONE, Cu^{2+}, I_2,ETC

This has been shown to be a general mechanism of antioxidant action in polymeric substrates when the rate of diffusion of oxygen into the system is limited and the oxygen concentration is therefore low [13]. The presently recognised antioxidant mechanisms are summarised in scheme 2 in the light of the presently accepted mechanism of oxidation by molecular oxygen [14]. Examples of CB-A antioxidants are listed in table 2.

SCHEME 2. ANTIOXIDANT MECHANISMS

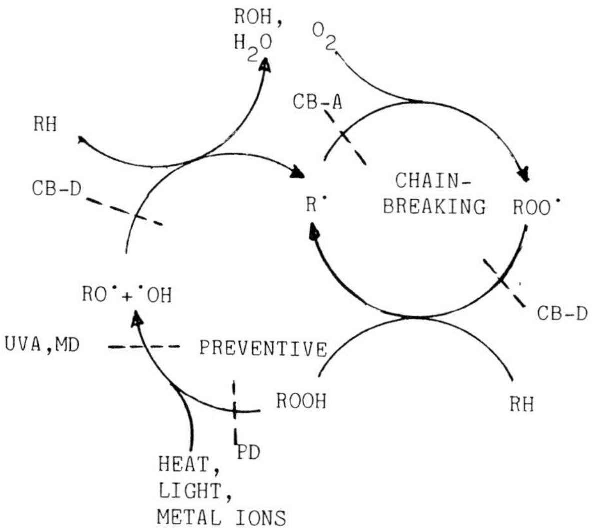

<u>Oxidation and stabilisation of biological substrates.</u>

The oxidation of biological substrates has been studied by chemists for almost as long as technological materials. The pure linoleate esters were frequently used as chemical models of more complex oils and fats and indeed the understanding of the chemistry of cross-linking of drying oils and the deterioration of food stuffs (rancidity) resulted from early observations of the effects of oxygen on the 1,4 unsaturated fatty acids and their esters [15,16].

TABLE 2. SOME CHAIN-BREAKING ACCEPTOR (CB-A) ANTIOXIDANTS

	ANTIOXIDANT STRUCTURE	CONDITIONS OF FORMATION AND USE
V	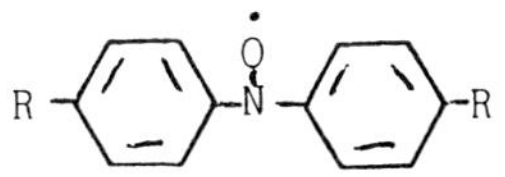	FORMED IN SITU FROM CORRESPONDING AMINE IN RUBBERS
VI	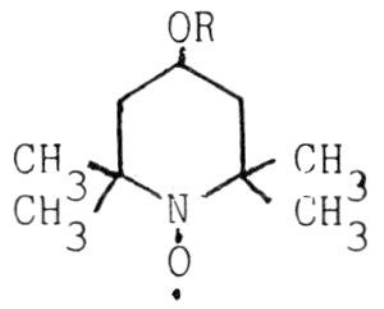	FORMED IN SITU FROM PARENT AMINE IN PLASTICS DURING PROCESSING AND ON EXPOSURE TO LIGHT CATALYTIC ANTIOXIDANT WITH THE DERIVED HYDROXYLAMINE
VII	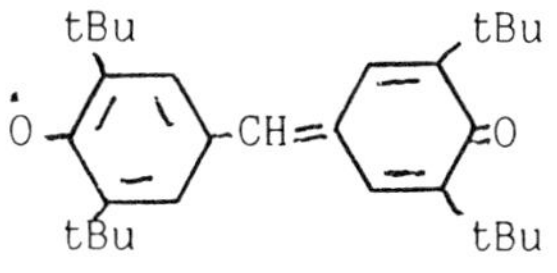 (G)	MELT STABILISER (MECHANO-ANTIOXIDANT) FOR POLYOLEFINS. CATALYTIC ANTIOXIDANT WITH THE DERIVED PHENOL
VIII	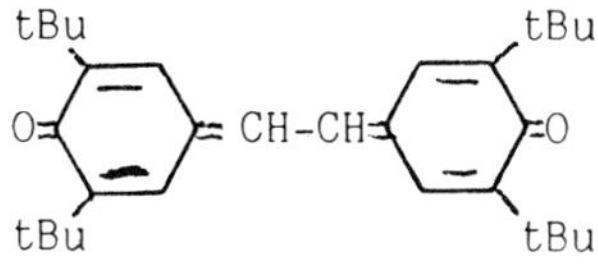 SQ	DITTO
	I_2 CUPRIC STEARATE	MELT STABILISERS FOR POLYOLEFINS CATALYTIC ANTIOXIDANTS WITH THE DERIVED REDUCED IONS

Not surprisingly, then, many of the early studies of antioxidant action were also carried out in the unsaturated fatty esters in vitro. A seminal discovery by Mattill and his co-workers in the 1930's [17,18], was the recognition that two different kinds of naturally occuring antioxidant, the tocopherols (VIII) (found in vegetable oils) and ascorbic acid (IX) (found in animal fats) when used together were much more effective than either alone.

TOCOPHEROLS	R_1	R_2	R_3
α	CH_3	CH_3	CH_3
β	CH_3	H	CH_3
γ	H	CH_3	CH_3
δ	H	H	CH_3

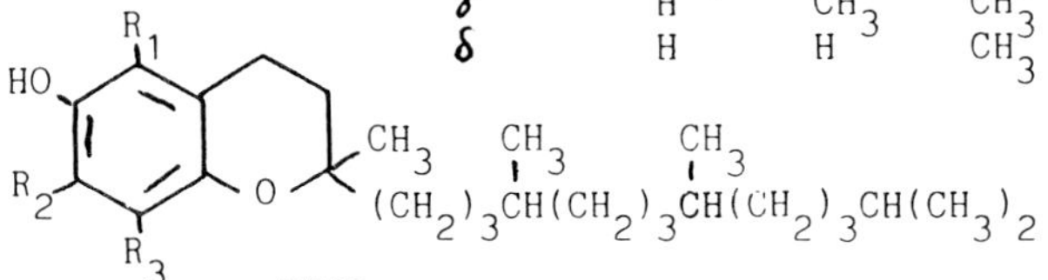

VIII

ASCORBIC ACID

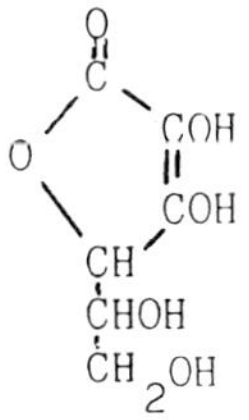

IX

This phenomenon which was named synergism was later suggested by Golumbic [19] to be due to the regeneration of the primary chain-breaking antioxidant from its oxidation product by ascorbic acid. Synergism has since been shown to be a practically important phenomenon in a variety of technological media and indeed, some very successful industrial polymers would probably never have been developed commercially had it not been for the discovery of multicomponent synergistic antioxidants.

In spite of the recognition by Mattill in 1936 [17] that biologically produced antioxidants were highly effective oxidation inhibitors in vitro for biologically important substrates, it was not until some twenty years later that Dann suggested [20] that prevention of oxidation could be a major function of the tocopherols in vivo. His views sparked off a controversy which was to continue among biochemists for the next twenty years. In 1965, Horwitt stated [21], "Nearly all biochemists would like to see α-tocopherol directly implicated in specific enzyme reaction that cannot be related to its antioxidant activity; to date, no such enzymic activity has been unequivocally demonstrated. In all cases where such activity has been claimed, some other chemically unrelated antioxidant could serve as a substitute for α-tocopherol". Since that time, the antioxidant role of α-tocopherol has widened and deepened and there are now probably few biochemists who would not admit the fundamental importance of Vitamin E as a biological antioxidant in the lipids in vivo.

Of equal importance, however, has been the recognition of synergism between α-tocopherol and other important constituents of the human diet, and in particular ascorbic acids, selenium and glutathione. Tappel [22] was able to link the antioxidant chemistry of these compounds with the synergism exhibited by these antioxidants in vitro. In particular he drew attention to the analogy between the function of selenocystine diselenoxide in vivo and organic disulphides in vitro. Tappel [23] also related Golumbic and Mattill's regenerative mechanism for synergism between α-tocopherol and ascorbic acid to their behaviour in living organisms. The mechanism of this process has recently been studied quantitatively by Ingold and his co-workers [24] in phospholipid bilayers confirming the mechanism originally proposed by Mattill and co-workers in unsaturated fatty esters almost fifty years ago. An important conclusion from this work is that in vivo ascorbic acid probably has to be looked at as a CB-D antioxidant in its own right. However, as has been observed in vitro, the efficiency of this process is likely to be very dependent on the concentration of other species (particularly oxidising and reducing agents) present in the system [25].

Mechanisms of Antioxidant Action

Chain-breaking mechanisms.

Over the years since the pioneering studies of Lowry [8] and Bolland [9] and their co-workers provided a sound mechanistic basis for the chain-breaking hydrogen donor (CB-D) antioxidants, the molecular requirements for high activity have been clearly defined [25]. In general terms, antioxidant activity in arylamines and phenols is increased by electron releasing groups in the aromatic ring, by delocalization of the unpaired electron in the derived radical $A^{\cdot}$ (reaction 5) and by steric protection of the derived phenoxyl. These requirements are met almost ideally by the hindered phenols I (see table 1) in which the

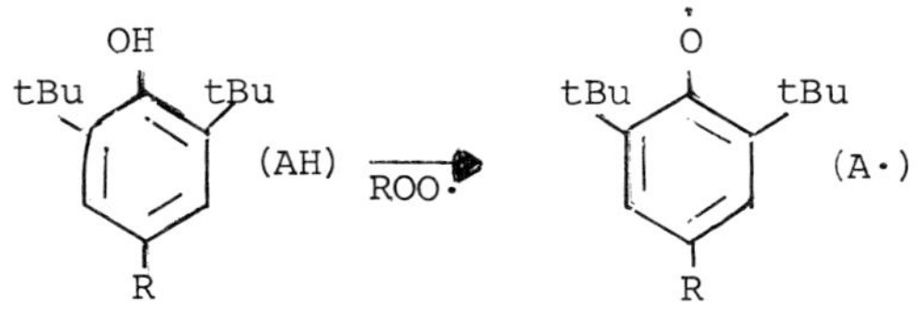

(5)

group R is electron releasing (e.g. alkyl, alkoxy and alkylamines). Further major improvements in this type of antioxidant over the structure shown in table 1 are probably not possible since reduction of critical oxidation reduction potential of the reducing system below a critical level, results in direct attack by dioxygen, thus limiting the effectiveness of the antioxidant. Similarly, more effective delocalization of the electron in the free radical, A$^{\bullet}$, although beneficial from the point of view of the unreactivity of the aryloxyl produced, also limits the oxidation stability of the parent phenol.

In spite of these general limitations, hindered phenols which give rise to very stable radicals (e.g. galvinoxyl, VII (table 2)) have been found to be highly effective antioxidants in polymers when the oxygen pressure is relatively low [13,26,27]. The evidence suggests that under those conditions, the parent phenol can be readily regenerated by reaction with a substrate alkyl radical [13] or at the expense of a more oxidisable species such as thiol or sulphuric acid present in the system [28].

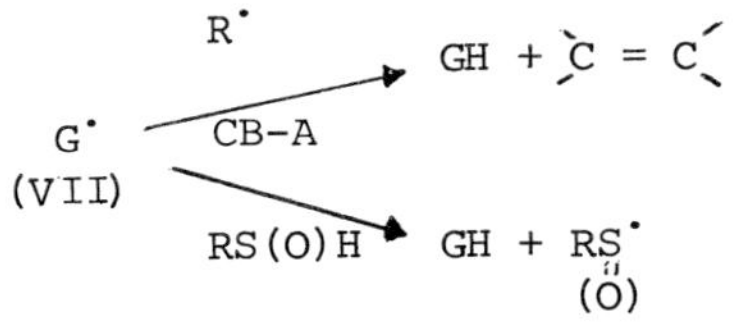

(9)

In the light of the above general "rules" for CB antioxidant activity, none of the tocopherols (VIII) would be expected to be particularly favoured structures. Although they all have an electron releasing alkoxy group in the 4-position, none has an effective hindering group in an ortho position. An acyclic analogue of α-tocopherol, 2,3,5,6-tetramethyl-4 methoxy phenol (XI) has a similar level of activity to the hindered analogue

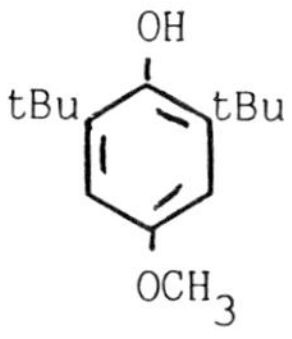

II(c)

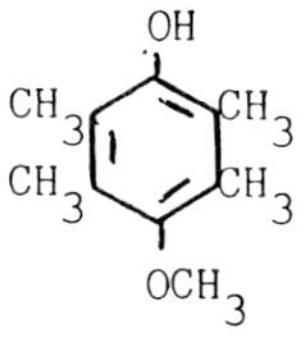

X

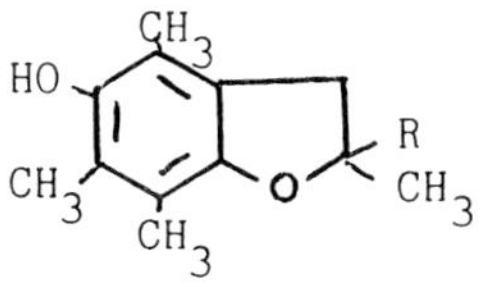

XI

II(C) [24] and both are much less effective than α-tocopherol as measured by the rate of reaction 5 and Ingold et al [24] have attributed this to stereoelectronic stabilisation of the derived phenoxyl radical in tocopherol by distortion of the C-O bond by the chroman ring. They confirmed this by synthesising a 5 membered furan anologue of α-tocopherol (Xl) in which the C-O bond is more distorted. This proved to be rather more effective than α-tocopherol. Stereoelectronic stabilisation of α-tocopheroxyl is thus an additional factor to be taken into account when considering its activity. However, whereas galvonoxyl is relatively stable in the presence of air, α-tocopheroxyl (XII) is not and readily transforms to the acyclic quinone (XIII) by

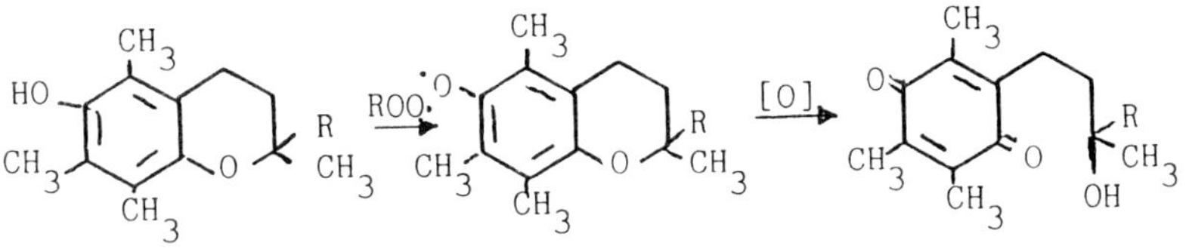

VIII XII XIII

breaking of the chroman ring. As has already been seen, reducing agents like ascorbic acid have the ability to regenerate the phenol from the aryloxyl and possibly even the semiquinone ion radical from the benzoquinone XIII. No evidence has been reported for the reduction of tocopheroxyl by stabilised alkenyl radicals formed as intermediates during the oxidation of 1,4-dienes but this could well occur in the absence of other reducing agents since the allylic C-H bond is relatively weak reaction 11

$$\dot{A} + -\dot{C}HCH{=}CHCH{=}CHCH_2- \longrightarrow AH + -CH{=}CHCH{=}CHCH{=}CH- \qquad (11)$$

A second class of antioxidant, which like galvinoxyl can participate in a cyclical CB mechanism, is the stable nitroxyls and their associated hydroxylamines [13,30-34]. Both the diaryl (IX) and the dialkyl (X) nitroxyls have been shown to be effective CB-A antioxidants when oxygen access is limited by diffusion. The diaryl nitroxyls (XIV) have the advantage that they can

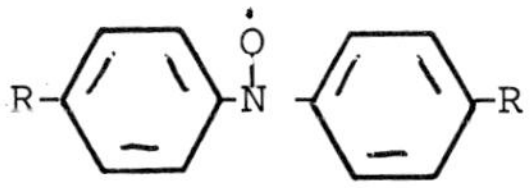

XIV

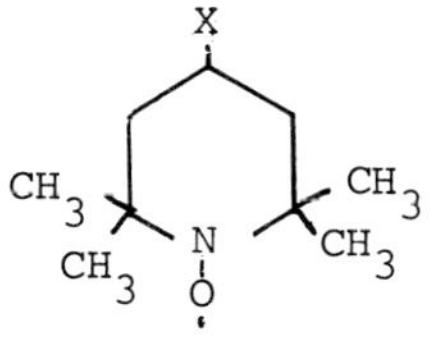

XV

trap both alkyl and alkylperoxyl radicals and XIV can participate in the regenerative cycle shown in scheme 1.

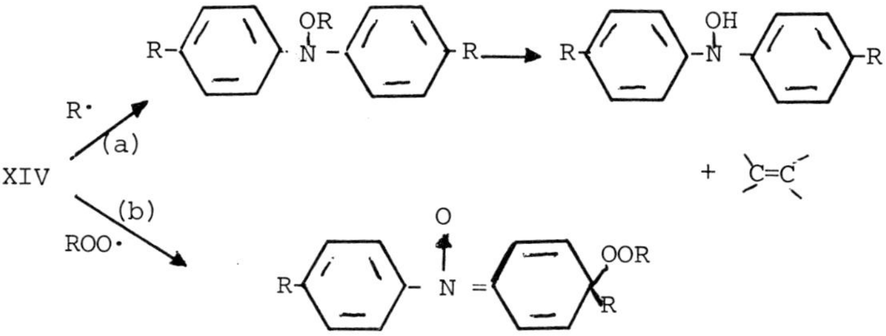

(12)

However, reaction 12(b) is a wasteful process since it leads to the irreversible destruction of the nitroxyl. The "hindered" piperidines, XV, are more stable and hence more effective in the catalytic cycle. The low oxygen pressures found in cells [29] and the relatively high stabilities of intermediate alkenyl radicals formed during autoxidation of the lipids encourage us to think that some of the catalytic systems identified in vitro might be adapted to biological oxidations if correctly targeted.

Preventive Mechanisms.

Remarkable progress has been made in recent years in the understanding of how the initiation process in autoxidation may be retarded or even prevented altogether. Combinations of chain-breaking and preventive antioxidants are now widely used in polymer technology to control oxidative degradation. The three most important classes of preventive antioxidant are

(1) the peroxide decomposers (PD)

(2) the metal deactivators (MD)

(3) the UV absorbers (UVA)

Since peroxides are the most important initiators in most autoxidising systems, the peroxidolytic antioxidants are potentially the most powerful group of antioxidants [10,11]. They range from the simple stoichiometric peroxide decomposers such as the phosphite esters (XVl), through the dialkyl sulphides (XVII) which act primarily by a catalytic peroxidolytic mechanism, to

$(RO)_3P$ $(ROCOCH_2CH_2)_2S$ $\left[R_2NC{\lesssim}^{S}_{S}\right]_n M$

XVI XVII XVIII

the sulphur metal complexes such as the dithiocarbamates (XVIII) which act by a combination of mechanisms, including chain-breaking donor (CB-D), catalytic peroxidolysis (PD-C) and in some cases UV screening (11). Most of the evidence now points to a predominantly ionic mechanism of peroxide decomposition by sulphur-containing antioxidants in vitro and in this respect they differ from glutathione peroxidase which appears to reduce hydroperoxides to alcohols in two coupled single electron steps in which the intermediate radicals are rapidly reduced further. Simple thiols related to glutathione (XIX) also destroy hydroperoxides catalytically [35] but the evidence suggests that they do so through ionic and not radical intermediates. The effective antioxidants are sulphur acids formed by oxidation of the thiols by hydroperoxides. In this respect they resemble the thiodipropionates (XVII) dithiocarbamates (XVIII) rather than

$$HOOC\underset{}{\overset{NH_2}{\overset{|}{C}H}}\,CH_2CH_2CONH\overset{CH_2SH}{\overset{|}{C}H}CONHCH_2COOH\cdot$$

XIX , GSH

glutathione. It appears that in glutathione peroxidase, as

for the biological antioxidant catalase, nature has developed a highly sophisticated system for deactivating otherwise damaging radicals within the environment of a complex molecule. It seems then that in the case of the peroxidolytic antioxidants man has much to learn from nature.

Similarly, chemical science has not so far succeeded in improving on the metal chelating ability of the porphoryns. Indeed, the difficulty in completely deactivating transition metal ions in autoxidising systems in vitro has led by empirical selection to the development of complexing agents with additional antioxidant function. A typical example of such an autosynergist is XX [36] in which the bis hydrazide group acts as a metal deactivating agent and the hindered phenol removes alkylperoxyl radicals by the CB-D mechanism.

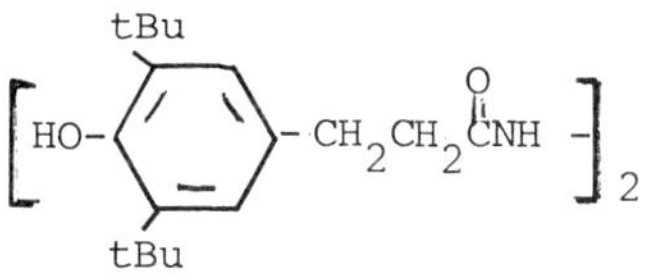

XX

Other metal deactivator autosynergists are known. For example, the zinc dialkyl dithiocarbamates (XVIII, M=Zn) readily undergo metathesis with transition metal ions in rubber which are thereby deactivated toward hydroperoxides [37]. As the dithiocarbamates are also peroxidolytic antioxidants, very complete inhibition of oxidation is achieved provided the dithiocarbamate is in excess. However, the sulphur ligand is destroyed during the course of its antioxidant action and this leads to very rapid oxidation at the end of the induction period [38]. Although synergism between two different molecules is common in biological substrates [39], autosynergism in which two antioxidant functions are contained in the same molecule does not seem to have been reported in vivo.

The simplest example of this in model systems is the CB-D antioxidant XXI which also contains a peroxidolytic (PD-C) function.

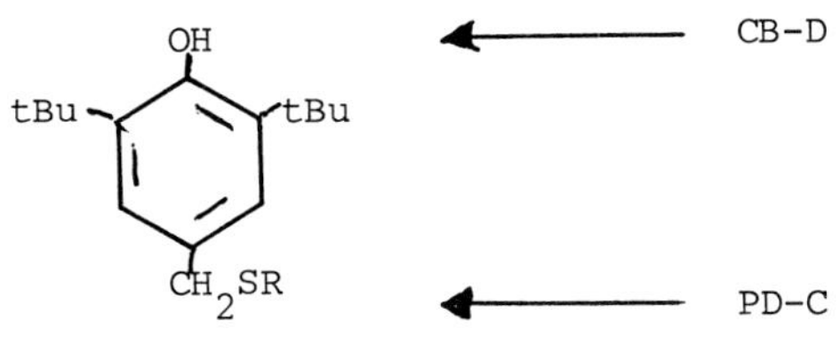

XXI

This molecule is approximately ten times more effective as an antioxidant on a molar basis than the analogous hindered phenol, BHT (IId)) [40], which is a widely used commercial antioxidant. It has been shown that XXI is a very powerful catalyst for peroxide decomposition; 1 mole of XV($R=C_{12}H_{25}$) can destroy 3×10^5 moles of hydroperoxide [41]. Autosynergists are also very effective in photo-stabilisation. The photo-antioxidants XXII and XXIII both have two antioxidant functions.

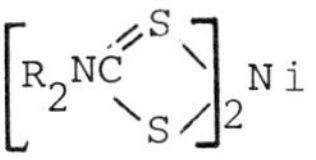

XXII

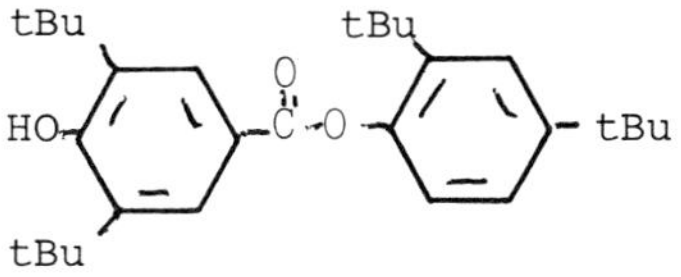

XXIII

Both have the ability to absorb UV light without themselves being destroyed. They thus protect the substrate from the damaging effects of UV irradiation. However, they also behave as antioxidants; XXII by decomposing hydroperoxides and XXIII by trapping alkylperoxyl radicals. The biological pigment melanin which screens the skin from damaging UV light, is formed by oxidation of tyrosine and many of the intermediates, notably 5,6-dihydroxyindole (XXIV) are powerful CB-D antioxidants. It seems likely that these are involved in synergistic action with the pigment itself.

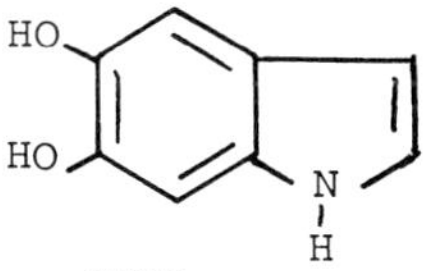

XXIV

It would seem that there is scope for the development of anti-oxidants in vivo containing more than one antioxidant function. Autosynergistic metal deactivators seem to be an obvious initial choice in the treatment of diseases associated with transition metal ions since it could lead to a reduction in the concentration of complexing agent necessary to achieve a desired effect. UV screening agents containing antioxidant functions would seem to be another viable objective.

Targeting antioxidants in vivo.

A major problem associated with the use of antioxidants is that to reach the site of oxidation they need to have a preference for that environment rather than for other substrates in the body. Nature has achieved this in the case of α-tocopherol by arranging that as a result of the length of the lipophilic phytyl tail, the antioxidant can become an integral part of the phospholipid bilayer [24]. However, it may well have done this at the expense of intrinsic antioxidant activity since as already discussed, the α-tocopherol molecule is almost certainly not the most efficient that could be devised. Furthermore, some synergists, notably ascorbic acid, are not physically compatible with the lipid substrate. The practical effect of this is that much of the synergist is wasted by extraneous chemical reactions or by excretion.

Targeting antioxidants to the site where they are most required has been the object of considerable research for the past twenty years in the polymer industry. Such techniques as co-polymerisation of the antioxidant molecule with the base monomer and adduct formation with reactive sites in the polymer have proved very successful in anchoring the antioxidant to the polymer backbone [42-44]. Scheme 3 shows how unsaturation in the

SCHEME 3. ATTACHMENT OF ANTIOXIDANTS TO AN UNSATURATED SUBSTRATE

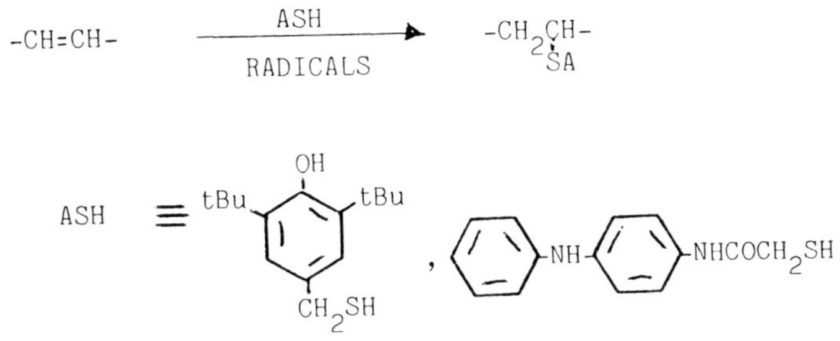

polymer (which is also the most oxidatively vulnerable segment) can be used to accept a sulphur adduct where A is a typical hindered phenol or aryl amine antioxidant group. This technique has been used successfully to locate antioxidants at high concentrations in the oxidatively vulnerable phase of rubber modified graft to co-polymers. In addition, because the reactive antioxidants attached to the polymer have the autosynergistic activity illustrated by compound XXI (where R is the substrate) the lifetime of the polymer can be increased using the same concentration of antioxidant by an order of magnitude [44,45]. In view of the very different natures of the biological chain-breaking antioxidants which contain a lipophilic tail and their synergists, notably ascorbic acid which is hydrophilic, the idea of attaching autosynergistic antioxidants to the lipid fatty acids seems to be particularly attractive.

Can in vivo antioxidants be improved ?

We have seen that most of the most important antioxidant mechanisms identified from studies in model compounds and technological substrates are also found in nature. Tables 3 and 4 summarise the present state of knowledge for the chain-breaking and preventive antioxidants and lists the types of compound which are active in vitro and in vivo.

TABLE 3. CHAIN-BREAKING ANTIOXIDANT MECHANISMS IN VITRO AND IN VIVO.

MECHANISM.	CHEMICAL CLASS. In vitro	In vivo
CB-D	Hindered phenols, aromatic amines	α-Tocopherol, thiols, ascorbic acid
CB-A	Stable phenoxyls and nitroxyls, quinones, quinonimes, α-tocopheroquinone	α-Tocoperoquinone (?) Vitamin K (?) Ubiquinone (?)
CB-C (CB-A/CB-D)	Phenoxyls/phenols Nitroxyls/hydroxylamines Cu^{2+}/Cu^{+}, $I\cdot/I^{-}$	Not so far recognised

TABLE 4. PREVENTIVE ANTIOXIDANTS IN VITRO AND IN VIVO.

MECHANISM	CHEMICAL CLASS In vitro	In vivo
PD-S	Phosphite esters	Not recognised
PD-C	Sulphur acids derived from organo sulphur compounds, cyclic phosphate esters.	Glutathione peroxidase catalase, horseradish peroxidase (?)
MD	Nitrogen, sulphur and oxygen chelating agents.	Porphoryns
UVA	Pigments, 2-hydroxybenzophenones, 2-hydroxy-benzotriazoles, nickel complexes	Melanin

Although the catalytic CB mechanism has not so far been identified in vivo it seems possible that both α-tocopheroxyl, α-tocopheroquinone and some naturally occuring quinones may function in this way since they are quite analogous to compounds which have been shown to have CB-A antioxidant activity in vitro. More detailed studies are required in biological substrates both in vitro and in vivo to elucidate these complementary mechanisms and to determine whether more intrinsically active structures and synergists may be developed. There is evidence that the α-tocopherol structure is not the optimum and there is potential for improving its participation in the CB-A/CB-D catalytic cycle and in its regeneration by reducing agents. In the latter context, targeting of both antioxidants and synergists by chemical attachment to the lipid substrate appears to offer the prospect of more efficient synergistic systems.

Although synergism has been recognised between α-tocopherol and ascorbic acid and α-tocopherol and selenium and possibly between α-tocopherol and glutathione peroxidase (see table 5) only one possible example of autosynergism has been observed (melanin).

TABLE 5. SYNERGISTS

MECHANISM	STABILISATION EFFECT	BIOLOGICAL EXAMPLES
CB-D/PD-C	Thermal oxidation	α-Tocopherol/glutathione peroxidase (?)
CB-D/UVA	Photo-oxidation	Melanin*
PD-C/UVA	Photo-oxidation	-
PD-C/MD	Metal catalysed oxidation	-
CB-A/MD	Metal catalysed oxidation	-
CB-A/CB-A	Thermal oxidation	α-Tocopherol/ascorbic acid. α-Tocopherol/selenium

* Melanin and its intermediates are potential CB-D and CB-A antioxidants.

There is clearly a potential advantage in having more than one type of antioxidant activity in the same molecule since this would make the targeting of such autosynergists to the site where they are intended to act very much easier. Autosynergists for metal catalysed oxidation (PD-C/MD, CB-A/MD) and photo-oxidation (CB-D/UVA) seem to be particularly attractive possibilities. CB-D/PD-C autosynergists for uncatalysed oxidation also have considerable potential in lipids.

REFERENCES.

1. G. Scott, Atmospheric Oxidation and Antioxidants, Elsevier (1965), p.2.
2. W. Ostwald, J. Soc. Chem. Ind., 32 (1913) 179.
3. H. L. J. Bäckström, J. Am. Chem. Soc., 49 (1927) 1460.
4. P. A. A. van der Beck and W. S. Jorissen, Rec. Trav. Chim. 45 (1926) 245.
5. G. Scott, Atmospheric Oxidation and Antioxidants, Elsevier (1965), p.16.
6. A. Genthe, Z. Angew, Chem. 19 (1906) 2087.
7. C. Moureau and Dufraisse, Chem. Rev., 3 (1926-7) 113.
8. C. D. Lowry, G. Egloff, J. C. Morrell and C. G. Dryer, Ind. Eng. Chem., 25 (1933) 804.
9. J. L. Bolland and P. ten Have, Trans. Farad. Soc., 43 (1947) 201.
10. G. Scott, Developments in Polymer Stabilisation-6, Ed. G. Scott, App. Sci. Pub., London, 1983, p.29.
11. S. Al-Malaika, K. B. Chakraborty and G. Scott, Developments in Polymer Stabilisation-6, Ed. G. Scott, App. Sci. Pub., London, 1983, p.73.

12. H. Berger, T. A. B. M. Bolman and D. M. Brouwer, Developments in Polymer Stabilisation, Ed. G. Scott, App. Sci., Pub., London, (1983) p.1.

13. G. Scott, Developments in Polymer Stabilisation-7, Ed. G. Scott, Elsevier App. Sci. Pub., London (1984), p.65.

14. N. Grassie and G. Scott, Polymer Degradation and Stabilisation, C.U.P., (1985), Chapter 5.

15. T. P. Hilditch, Nature, 160 (1950) 558.

16. see Ref. 1, p. 19 et seq for a review of 1,4-diene oxidation up to 1965.

17. H. S. Olcott and H. A. Mattill, J. Am. Chem. Soc., 58 (1936) 1627, 2204.

18. H. S. Olcott and H. A. Mattill, Chem. Rev., 29 (1941) 257.

19. C. Golumbic, Oil and Soap, 23 (1946) 184.

20. H. Dann, Pharmacol. Revs., 9 (1957) 1.

21. M. K. Horwitt, Fed. Proc., 24 (1965) 68.

22. A. L. Tappel, Fed. Proc., 24 (1965) 73.

23. A. L. Tappel, Geriatrics, 23 (1968) 97.

24. C. W. Burton, D. V. Foster, B. Perly, T. F. Slater, I.C.P. Smith and K. V. Ingold, Trans. Roy. Soc. B 3111 (1985) 565.

25. Ref. 1, Chapters 4, 5.

26. K. B. Chakraborty and G. Scott, J. Polym. Sci., Polym. Lett. Ed., 22 (1984) 553.

27. R. Bagheri, K. B. Chakraborty and G. Scott, Polym. Deg.

and Stab., 5 (1983) 145.

28. C. Armstrong and G. Scott, J. Chem. Soc. (B) (1971) 1747.

29. L. Stryer, Biochemistry, Freeman, San Francisco,(1975) p.72.

30. A. A. Katbab and G. Scott, Europ. Polym. J., 17 (1981) 559.

31. R. Bagheri, K. B. Chakraborty and G. Scott, J. Polym. Sci., Polym. Chem. Ed., 22 (1984) 1573.

32. H. S. Dweik and G. Scott, Rubb. Chem. Tech., 57 (1984) 735.

33. S. Al-Malaika, E. O. Omikorede and G. Scott, Polymer Communications, 27 (1986) 173.

34. S. Al-Malaika, E. O. Omikorede and G. Scott, J. App. Polym. Sci., 33 (1987) 703.

35. J. D. Holdsworth, G. Scott and D. Williams, J. Chem. Soc., (1964) 4692.

36. Z. Osawa, Developments in Polymer Stabilisation, Ed. G. Scott, Elsevier App. Sci. Pub., London (1984), p.193.

37. B. N. Leyland and R. L. Stafford, Trans. I.R.I., 35 (1959) 25.

38. B. N. Leyland and R. L. Stafford, Chem. Canada, Nov. (1959) 45.

39. Ref. 1, p.361 et seq.

40. G. Scott and M. F. Yusoff, Europ. Polym. J., 16 (1980) 497.

41. V. M. Farzaliev, W. S. E. Fernando and G. Scott, Europ. Polym. J., 14 (1978) 785.

42. G. Scott, Developments in Polymer Stabilisation-1, Ed. G. Scott, App. Sci. Pub., London, 1979, Chap. 9.

43. G. Scott, Developments in Polymer Stabilisation-4, Ed. G. Scott, App. Sci. Pub., London, 1981, Chap. 6.

44. G. Scott, Developments in Polymer Stabilisation-8, Ed. G. Scott, Elsevier App. Sci. Pub., London, 1987, Chap. 2.

45. M. Ghaemy and G. Scott, Polym. Deg. and Stab., 3 (1980-1) 341.

FREE RADICALS: CHEMISTRY, PATHOLOGY AND MEDICINE
edited by Catherine Rice-Evans & Thomas Dormandy,

Antioxidant Status and Neurological Function in the Vitamin E Deficient Rat.

MA Goss-Sampson, CJ MacEvilly, DPR Muller.
Institute of Child Health, 30, Guilford Street,
London WC1N 1EH.

Abstract.

Vitamin E (alpha-tocopherol) concentrations were measured longitudinally over a period of greater than 1 year in vitamin E deficient and control rats. Concentrations in serum, liver and adipose tissue fell rapidly and were undetectable by 24 weeks. Nervous tissue (brain, cord and nerve) of deficient animals retained a greater percentage of the vitamin, and by 52 weeks concentrations were between 4-5% of control values. After 55 weeks of deficiency the activities of glutathione peroxidase and superoxide dismutase in the various brain regions were not significantly different from that of controls.

Fast axonal transport of acetylcholinesterase (anterograde and retrograde) was studied and by 52

weeks a significant reduction of both anterograde ($p<0.02$) and retrograde ($p<0.05$) transport was observed in the deficient rats.
Electrophysiological measurements, somatosensory evoked potentials (SEP) and electromyography (EMG) were made in 40-42 week deficient and control animals. No differences in peripheral conduction (lumbar SEP) were observed between the two groups, whereas central conduction (cortical SEP) was significantly delayed ($p<0.001$) in the deficient rats. All of the vitamin E deficient rats ($n=12$) showed abnormal EMG indicative of chronic partial denervation, whereas none of the controls ($n=8$) showed any abnormality.

Introduction.

It is now well established that vitamin E (alpha-tocopherol) is important for the maintenance of normal neurological structure and function in man and experimental animals. In man a severe and prolonged deficiency of the vitamin results in a characteristic and progressive neurological syndrome characterised by ataxia of limbs and gait, areflexia, proprioceptive loss, ophthalmoplegia, retinal pigmentation and general muscle weakness (1). The neuropathological lesions associated with vitamin E deficiency are similar in man and experimental animals (eg. rat and monkey); there is degeneration of axons of the gracile and cuneate nuclei in the brainstem, the posterior columns of the spinal cord and in the peripheral nerves with a selective loss of large calibre myelinated fibres (2).

Alpha-tocopherol appears to be the only significant lipid soluble chain-breaking antioxidant *in vivo* (3), and this is presumed to be its principal role. It is, however, not known why the nervous system should be particularly susceptible to a deficiency of this vitamin. In order to further understand the role of alpha-tocopherol in neurological tissue, aspects of the neurobiology of vitamin E and some other antioxidant systems were studied, and correlated with the functional studies of axonal transport and neuro-electrophysiology.

Materials and methods.

180 weanling male Wistar rats were obtained from Charles River Ltd., UK. Half of the rats were placed on a vitamin E deficient diet (Machlin/Draper-HLR 814), supplied by Dyets, Pennsylvania, USA. The remainder (controls) received the same diet to which alpha-tocopheryl acetate (100mg/kg) had been added. Rats from each group were sacrificed at fixed times over 55 weeks for the different investigations. The various tissues and neuro-anatomical regions were obtained following perfusion with ice-cold saline as previously described (4).

Alpha-tocopherol concentrations were determined by high performance liquid chromatography/fluorimetry using a modification (5) of the method of Buttriss and Diplock (6). Concentrations of alpha-tocopherol were routinely expressed as μg/g lipid. Total lipid concentrations were determined by the hydroxamic acid method of Snyder and Stevens (7) after extraction with 2:1 chloroform-methanol (8). The activities of superoxide dismutase and glutathione peroxidase were determined by the methods of Crapo *et al* (9) and Beutler *et al* (10) respectively, and were expressed as units/mg protein. Protein concentrations were determined by the method of Lowry *et al* (11).

Fast axonal transport of acetylcholinesterase (anterograde and retrograde) was measured in the sciatic nerve using a double ligation/accumulation technique similar to that described by Oikarinen and Kalimo (12). Results are expressed as a percentage increase of acetylcholinesterase activity at the relevant ligature site after 3 hours accumulation.

Electromyographic recordings, lumbar and cortical SEP's and their respective conduction velocities, were determined after 40-42 weeks of deficiency.

Results are expressed throughout as mean $\pm$ 1SD, and the significance of difference between mean values was calculated by the Student's t-test.

Results.

Alpha-tocopherol was undetectable in serum, liver and adipose tissue after 16-24 of weeks deficiency. Nervous tissues (brain, cord and nerve) of vitamin E deficient animals retained a greater percentage of alpha-tocopherol and by 52 weeks concentrations were still between 4-5% of control values. The concentrations of alpha-tocopherol from weaning to 24 weeks in all the tissues studied have been reported previously (13). Fig.1 shows the

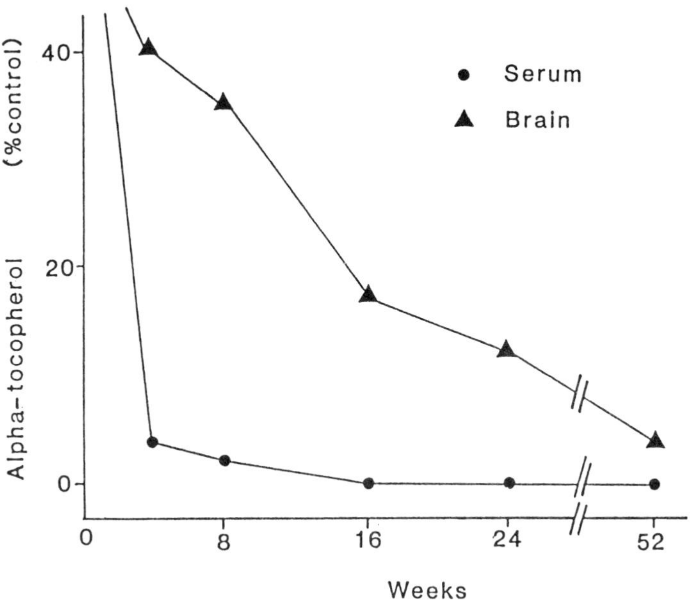

Fig 1. Concentration of alpha-tocopherol (μg/g lipid)in serum and brain of vitamin E deficient rats expressed as a percentage of control values at the same time point (n=3).

alpha-tocopherol concentrations (expressed as a percentage of control values) in serum and brain (representative of peripheral and nervous tissues) in deficient animals from weaning to 52 weeks. The concentrations of alpha-tocopherol in the various regions of brain and cord of deficient and control animals at 52 weeks are shown in Fig 2 .

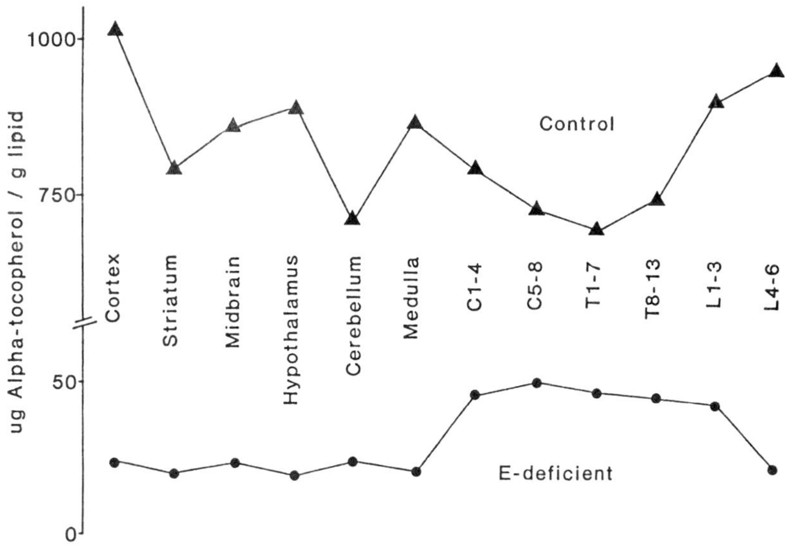

Fig.2 Concentration of alpha-tocopherol in regions of brain and spinal cord of control and deficient rats after 52 weeks (n=3). C=cervical, T=thoracic, L=lumbar regions of spinal cord.

In the control rats there was a rostral to caudal decrease of alpha-tocopherol in the brain; the highest concentration being in the cortex (1032 ± 196 μg/g lipid) and the lowest in the cerebellum (704 ± 69 μg/g lipid). Conversely there tended to be a rostral to caudal increase along the spinal cord. Alpha-tocopherol concentrations were similar along all regions of sciatic and tibial nerve (775 ± 92 μg/g lipid) . After 52 weeks of deficiency the concentrations of alpha-tocopherol were similar (20-24 μg/g lipid) in all the brain regions with no region exhibiting selective retention or loss. The concentrations of alpha-tocopherol in the various regions of the spinal cord were approximately twice that found in the brain (43-50 μg/g lipid). There was no selective decrease in the various nerve regions with concentrations varying between 23-47 μg/g lipid.

The activities of superoxide dismutase and glutathione peroxidase were estimated longitudinally in the various brain regions of control and deficient animals. At no time was there any significant differences between the activities of either enzyme in the two groups of animals. The activities found in whole brains at 52 weeks are shown in Fig 3.

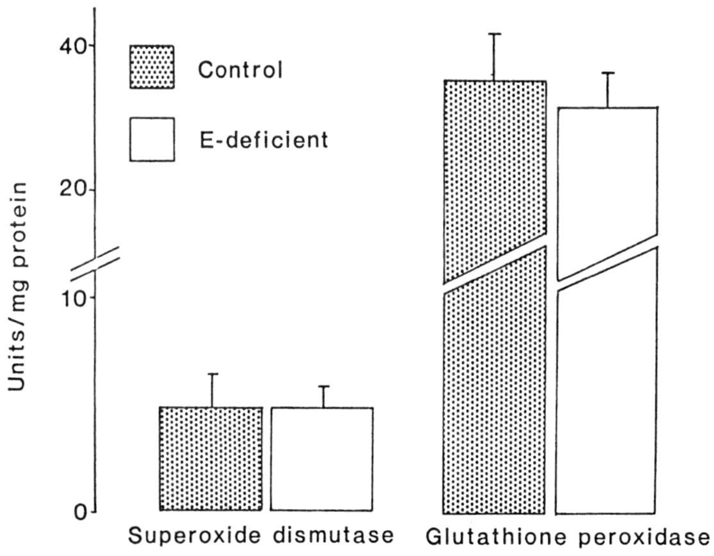

Fig.3 Activities of superoxide dismutase and glutathione peroxidase (mean $\pm$1SD) in whole brains of control and deficient rats after 55 weeks (n=3).

Anterograde and retrograde fast axonal transport of acetylcholinesterase were determined from 8-52 weeks. The results at 52 weeks are shown in fig 4. The deficient animals exhibited a 25% reduction in anterograde accumulation ($p<0.02$) and a 20% reduction in retrograde accumulation ($p<0.05$). No significant differences in either anterograde or retrograde transport were found at earlier time points.

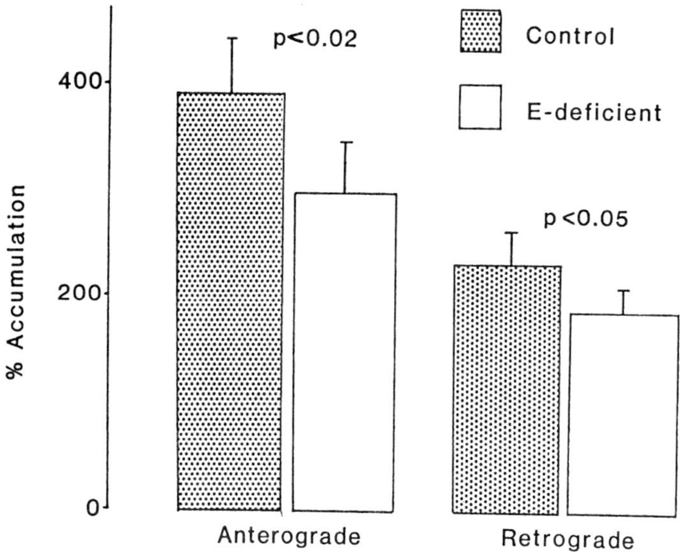

Fig.4 Fast anterograde and retrograde axonal transport of acetylcholinesterase in sciatic nerves of control and deficient rats after 52 weeks (n=6).

After 40-42 weeks of deficiency no differences in peripheral conduction (lumbar SEP) were observed between the deficient and control animals (Fig 5). However, central conduction (cortical SEP) in the deficient animals was found to be significantly ($p<0.001$) slowed (Fig 5). EMG studies of distal muscle (gastrocnemius) showed abnormalities (presence of spontaneous fibrillation, positive sharp waves or polyphasic groups) in all the

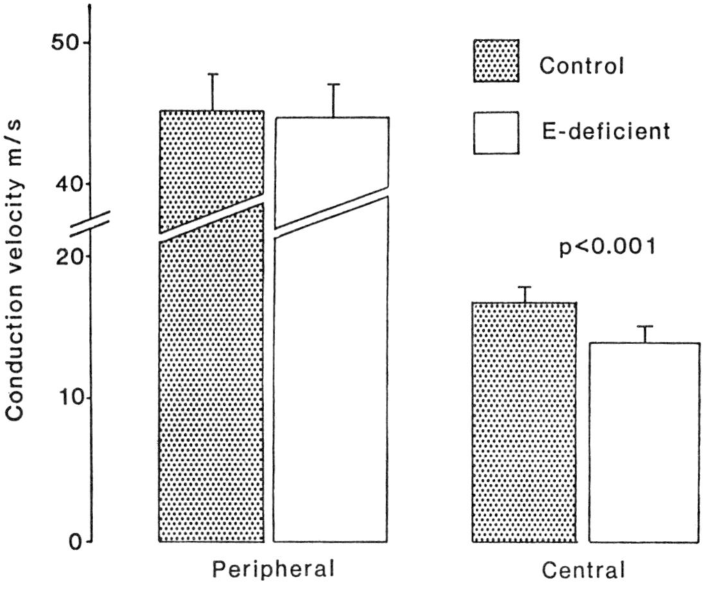

Fig.5 Peripheral (lumbar SEP) and central (cortical SEP) sensory conduction velocities in control (n=8) and deficient (n=12) rats after 40-42 weeks.

deficient animals studied (n=12). None of these abnormalities were found in any control rat (n=8).

Discussion.

There have been a number of studies on the effects of chronic vitamin E deficiency in the rat (14,15), and the concentration of vitamin E in various non-neurological tissues throughout the deficiency has been documented (16-18). There have not, however, been any previous detailed longitudinal studies on the neurobiology of vitamin E and other antioxidant systems during long-term deficiency in the rat, although there is some information on the vitamin E deficient mouse (19). This is also the first study which has attempted to correlate the concentrations of antioxidants with neurological function during vitamin E deficiency.

The rapid decrease in alpha-tocopherol concentrations in serum, liver and adipose tissue observed in this study has been reported by others (16-18). The decrease in concentration in brain, cord and nerve was less rapid and the nervous tissues still retained 4-5% of the control concentrations after 52 weeks of deficiency. Similar observations have been reported in the vitamin E deficient mouse (19). These findings, together with those of Burton and Ingold (20), who showed, using deuterated tocopheryl acetate, that the half-life of alpha-tocopherol was increased in the brain compared to other tissues, strongly suggest that neurological tissues preferentially conserve vitamin E and that its turnover may be reduced.

The distribution of alpha-tocopherol in the brain and cord regions of control rats agrees with that previously reported (21); the lowest concentrations being found in the cerebellum and the rostral end of the spinal cord. This is of interest since in man, a severe deficiency of vitamin E results in a characteristic spinocerebellar syndrome (1). In the deficient animals, the concentrations of alpha-tocopherol were similar in the various brain regions and along the cord. On a percentage basis therefore, the cerebellum lost the least tocopherol which contrasts with a previous report in the vitamin E deficient mouse (19), where the cerebellum lost the greatest percentage of alpha-tocopherol. After 55 weeks, no differences in the activities of either glutathione peroxidase or superoxide dismutase were observed between the vitamin E deficient and control groups. This indicates that these two water soluble antioxidant enzyme systems do not compensate for the reduced concentrations of the lipid-soluble antioxidant, vitamin E. In a recent study by Nelson (22), the effects of the water-soluble free radical scavengers ethoxyquin and promethazine on the neuropathology of vitamin E deficiency in rats were studied. Results showed that the rats which received vitamin E deficient diets supplemented with either ethoxyquin or promethazine for 36-39 weeks had no evidence of the neuropathological lesions which were present in the unsupplemented deficient rats. There was also no evidence that alpha-tocopherol was spared in the rats receiving the ethoxyquin or promethazine. This study therefore confirms the assumption that free radical damage is involved in the pathogenesis of the neurological lesions resulting from a deficiency of vitamin E.

The characteristic pathological features of the neurological lesions are a progressive primary axonopathy with a secondary demyelination which predominantly affects the centrally directed primary sensory neurons (2). This is consistent with the electrophysiological findings after 40-42 weeks of deficiency which showed a significantly slowed conduction within the central somatosensory pathway but normal conduction in the peripheral sensory pathway of the hind-limbs. These findings are similar to those previously reported in man with severe vitamin E deficiency (24,25).

A reduction in anterograde and retrograde fast axonal transport of acetylcholinesterase was observed after 52 weeks of deficiency but not before. This is in agreement with the finding that if alterations in axonal transport occur they generally arise late in the course of nerve degeneration . Ultrastructural studies of sensory nerve endings (15,26) show an accumulation of organelles, multivesicular bodies, and tubulovesicular structures. These observations, along with the reduced axonal transport suggest there may be a defect in "turnaround" at the axon terminal.

Histological changes compatible with a necrotising myopathy are observed in vitamin E deficient rats from as early as 12 weeks of deficiency (27). However, the abnormal EMG results obtained in the deficient animals after 40-42 weeks indicate that a process of chronic partial denervation is also occurring. Similar abnormalities of EMG have been reported in vitamin E deficient states in man (23,24). Further studies are therefore required to delineate the precise effect of vitamin E deficiency on nerve-muscle interaction.

The abnormal EMG and SEP findings in experimental vitamin E deficiency do not appear to have been previously reported. Measurement of the cortical SEP provides for the first time a sensitive and objective method of assessing the neurological effects of vitamin E deficiency in longitudinal studies in the same rat.

Acknowledgements.

We thank Hoffmann La Roche and Co. and the Friedreich's Ataxia Group for financial support. We also thank Prof. PK Thomas (Royal Free Hospital, London) and Dr A Kriss (The Hospitals for Sick Children, Great Ormond Street, London) for help with the electrophysiological studies.

References.

1. Muller DPR, Lloyd JK, Wolff OH. Lancet i. (1983) 225.
2. Nelson JS, Fitch CD, Fischer VW, Broun GO, Chou AC. J. Neuropath. Exp. Neurol. 40 (1981). 166.
3. Burton GW, Joyce A, Ingold KU. Arch. Biochem. Biophys. 221. (1983). 281.
4. Goss-Sampson MA, Muller DPR. Neuropath. Appl. Neurobiol. (1987). In Press.
5. Metcalfe T, Muller DPR, Brooksbank BWL. IRCS Med Sci. 12. (1984). 121.
6. Buttriss JL, Diplock AT. Methods Enzymol. 105. (1985). 131.

7. Snyder F, Stevens NA. Biochem. Biophys. Acta 34. (1958). 244.

8. Folch J, Lees M, Sloane SGM. J. Biol. Chem. 226 (1957). 497.

9. Crapo JD, McCord JM, Fridovich I. Methods Enzymol. 53. (1978). 382.

10. Beutler E, Blume KG, Kaplan JL, Lohr GW, Ramot B, Valentine WM. Br. J. Haematol. 35. (1977). 331.

11. Lowry OH, Rosebrough NJ, Farr AL, Randall RJ. J. Biol. Chem. 193. (1951). 265.

12. Oikarinen R, Kalimo H. Neuropathol. Appl. Neurobiol. 10. (1984). 163.

13. Muller DPR, Goss-Sampson MA. In: International Symposium of Vitamin E - Nutritional and Medical Aspects. Elsevier. (1987). In Press.

14. Machlin LJ, Filipski R, Nelson J, Horn LR, Brin M. J. Nutr. 107 (1977) 1200.

15. Towfighi J. Acta Neuropath. 54 (1981) 261.

16. Century B, Horwitt MK. J. Nutr. 72 (1960) 357.

17. Bieri JG. Ann. NY. Acad. Sci. 203 (1972) 181.

18. Machlin LJ. In: Vitamin E : A Comprehensive

Treatise. Dekker (NY). (1980).

19. Vatassery GT, Angerhofer CK, Peterson FJ. J. Neurochem. 42 (1984) 554.

20. Burton GW, Ingold KU. Acc.Chem.Res. 19 (1986) 194.

21. Vatassery GT, Angerhofer CK, Knox CA, Deshmukh DS. Biochem. Biophys. Acta. 792 (1984) 118.

22. Nelson JS. In: International Symposium of Vitamin E - Nutritional and Medical Aspects. Elsevier (1987) In Press.

23. Wichman A, Buchthal F, Pezeshkpour GH, Gregg RE. Neurology 35 (1985) 1279.

24. Brin MF, Pedley TA, Lovelace RE, Emerson RG, Gouras P, Mackay C, Kayden HJ, Levy J, Baker H. Neurology 36 (1986) 669.

25. Satya-Murti S, Howard L, Krohel G, Wolf B. Neurology 36 (1986) 917.

26. Bradley DJ, Muddle JR, Southam E, Thomas PK. J. Physiol. 376 (1986) 34p.

27. Nelson JS, Machlin LJ, Welsh R. Fed. Proc. 37 (1978) 758.

FREE RADICALS: CHEMISTRY, PATHOLOGY AND MEDICINE
edited by Catherine Rice-Evans & Thomas Dormandy, © 1988 Richelieu Press, London

The influence of dietary vitamin E and selenium on the production of free malondialdehyde (MDA) in rat liver measured by high performance liquid chromatography (HPLC)

Tokarz A[1,2], McCarthy P T[2], Diplock A T[2]

[1]Department of Bromatology, Medical Academy, Warsaw, Poland
[2]Department of Biochemistry, United Medical and Dental Schools of Guy's and St Thomas's Hospitals, London

Weanling male Wistar rats, aged 21 days, were fed a Torula yeast based diet for 28 days, which was deficient in both vitamin E and selenium. The food intake of some animals was supplemented with all-rac-α-tocopherol acetate (100 or 1000 mg/kg) and selenium was added to the drinking water at a level of 0.1 or 1.5 mg/litre. The concentration of MDA in hepatic extracts was measured by HPLC with UV detection using a Spherogel-TSK G1000 PW size-exclusion column with a mobile phase of 0.1M sodium phosphate buffer, pH 7,0 and a detection wavelength of 267nm. The results showed that the concentration of MDA was inversely related to the levels of vitamin E and selenium in the diet.

Introduction

Both vitamin E and selenium reduce peroxidation of polyunsaturated fatty acids (PUFA) in cell membranes and influence PUFA breakdown induced by carcinogenic chemicals. To explain the effects of vitamin E and selenium most attention has been devoted to their antioxidant properties but some recent evidence suggests that they may also have more subtle modes of action *in vivo*[1,2]. For example selenium is not an antioxidant *per se*[1] but may be incorporated into complexes or compounds with antioxidant properties[3]. These may inhibit lipid peroxidation, accelerate the decomposition of peroxides, scavenge free radicals and repair damage to molecules. An example of such a cellular component is the seleno-enzyme glutathione peroxidase[4]. In addition to its antioxidant properties vitamin E also acts as a stabiliser of lipid-lipid and protein-lipid bonds[3,5].

One of the most intensively studied products of lipid peroxidation is malondialdehyde (MDA). It is thought to cause cell damage, the formation of lipofuscin age pigments[6] and to react with amino acids[7], proteins[8] and nucleic acid[9]. Other studies indicate that MDA is carcinogenic when applied to the skin of mice[10] mutagenic in several strains of *Salmonella Typhimurium*[11], and it is reported to be highly mutagenic and carcinogenic in cultured mammalian cells[1,2].

The amount of MDA in tissue extracts is used as an index of the degree of lipid peroxidation in vivo. Usually the MDA is measured by reacting it with thiobarbituric acid. However, this reaction lacks specificity resulting in overestimates of the amount of MDA present[13]. We have used, therefore, an HPLC[15] method to separate and quantitate free MDA in order to obtain an accurate measurement of the influence of dietary vitamin E and selenium on the production of MDA in rat liver.

Materials and Methods

Nine groups of six weanling male Wistar rats, aged 21 days were fed a Torula yeast based diet deficient in vitamin E and selenium ad libitum. Some animals were supplemented with all-rac-α-tocopherol acetate at a concentration of 100 and 1000mg/kg diet and selenium was added to the drinking water at a level of 0.1 and 1.5mg/litre. After 28 days the animals were killed, the liver excised, frozen in liquid nitrogen and stored at -70°C.

Livers were homogenised in ice cold 0.01M Na_3PO_4(pH8.00) using a Polytron homogeniser and a 10%(W/V) solution prepared. To study the effects of in vivo autooxidation on the levels of hepatic free MDA, aliquots of liver were homogenised in a buffer containing 100μg α-tocopherol. After centrifugation at 2000g for 5 mins, the supernatant was filtered through an Amicon cell equipped with a UM05 membrane ultra filter. (Amicon Corporation, Massachusetts, U S A).

Analyses were performed on a Spherogel TSK 1000 PW size exclusion column (30cmx7,5mm id; Beckmann, UK) using a mobile phase of 0.1M Na_3Po_4 buffer (pH 7.00) delivered at a flow rate of 0.6ml/min using a Knauer model 64 pump (Biotech.Instruments Bedfordshire). Free MDA was detected at 267nm using a Varian UV-50 variable wavelength UV detector at a detector sensitivity of 0.01 aufs. A 100μl aliquot of the clear supernatant was injected using a Rheodyne syringe loading injector fitted with a 100μl loop.

Results and Discussion

Typical chromatograms of free MDA in the liver of rats given 1000, 100 and 0 mg/kg dietary α-tocopherol acetate are shown in figure 1.

The concentration of MDA in the liver of rats fed on the different diets is shown in table 1 and figure 2.

The highest level of MDA was found in liver from animals given a diet deficient in both selenium and vitamin E (figure 1C; group 1). In the absence of dietary α-tocopherol, liver MDA levels were substantially reduced by the inclusion of selenium in the drinking water. The highest doses of selenium (group 3) reduced hepatic MDA levels to that found in the liver of animals given optimal amounts of α-tocopherol and selenium together (group 5; figure 2)

Figure 1

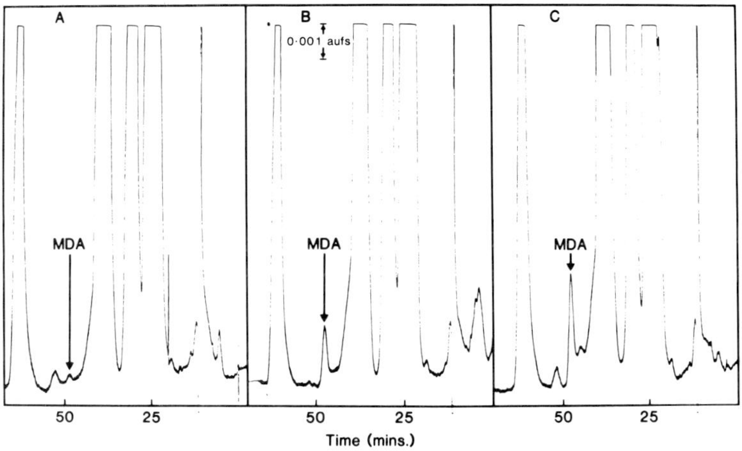

HPLC separation of different concentrations of free MDA in rat liver

A - 85ng/g: B - 500ng/g: C - 860ng/g

High levels of α-tocopherol (group 7) reduced hepatic MDA levels to about 33% of those found in the liver of rats given the control diet. When high levels of selenium were also given (group 9), MDA levels were still further reduced but this difference was not statistically significant.

When α-tocopherol was added to the extract during its preparation for chromotography, the level of free MDA which could be measured in the extract was significantly reduced (Table 2).

Table 1

Selenium (mg/litre)	Vitamin E (mg/kg)		
	0	100	1000
0	0.72 ± 0.17**	0.26 ± 0.05	0.17 ± 0.06*
0.1	0.53 ± 0.10**	0.28 ± 0.12	0.08 ± 0.07**
1.5	0.28 ± 0.09	0.18 ± 0.13	0.06 ± 0.02**

Concentration of MDA in rat liver (μg/g)

(Values are mean ± sd of six animals. * $p<0.05$; ** $p<0.001$)

Table 2

Selenium (mg/litre)	Vitamin E (mg/kg)		
	0	100	1000
0	0.33 ± 0.06**	0.09 ± 0.01	0.04 ± 0.01**
0.1	0.21 ± 0.06**	0.09 ± 0.01	0.04 ± 0.01**
1.5	0.20 ± 0.05**	0.07 ± 0.01**	0.04 ± 0.01**

Concentration of MDA in rat liver after addition of 100μg α-tocopherol added during homogenisation (μg/g).

(Values are mean ± sd of six animals. * $p<0.05$; ** $p<0.001$)

Figure 2

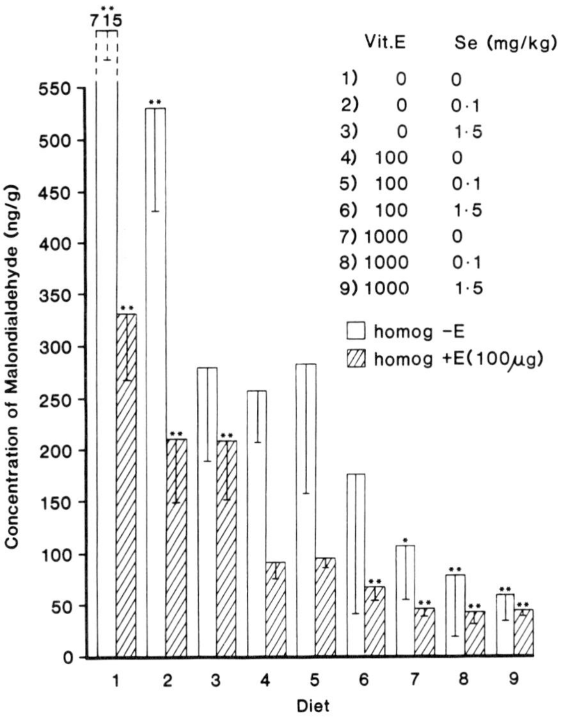

Concentration of free MDA in the liver of rats given different amounts of α-tocopherol and sclcnium

(.p<0.05 vs group 5;   ..p<0.001 vs group 5)

In the absence of an antioxidant in the extraction medium MDA levels were considerably higher than in its presence; possibly because α-tocopherol reduced lipid peroxidation during homogenisation, or alternatively the presence of α-tocopherol could inhibit the release of bound MDA. According to Hye-Sung Lee and Csallany[16] the concentration of bound MDA in rat liver is 10 fold that of free MDA. It is necessary, therefore, to include an antioxidant in the extraction medium to prevent this spurious production of MDA.

In order to resolve MDA in rat liver using this method we found that the pH of the eluting buffer was critical. In contrast to Csallany[15] who used phosphate buffer (pH 8.0) we obtained poor separation from UV absorbing containments at this pH, but baseline resolution was achieved at pH 7.0.

The results showed that both selenium and vitamin E in the diet of rats reduced hepatic MDA levels and that the effect of these substances is synergistic and dose-dependent.

References

1. Hoekstra W G. (1975) Fed Proc. 34;2083
2. Tappel A L (1980) Ann. N Y Acad. Sci 355;18
3. Combs G F Jnr. Noguchi T, Scott M L. (1975) Fed Proc. 34;2090
4. Becking G C. (1978). In Nutrition and Drug Interrelations. ed. J H Hathcock, J Coon. pp 371-408, New York Academic

5. Stohs S J, Hassan M Q, Murray W J. (1984) Xenobiotica 14; 533
6. Schemberger R J, Andreone T L, Willis C E.(1977) J Natl. Cancer Inst. 53, 1771
7. Mukai F M, Goldstein B D (1976) Science (Washington D C) 191, 868
8. You T M (1979) Mech. Aging Dev. 11, 137
9. Gardner W M (1979) J Agr Food Chem. 27, 220
10. Chio K S, Tappel A L (1968) Biochem. 8, 2821
11. Chiba M, Doi M, Yoshikawa M, Sugimoto E. (1972) Agric.Biol. Chem 40, 1001
12. Reiss U, Tappel A L, Chio K S (1972) Biochem Biophys Res Commun. 48;921
13. Esterbauer H, Cheeseman K H, Dianzani M U, Poli G, Slater T F (1982) Biochem J. 208, 129
14. Witz G, Zaccaria A, Lawrie N, Ferran H E, Goldstein B D (1985). Toxicologist 5; 919
15. Csallany A S, Der Guan M, Manwaring J D, Addis P B. (1987) Anal Biochem. 142, 277
16. Hye-Sung Lee, Csallany A S. (1987) Lipids 22,2, 104

FREE RADICALS: CHEMISTRY, PATHOLOGY AND MEDICINE
edited by Catherine Rice-Evans & Thomas Dormandy, © 1988 Richelieu Press, London

Chain-breaking antioxidants in the blood of malaria-infected Nigerian children

D.I. Thurnham*, Surangchit Koottathep*, D.A. Adelekan+.

* Nutrition Section, Clinical Investigation Unit, Dudley Road Hospital, Birmingham B18 7QH, UK. & Wolfson Research Laboratories, Queen Elizabeth Medical Centre, Birmingham B15 2TH.

\+ Department of Pediatrics, University of Ife, Ile-Ife, Nigeria.

The peroxyl radical trapping properties (TRAPexp) of plasma from 24 Nigerian children was measured. The subjects were selected in the Hospital Outpatient Department of Ife University or Health Clinics nearby: 13 had malaria, 6 mild respiratory disorders and 5 had previously been in-patients and were now apparently well and attending for follow-up examinations. TRAPexp measurements for all children were lower than those of caucasian control subjects and, in approximately half, TRAPexp was not detectable. TRAPexp measurements of the malaria patients were not different from the other two groups of children. Measurements of individual chain-breaking antioxidants showed that urate and protein sulphydryl were within the normal range and, although tocopherol concentrations were low and ascorbate negligible, theoretical radical trapping potential (TRAPcalc) was well within the limits of detection for the assay. Experiments suggested that substances were present in all samples depressing radical-trapping properties and these may be associated with the high level of infection present in this community. The results support the hypothesis that children in areas where the prevalence of infection and malnutrition is high, have impaired antioxidant defenses.

The chain-breaking, antioxidant vitamins alpha-tocopherol, ascorbate and beta-carotene are currently exciting considerable interest as protective agents against cancer (1) and heart disease (2). In malaria too, the antioxidant status of the host may be important in influencing the pathology of the disease. The parasite has been shown to create oxidant stress within the erythrocyte and to be responsible for producing peroxide (3). Extracellularly, the host's immune response generates reactive oxygen species (ROS; 4,5) and haemolysis and extensive damage to parenchymatous organs and endothelial tissues is a normal accompaniment of the disease. In experimental animals, deficiencies of tocopherol (6), riboflavin (7) and ascorbate (8) suppressed parasite development, possibly by causing premature release of immature parasites by oxidant-sensitive red cells (9). Paradoxically, however, host mortality in these nutritionallydeficient, malaria-infected animals was usually as high as in those animals without underlying deficiency possibly because the low antioxidant status may have also depressed the tissues' defenses against extra-erythrocytically-generated ROS.

In experimental malaria, changes in the antioxidant status of the host in response to the malaria both within the red cell and the plasma have been reported. In heavily infected mice, erythrocyte tocopherol and ascorbate concentrations were doubled and plasma ascorbate and urate increased but plasma protein decreased (10). The same workers have also reported that 'bleomycin-iron' (which could be capable of catalysing the Fenton reaction) was increased almost a 100 fold in heavily infected mice (11).

In man also there is evidence that antioxidant status can influence the disease. It has been suggested that vitamin E deficiencies limit the severity of malaria in herding peoples in Eastern Niger where the diet consists of defatted and rancid milk derivatives (12) and following the provision of grain, which would increase vitamin E intake, there was a recrudescence of latent malaria (13). Marginal riboflavin status is common throughout the developing world and has been associated with protective effects against malaria in Papua New Guinea (14) and, in a more recent study, children who were supplemented with riboflavin, ascorbate, thiamin and iron in the Gambia, showed greater parasite densities (15). Furthermore, iron administered to infants i.m. in Papua New

Guinea was associated with a doubling of the prevalence of malaria (16) possibly indicating an increased susceptibility to the infection, although this has recently been questioned (17).

There is much evidence that nutritional status is influenced by disease (18-20) but very little work would seem to have been done concerning malaria (12-14). In this study we used the 'TRAP' assay (21,22) to measure the ability of the plasma samples from malaria-infected children to quench an *in vitro* source of peroxyl radicals as a measure of the body's defense against ROS. In addition, we measured plasma tocopherol and ascorbate concentrations in the samples together with the other chain-breaking antioxidants, urate and protein sulphydryl to compare potential radical trapping capacity with observed measurements.

Experimental Details

Subjects

The malaria patients were children (14 to 18 m) presenting in the Outpatient Department of Ife University Hospital or at Health Clinics in the surrounding areas with a fever of rapid onset, intermittant and sometimes with vomiting. Control patients were children in the same age range presenting at the same locations with no history of recent fever but with mild respiratory disorders although otherwise stable and not in distress. Samples were also obtained from children aged 5 to 12 y who had previously been in-patients at the Teaching Hospital and were attending Outpatient Clinics for follow-up investigations. None had a fever at the time of examination. The investigations had the approval of the Ethics Committee, University of Ife.

One or two mL K_2EDTA blood was obtained from all children and transported at ambient temperatures to the laboratory; a journey which could take up to 2 h. Blood was centrifuged to separate plasma which was then stored at -70°C. From a random sample of specimens where sufficient plasma was available, 0.1 mL was mixed with 0.4 mL 5 % trichloracetic acid. The protein-free extract was also stored at -70°C for total ascorbate measurements in the U.K.

The 'TRAP' Assay

The 'TRAP' assay is based on that described by Wayner and colleagues (21,22) and we have made only minor modifications (23). Reagents were obtained from BDH (Poole, Dorset) unless otherwise indicated. The system comprises a water-soluble, thermolabile, free-radical initiator 2,2'-azo-bis-(2-amidinopropane hydrochloride) (ABAP; Polysciences, Warrington, PA, USA) and the peroxidation of linoleic acid dispersed in phosphate-buffered saline (PBS, pH 8) is measured using a Clarke-type oxygen electrode (Yellow Springs Instrument Co., Yellow Springs, Ohio; Model 5331). In the routine assay, 250 uL serum or plasma is vortexed 30 to 45 secs with 10 uL linoleic acid (Sigma, Poole, Dorset), 50 uL of the mixture was added to 3 mL PBS in the oxygen electrode chamber and the reaction started by addition of 30 uL 0.4 mol/L ABAP. Oxygen uptake was monitored for approximately 40 min until the rate of uptake reached its maximum. At approximately 50 % oxygen saturation in the chamber, 10 nmol 6-hydroxy-2,5,7,8-tetramethylchroman-2-carboxylic acid (Trolox, 25 uL 0.4 mmol/L; Hoffmann-La Roche, Nutley, NJ) was added to calibrate the system. Addition of serum and Trolox retard oxygen uptake until all antioxidant is consumed in quenching peroxyl radicals. Induction times are measured from the recorder trace. The total assay time for a normal serum sample is approximately 90 min. 'TRAP' is defined as the number of umol of peroxyl radicals trapped per litre of fluid and is calculated from the following equation.

TRAPexp (umol/L) = n[Trolox] x 20 x 'T'serum / 'T'Trolox

where

n	=	stoichiometric factor for Trolox which equals 2 (21,22).
'T'	=	Induction time i.e. time taken for maximum oxygen uptake to be achieved following addition of sample or Trolox and measured from the recorder trace.
[Trolox]	=	10 nmol

Stoichiometric Factors (n)

In this study we have used the following values for n of 2, 1.7

and 1.3 for tocopherol, ascorbate and urate respectively (22). For protein sulphydryl, we used a factor of 0.2 (23).

Measurements of Individual Antioxidants

Fat-soluble vitamins - A heptane extract was prepared from 0.1 mL plasma (24), evaporated to dryness, reconstituted in 0.1 mL mobile phase (acetonitrile:methanol:chloroform, 47:47:6) and 25 uL were injected onto a 100 x 4.5 mm, 3 um Sperisorb ODS-2 cartridge column (LKB) using reverse phase chromatography. Tocopherol acetate was used as internal standard and tocopherol, retinol and carotenes were monitored at 292, 325 and 450 nm respectively (25). Total ascorbate in trichloracetic acid extracts prepared at the time of blood collection was measured colorimetrically following reaction with 2,4-dinitrophenylhydrazine (26).

Urate and total ascorbate in plasma were measured using a modified HPLC method (22). 0.1 mL plasma was mixed with 25 uL 40 mmol/L dithiothreitol (DTT) and stood at room temperature for 5 min to convert all dehydroascorbate to ascorbate. (Substituting water for DTT enables ascorbate only to be measured.) Metaphosphoric acid (0.1 mL 10 %) was added to stablise the sample followed by 0.4 mL methanol, 20 uL 3,4-dihydroxybenzylamine hydrobromide as internal standard (DHBA, Aldrich Chemical Co. Ltd, Gillingham, Dorset) and 0.2 mL 5 mol/L NaH_2PO_4 adjusted to pH 5 with NaOH. Samples were then centrifuged and supernatants could then be stored for at least 3 weeks at -50°C prior to injecting 50 uL onto a 10 um Lichrosorb-NH_2. column (Technicol Ltd, Stockport, Cheshire) and eluted at 2.0 mL/min with acetonitrile:25 mmol/L NaH_2PO_4 (pH 4.8) mixed in the ratio 78.6:21.4. Absorbances of both urate and ascorbate were measured at 254 nm and 0.01 AUFS using a Spectromonitor III (LDC Ltd, Stone, Staffordshire).

Serum sulphydryl concentrations were measured using DTNB reagent at pH 7.4 (27).

Statistical Analysis

The data were analysed with the assistance of SPP (28) and the tests used are indicated in the text.

Results

TRAP measurements made on plasma from 13 children with malaria and 11 from the control groups are shown in Figure 1. All measurements of TRAPexp were made within 4 to 6 weeks of sample collection but, TRAPexp was not detectable in blood from 7 malaria patients and 6 of the controls. A typical recording of the oxygen consumption produced by one of the unmeasurable samples compared with that from a normal caucasian sample is shown in Figure 2. Induction times produced by normal caucasian plasma or serum can vary from 25 to 35 min and that following the addition of 10 nmol Trolox is usually very close to 16 min. In contrast, induction

FIGURE 1

MEASUREMENTS OF TRAPEXP IN NIGERIAN CHILDREN AND COMPARED WITH CAUCASIAN ADULTS

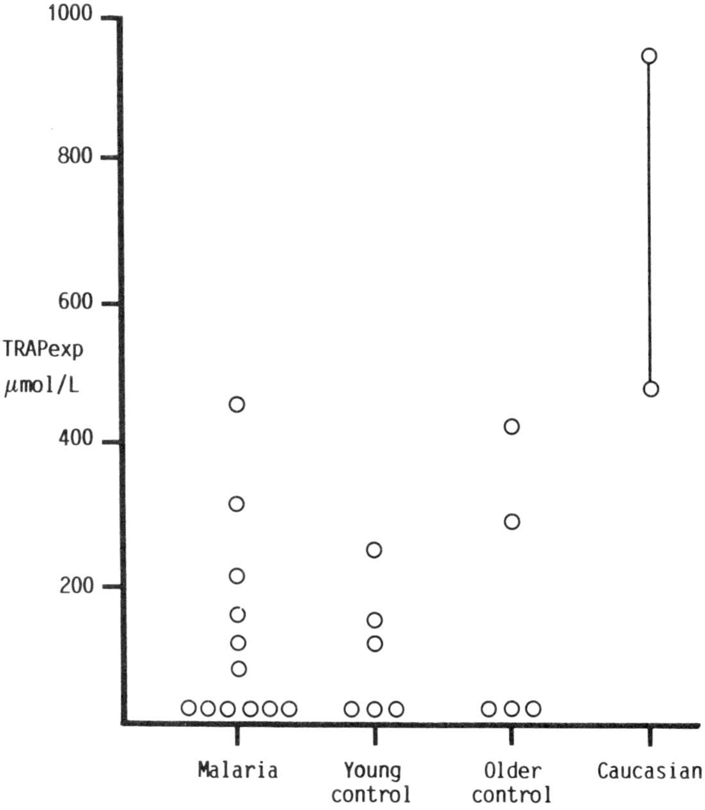

FIGURE 2

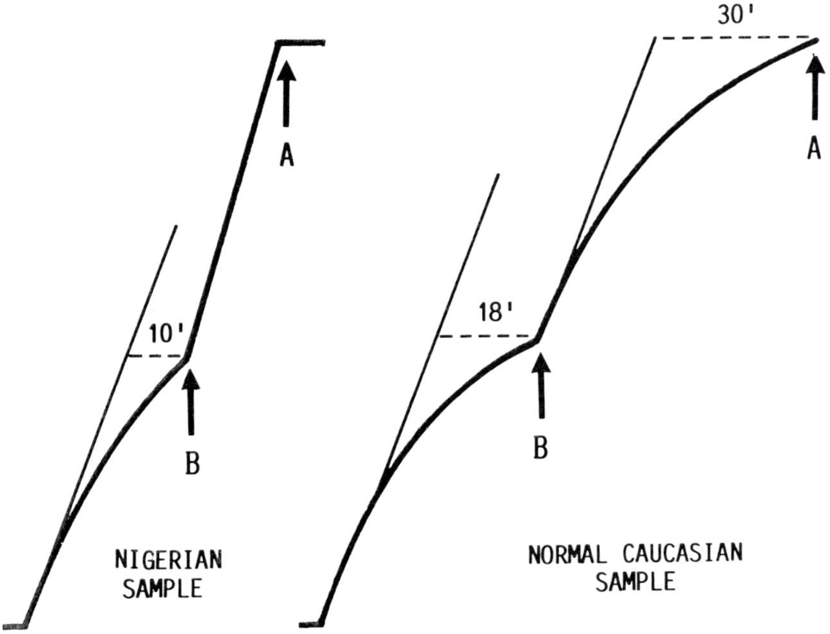

Measurement of TRAP in plasma from a normal caucasian adult and a Nigerian child with no apparent peroxyl-radical trapping activity. Samples are added to the oxygen electrode cell at A and Trolox is added at B.

Induction times ('T') of 30 min (plasma) and 18 min (Trolox) were obtained for the caucasian sample but there was no 'T' for the plasma and a reduced 'T' for Trolox in the presence of the Nigerian sample.

times produced by both malaria and control Nigerian plasma varied from indetectable to little more than 7 min except one 9 y child in the control group whose induction time was 15 min. Furthermore, induction times produced by addition of Trolox to the Nigerian samples were also much shorter than those produced with caucasian samples (Figure 2).

Four possibilities were considered to explain the results.

1. Deterioration of samples between collection and the time of arrival in the UK.
2. Presence in the samples of substances able to accelerate the generation of peroxyl radicals i.e. the breakdown of ABAP.
3. Insufficient chain-breaking antioxidants in the samples to produce measureable TRAPexp values.
4. Impaired quenching properties of the individual antioxidants in the plasma.

Deterioration of Samples

We have previously reported that TRAPexp values obtained from serum stored below $-50^{\circ}C$ remained constant for up to 2 months (23) and since all measurements reported here were done within 6 weeks of blood collection, the results should still have been reliable. Furthermore, there was no obvious relationship between the length of time the Nigerian samples had been stored and the TRAPexp values obtained.

Most samples were brought in from clinics in outlying districts, thus there was up to a 2 h delay at ambient temperatures before samples reached the laboratory. However there was no reduction in measurements of TRAPexp made on plasma from a healthy caucasian heparinised blood sample which was incubated at 37°C for up to 4 h. In addition none of the samples were visibly haemolysed which we know will reduce TRAPexp values.

There is evidence however that some deterioration may have occurred. Ascorbate is a particularly labile antioxidant and while we have found that ascorbate in plasma stored at -50°C or below is fairly stable for 6 to 8 weeks, most of the Nigerian samples were found to contain no detectable ascorbate in the plasma samples at the time of measuring TRAP in the UK. In

Table 1

Total Plasma Ascorbate* (umol/L)

Samples	n	Median	Minimum	Maximum
Malaria cases	23	21.4	10	62
< 11.4 umol/L+	1			
Control cases	13	17	6	44
< 11.4 umol/L	5			

* Measurements made in UK on TCA extracts of fresh plasma prepared in Nigeria.

\+ Threshold of vitamin C deficiency

contrast, Table 1 shows ascorbate concentrations in plasma TCA extracts prepared at the time of blood collection in Nigeria and measured on arrival in the UK. The concentrations are in the lower normal range, with 6 samples (1 malaria patient and 5 control children) suggesting a risk of deficiency (< 11.4 umol/L; ref 28). For 11 specimens on which ascorbate results were available at both times (i.e. time of collection in Nigeria and of TRAP analysis in the UK), only 2 showed ascorbate concentrations which had not decreased while in 2 it had decreased by 75 % and in the rest it was not detectable. Two factors may have contributed to the loss of ascorbate. During transport from Nigeria to the UK, the plasma samples were carried on ice, i.e. 4^oC, which will have allowed some ascorbate to deteriorate. In addition, most Nigerian samples contain elevated caeruloplasmin concentrations (Table 6). High caeruloplasmin concentrations may also accelerate ascorbate deterioration but this will be discussed later.

Accelerated breakdown of ABAP

The rate of oxygen consumption was calculated as % O_2 removed min^{-1} from the maximum oxygen consumption curve. The rate of oxygen consumption was significantly greater in the control Nigerian children than those with malaria (Table 2; $P < 0.01$) but the difference was too small to account for the depressed TRAPexp values. Furthermore, there was no difference between the oxygen

Table 2

Maximum Rate of Oxygen Consumption when Measuring TRAPexp of Nigerian Plasma Samples

Group	n	Median	Minimum	Maximum	P <*
		% O_2 removed min^{-1}			
Malaria cases	13	2.02	1.61	2.46	0.01
All Controls	11	2.46	1.92	3.08	
TRAPexp > zero	11	2.13	1.90	2.63	NS
TRAPexp = zero	13	2.34	1.61	3.08	

* Mann-Whitney

comsumption of the malaria-infected children and that of healthy caucasian adults (data not shown) nor any difference in the rate of oxygen consumption of the Nigerian children when grouped according to the presence or absence of a detectable TRAPexp measurement (Table 2).

Chain-Breaking Antioxidant Concentrations in the Plasma

Burton and colleagues have concluded that the 4 main chain-breaking antioxidants in plasma are alpha-tocopherol, ascorbate, urate and protein sulphydryl (22) and the concentrations of these substances in plasma of malaria-infected and control children are shown in Table 3. The only statistically significant difference between the two groups was in the ascorbate results. That is, in the plasma samples which we investigated by the 'TRAP' assay, ascorbate was only detected in 1/10 of the control children and 5/11 malaria cases. The absence of ascorbate however, was not by itself responsible for the zero TRAPexp values since 3 children with such values had detectable ascorbate concentrations in their plasma of 20.2, 16.4 and 12.2 umol/L and theoretical radical trapping capacities (TRAPcalc) of 489, 618 and 244 umol/L respectively.

Since there was no major difference between antioxidant concentrations in the plasma of malaria-infected and control children, the data was also re-examined according to whether a TRAPexp measurement was detectable or not (Table 3). The median TRAPcalc measurement for the group with TRAPexp values of zero was lower than that of the other group but the difference was not significant ($P = 0.055$). The concentration of urate however was lower in the group with no detectable TRAPexp values ($P = 0.02$).

Alpha-tocopherol concentrations were also low in the Nigerian children and 4/10 control and 7/11 malaria-infected children had deficient status (< 11.6 umol/L) but there was no association between low tocopherol concentrations and the zero TRAPexp values. Measurements of tocopherol:cholesterol ratios however were all greater than 2.22 umol/mmol (30) suggesting vitamin E status was adequate.

Table 3

Chain-Breaking Antioxidants, Calculated TRAP and Experimentally-Determined TRAP (umol/L) in Plasma from Nigerian Children and Caucasian Control Subjects

Groups	Tocopherol	Ascorbate	Urate	Sulphydryl	TRAPcalc@	TRAPexp
Malaria Cases (11)**	11.1* 5.3 - 17.1	2.1^{+} 0 - 127.5	240 65 - 538	585 346 - 868	463 244 - 1046	37.5 0 - 455
All Nigerian Controls (10)	13.7 3.4 - 47.6	0^{+} 0 - 20.2	210 150 - 287	551 419 - 971	424 322 - 560	0 0 - 421
Adult Caucasian Controls (20)	25.1 16.0 - 36.0	69.5 22.5 - 96.5	291 207 - 468	574 353 - 684	635 480 - 970	667 480 - 960
All Nigerian Cases						
TRAPexp > zero (10)	13.1 5.3 - 44.2	0 0 - 127.5	247^{++} 150 - 538	535 409 - 677	466 367 - 1046	207 74 - 455
TRAPexp = 0 (11)	9.7 3.4 - 47.6	0 0 - 20.0	176^{++} 65 - 307	609 346 - 971	390 244 - 618	not detectable

* Values shown are medians and ranges.

** Number of cases shown in parentheses.

+,++ Mann-Whitney; '+' $P < 0.05$, '++' $P < 0.02$.

@ Sum of (n x umol/L) for each antioxidant using the appropriate stoichiometric factors.

Alterations in the Quenching Properties of the Individual Antioxidants

The theoretical (TRAPcalc) radical-trapping activity of the group with non-detectable TRAPexp values ranged from 244-618 umol/L. Such values are well within the measurement sensitivity of the TRAP assay which was linear down to 100 umol/L in an experiment with diluted healthy caucasian blood (S. Koottathep unpublished). We therefore examined the possiblity that the radical quenching efficiencies of the different antioxidants in the plasma of the Nigerian children were reduced. It has already been mentioned that induction times following the addition of Trolox to the Nigerian samples were lower than we normally obtain for caucasian samples (Figure 2). Table 4 shows that where TRAPexp was zero, the Trolox induction times were shorter than those obtained where TRAPexp was measureable ($P < 0.01$).

Table 4

Induction Times (min) Following Addition of Trolox in the Measurement of TRAPexp on Nigerian Plasma Samples

Samples	n	Median	Maximum	Maximum	P =*
TRAP > zero	10	11.3	5.7	15.0	0.007
TRAP = zero	11	9.7	3.5	12.5	

* Mann-Whitney

In addition, we have measured the increase in induction times produced by urate and ascorbate added to 2 of the Nigerian samples. Both samples were from the control group viz a boy (5 y) with a TRAPexp of 280 umol/L and a girl (6 y) with a TRAPexp of zero. Plasma from a control child had to be used as there was not sufficient from any of the malaria patients. The increased induction times are shown in Table 5 and compared with those obtained when added to a normal caucasian sample. In the sample with no detectable TRAPexp, there was was a 72 to 85 % reduction in response to the added antioxidants. However, in the sample where TRAPexp was measureable, the response to the addition of urate was almost normal while that produced by ascorbate was only reduced by 30 %. That is something present in the sample reduced the quenching efficiency of the added antioxidants and more of this material was present in those plasma with the lowest TRAPexp values.

Table 5

Increased Induction Times (min) Produced by Addition of Urate and Ascorbate to Caucasian and Nigerian Plasma

Antioxidant Added	Caucasian Control	Nigerian Samples TRAPexp = 0	Nigerian Samples TRAPexp = 280
Urate 8.4 nmol cell^{-1}	10	1.5 (85 %)*	9.5 (5 %)
Ascorbate 14.4 nmol cell^{-1}	18	5.0 (72 %)	13.0 (28 %)

* Percentage inhibition shown in parentheses.

Table 6

Plasma Caeruloplasmin* (g/L)

Samples	n	Median	Minimum	Maximum
Malaria Cases	13	0.444	0.293	0.592
Nigerian Controls	10	0.503	0.355	0.566
Caucasian Controls	7	0.293**	0.230	0.341

* NOR-Partigen immunodiffusion plates (OSLG 03) using protein standard (OTFI 07; Behring-Hoechst UK Ltd, Hounslow, Middlesex)
** $P < 0.001$ Kruskall-Wallis

Caeruloplasmin

We have previously suggested that raised caeruloplasmin levels may be responsible for lower TRAPexp values in rheumatoid patients through copper-stimulated oxidation of ascorbate (23). Ascorbate concentrations were low to negligible in all but one of the Nigerian plasma samples (Table 3) and the lower induction time following the addition of ascorbate to the sample where urate gave a normal response (Table 5), led us to examine caeruloplasmin in the Nigerian samples. Table 6 shows that caeruloplasmin concentrations are significantly higher in the Nigerian samples than the caucasian controls. However, there were no differences in caeruloplasmin concentrations between:

1. the malaria cases and the local control children,
2. those samples with or without measureable TRAPexp values or
3. those samples with or without ascorbate in the plasma.

Both samples of plasma used to measure the induction times of added antioxidants (Table 5) had the same high caeruloplasmin concentration (0.524 g/L) which may partly explain the reduced induction times following the addition of ascorbate to the Nigerian samples in comparison to that obtained with caucasian blood.

Discussion

The peroxyl radical trapping capacity in all the Nigerian childrens' plasma samples was either not detectable or significantly lower than that found in normal adult caucasian samples. Theoretical radical-trapping capacities (TRAPcalc) of all plasma samples were well above the limit of detection of the TRAP assay and experiments suggested that the inability to detect TRAPexp was due to impaired quenching efficiencies of urate, ascorbate (Table 5) and possibly tocopherol as indicated by the results with Trolox (Table 2). Although raised caeruloplasmin concentrations may lower the trapping efficiency of ascorbate (23), the low TRAPexp values obtained in this study could not be linked with caeruloplasmin as plasma ascorbate concentrations were negligible when the samples were studied. However, the raised caeruloplasmin concentrations may have been responsible for the impaired trapping efficiency of the added ascorbate (Table 5).

We have not yet investigated what other substances present in the Nigerian plasma samples may be responsible for impairing the trapping efficiencies of the chain-breaking antioxidants but there are several possibilities. Aldehydes generated during lipid peroxidation can bind thiol groups (31) while cations may be bound to urate (32). Plasma tocopherol concentrations were low in many of the samples studied and, while vitamin E status may be adequate, the potential contribution to the radical trapping efficiency was very low.

Vitamin C status, on the basis of the TCA extracts prepared at the time of collecting the blood, was significantly better in the malaria-infected than in the control children (Table 1). We do not know whether this observation bears any relationship with that reported in murine malaria where both plasma and red cell ascorbate were raised (10). The mouse can synthesise its ascorbate whereas in man, plasma ascorbate concentrations reflect dietary intake. However, it is conceivable that the pyrexia will be accompanied by increased tissue breakdown releasing ascorbate from the tissues.

Caeruloplasmin is an acute phase protein and plasma concentrations are raised during infections (33). High caeruloplasmin concentrations were present in most of the children in this study, even the controls, suggesting that all were recently exposed to or still suffering from infections of various kinds. In addition, in other groups of children in this area we have found raised concentrations of 2 other acute phase proteins viz ferritin and transferrin (DA Adelekan, unpublished observations). We have not identified any specific feature in the chain-breaking antioxidants which is characteristic of malaria-infected children but we have revealed poorly functional antioxidant defenses in all the children. These children are living in an area where kwashiorkor, the end result of severe malnutrition, is a persistent problem. Golden and colleagues have suggested that malnutrition, toxins in the diet and a background of heavy or repeated infection may all combine to impair the tissues' antioxidant defense against free radical damage and this may precipitate kwashiokor in those who are worst affected (34). The impaired chain-breaking antioxidant capacity found in the plasma of the Nigerian children studied in this report is consistent with this hypothesis.

Acknowledgements

We are grateful to the Nestle Foundation for a grant to support this research and the travelling expenses of DAA. DIT is grateful for continued support by the Department of Health & Social Security and SK is in receipt of a British Council Grant. DIT would also like to thank Drs Burton & Ingold for the opportunity to visit the National Research Council Laboratories, Ottawa, Canada where some of the development work on the combined ascorbate and urate analysis was done.

References

1. Kromhout D.:Amer. J.Clin. Nutr. 45, (1987) 1361-1367.
2. Gey K.F., Brubacher G.B., Staehelin B.: Amer. J. Clin. Nutr. 45, (1987) 1368-1377.
3. Etkin N.L., Eaton J.W. in: Erythrocyte Structure and Function (Liss: New York 1975) pp 219-232.
4. Clarke I.A., Hunt N.H.: Infect. Immun. 39, (1983) 1-6.
5. Descamps-Latscha B., Lunel-Fabiani F., Kara-Binis A., Druilhe P.: Parasit. Immunol. 9, (1987) 275-279.
6. Eaton J.W., Eckman J.R., Bergman E., Jacob H.S.: Nature 264, (1976) 758-760.
7. Kaikai P., Thurnham D.I.: Trans. R. Soc. Trop. Med. Hyg. 77, (1983) 680-686.
8. Geiman Q.M., McKee R.W.: Proc. Soc. Exptl Biol. Med. 63, (1946) 313-315.
9. Thurnham D.I. in: Proc. XIII Int. Cong. Nutr. (J. Libbey: London, 1986) pp 129-131.
10. Stocker R., Hunt N.H., Weideman M.J., Clark I.A.: Proc. Nutr. Soc. 45, (1986) 70A.
11. Buffington G.D., Cowden W.B., Hunt N.H., Clarke I.A.: Febs Lett. 195, (1986) 65-67.
12. Eckman J.R., Eaton J.W., Berger E., Jacob H.S.: Trans. Assoc. Amer. Physic. 89, (1976) 105-113.

13. Murray M.J., Murray A.B., Murray M.B., Murray C.J.: Lancet i, (1976) 1283-1285.

14. Thurnham D.I., Oppenheimer S.J., Bull R.: Trans. R. Soc. Trop. Med. Hyg. 77, (1983) 423-424.

15. Bates C.J., Powers H.J., Lamb W.H. et al.: Trans. R. Soc. Trop. Med. Hyg. in the press.

16. Oppenheimer S.J., Gibson E.D., Macfarlane S.B. et al.: Lancet i, (1984) 389-390.

17. Harvey P., Heywood M.C., Nesheim J-P. et al.: Fed. Proc. 46, (1987) 1161 4920A.

18. Bamji M.S., Bhaskaram P., Jacob C.M.: Ann. Nutr. Metab 31, (1987) 191-196.

19. Sommer A., Tarwotjo I., Katz J.: Amer. J. Clin. Nutr. 45, (1987) 977-980.

20. Scrimshaw N.: Amer. J. Clin. Nutr. 30, (1977) 1536-1544.

21. Wayner D.D.M., Burton G.W., Ingold K.U., Locke S.J.: Febs Lett. 187, (1985) 33-37.

22. Wayner D.D.M., Burton G.W., Ingold K.U., Barclay L.R.C., Locke S.J.: Biochim. Biophys. Acta in the press.

23. Thurnham D.I., Situnayake R.D., Koottathep S., McConkey B., Davis M. in: Free Radicals, Oxidant Stress & Drug Action (Richelieu Press: London 1987) pp 169-192.

24. Burton G.W., Webb A., Ingold K.U.: Lipids 20, (1985) 29-39.

25. Thurnham D.I., Smith E., Flora P.S. in: 8th Int. Sym.

Carotenoids 1987 P18.

26. Thurnham D.I., Stephen J.M.L. in: Nutrition & Health in Old Age, Rep. Hlth Soc. Subj. no 16 (HMSO: London, 1979) pp 191-197.

27. Koster J.F., Biemond P., Swaak A.J.B.: Ann. Rheumat. Dis. 45, (1986) 44-46.

28. Royson P.: SPP - A statistical package for personal computors (Timberlake Clarke Ltd, 40 B Royal Hill. London, 1985).

29. Thurnham D.I. in: Vitamin Deficiency in the Elderly (Pergamon Press: London 1985) pp 46-67.

30. Thurnham D.I., Davies J.A., Crump B.J., Situnayake R.D., Davis M.: Ann. Clin. Biochem. 23, (1986) 514-520.

31. Clark I.A., Butcher G.A., Buffinton G.D., Hunt N.H., Cowden W.B.: Biochem Pharmacol. 36, (1987) 543-546.

32. Lam K-W., Fong D., Lee A., Ming K., Liu D.: J. Inorg. Biochem. 22, (1984) 241-248.

33. Markowitz H., Gubler C.J., Mahoney J.P., Cartwright G.E., Wintrobe M.M.: J. Clin. Invest. 34, (1955) 1498-1508.

34. Golden H.M.N., Ramdath D.: Proc. Nutr. Soc. 46, (1987) 53-68.

FREE RADICALS: CHEMISTRY, PATHOLOGY AND MEDICINE
edited by Catherine Rice-Evans & Thomas Dormandy, © 1988 Richelieu Press, London

Increases in Antioxidants After Endotoxin in Mice

J.M. Stark, S.K. Jackson and Joan Parton

Department of Medical Microbiology, University of Wales College of Medicine, Heath Park, Cardiff CF4 4XN, Wales, U.K.

Abstract

Endotoxin significantly increased the amounts of superoxide dismutase and catalase in the livers and spleens of mice sensitized with BCG. The increase in these enzymes was maximal at 6 hours after endotoxin *in vivo*, which corresponded to a decrease in the measured TBA-reactive products after oxidation *in vitro*. Metallothionein was also significantly increased after endotoxin, reaching maximum amounts after 24 hours. The results suggest that endotoxin initiates oxidative, free radical damage which induces a rise in antioxidants. The final outcome of the endotoxic crisis may depend on factors which influence this balance between oxidation and antioxidants.

INTRODUCTION

Bacterial lipopolysaccharides (endotoxins) are macromolecular glycolipids that form an integral part of the outer cell wall of Gram-negative bacteria. The incidence of septicaemias due to Gram-negative organisms is increasing, particularly in surgical and immune-compromised hospital patients (1). The morbidity and mortality associated with these infections are considered to be due to the release of endotoxin from the bacteria, and antibiotic therapy may sometimes accelerate this process (2).

Injection of nanogram or microgram amounts of purified endotoxin preparations into experimental animals causes a range of non-specific acute effects such as fever, hypotension, disseminated intravascular coagulation and, with higher doses, irreversible shock and death (3).

Despite a wealth of accumulated experimental information, little is known about the intimate events initiating these diverse reactions which typify the endotoxic crisis. Certain facts have, however, been established. First, in experiments with endotoxins from different bacterial mutants which produce endotoxins of differing polysaccharide composition, it is the lipid region of the molecules (termed 'lipid A') which elicits most, if not all, of their biological activity (4). Secondly, it has become clear that a large proportion of the pathophysiological effects of endotoxin are due to the presence of mediators, many released by cells of the macrophage/monocyte line. Examples of mediators that have been implicated in the progression of endotoxic shock include tumour necrosis factor (5), platelet activating factor (6) prostaglandins (7) and leukotrienes (8). The macrophage has emerged as the most important cell type for the development of endotoxic shock; even the unresponsive C3H/HeJ mice can be made sensitive to endotoxin by transfer of macrophages from 'normal' endotoxin-responsive animals (9).

Work in our laboratory has been concerned with the development of sensitivity to endotoxin; it is not clear, for example, why certain species are sensitive to minute amounts of endotoxin (rabbits, man) when others are fairly resistant (rats, mice).

The sensitivity may be modified however, by a sensitizing dose of endotoxin itself, by other infections (bacterial, mycobacterial, viral), and by killed organisms (10,11). These sensitized states are associated with a profound and prolonged state of macrophage activation, with greatly increased numbers of phagocytes in the liver and spleen (12). Activated macrophages can be stimulated to undergo a 'respiratory burst', producing reactive oxygen species such as superoxide anion and hydrogen peroxide (13). One hypothesis for the induction of sensitivity to endotoxin could thus be that the increased phagocytic state of the animal results in an explosive triggering of the production of various mediators and of oxygen-derived free radicals which may cause extensive tissue damage. Recent investigations have shown that macrophages from the spleens of mice sensitised with Bacille-Calmette-Guèrin (BCG) produce oxygen-derived radicals in vitro after stimulation with endotoxin. The blood and ileum from these mice showed an increase in lipid peroxidation products whereas the spleen and liver showed a decrease in such products (14). Splenectomy of sensitised animals reduced the sensitivity to endotoxin. The present study was carried out to investigate the antioxidant changes associated with endotoxin in sensitised animals to see if the pattern of tissue oxidation obtained could be explained by an increase in antioxidant status.

Materials and Methods

Mice: Mature adult male 25-30 mg TO mice were used.

Bacterial preparations and sensitization to endotoxin: BCG-sensitization to endotoxin was induced by giving intravenously approximately 10^7 organisms "Percutaneous" BCG (Glaxo) to warmed, vasodilated mice 14-18 days before endotoxin. Endotoxin (Westphal preparation of the lipopolysaccharide of *Escherichia coli* 026:B6 from Difco) was given in 10 ug amounts intraperitoneally in 0.4 ml phosphate buffered saline (PBS).

Measurement of tissue oxidation: The animals were killed by cervical dislocation at intervals (0,3,6,24,48 hours) after the giving of endotoxin and the spleens and livers removed. After being weighed the tissues were kept at -20°C until processed, not later than seven days later. Tissue oxidation was estimated from the thiobarbituric acid (TBA) reactive material measured in homogenates by a method modified from that of Heys and Dormandy (15). Briefly, a 1:9 dilution of 10% (w/v) tissue homogenate was prepared and aliquoted in 0.5 ml amounts. PBS (2.5 ml) was added to each aliquot. Preparations were allowed to oxidize *in vitro* at 37°C in a shaking water bath for 90 minutes. To the preparations 1.5 ml trichloroacetic acid (1.7M) was added and the precipitated protein removed by centrifugation. To 3 ml of super natant, 1 ml of freshly prepared TBA (69mMol in 0.05 M NaOH) was added. The mixture was boiled for 15 minutes, cooled, and the chromogen extracted in 3 ml acidified (5% HCl) butanol. After centrifugation, the light emission of the butanol layer was measured in arbitrary units in a Perkin-Elmer LS-3 fluorimeter at a wavelength of 548 nm after excitation at 528 nm.

Protein measurement: Tissue protein was measured by the Coomassie Blue dye absorption method (16).

Measurement of antioxidant enzymes: Antioxidant enzymes were measured in BCG-primed liver and spleen homogenates diluted with appropriate buffer. Measurements were made on tissues removed at various times after endotoxin challenge *in vivo*. All chemicals and enzymes were purchased from the Sigma Chemical Company. Enzymes were assayed by accepted methods outlined as follows:

Superoxide dismutase (17) Competition of superoxide dismutase with NBT for O_2^- generated by the xanthine/xanthine oxidase system was measured. The activity of superoxide dismutase in tissue homogenates was calculated by comparison with the inhibition of NBT reduction (A 540 nm) by solutions of pure superoxide dismutase.

Catalase (18) The decomposition of hydrogen peroxide catalysed by catalase was followed at 240 nm. The source of hydrogen peroxide was 30% analytical grade.

Glutathione Peroxidase (19) Sample tissue homogenates were added to solutions of reduced glutathione, glutathione reductase and NADPH. The reaction was initiated by addition of t-butyl hydroperoxide. Glutathione peroxidase activity was calculated from the consumption of NADPH by measuring the decrease in absorbance at 340 nm.

Glutathione s-transferase (20) Reaction of chlorodinitrobenzene with reduced glutathione catalysed by glutathione-S-transferase was monitored at 340 nm.

Metallothionein: Metallothionein was assayed by a competitive ELISA technique.

Vitamin E: Vitamin E was measured by the specific fluorescence at 330 nm after excitation at 286 nm. (21).

RESULTS

The pattern of tissue oxidation obtained *in vitro* in livers and spleens after various times of endotoxin *in vivo* is shown in Fig. 1. The results are expressed as a percentage of the values obtained for tissues from animals without endotoxin

Figure 1. Measurement of TBA-reactive substances (TBARS) in incubated murine liver and spleen homogenates at various times after endotoxin _in vivo_. Values of TBARS are expressed as a percentage of the values obtained without endotoxin.

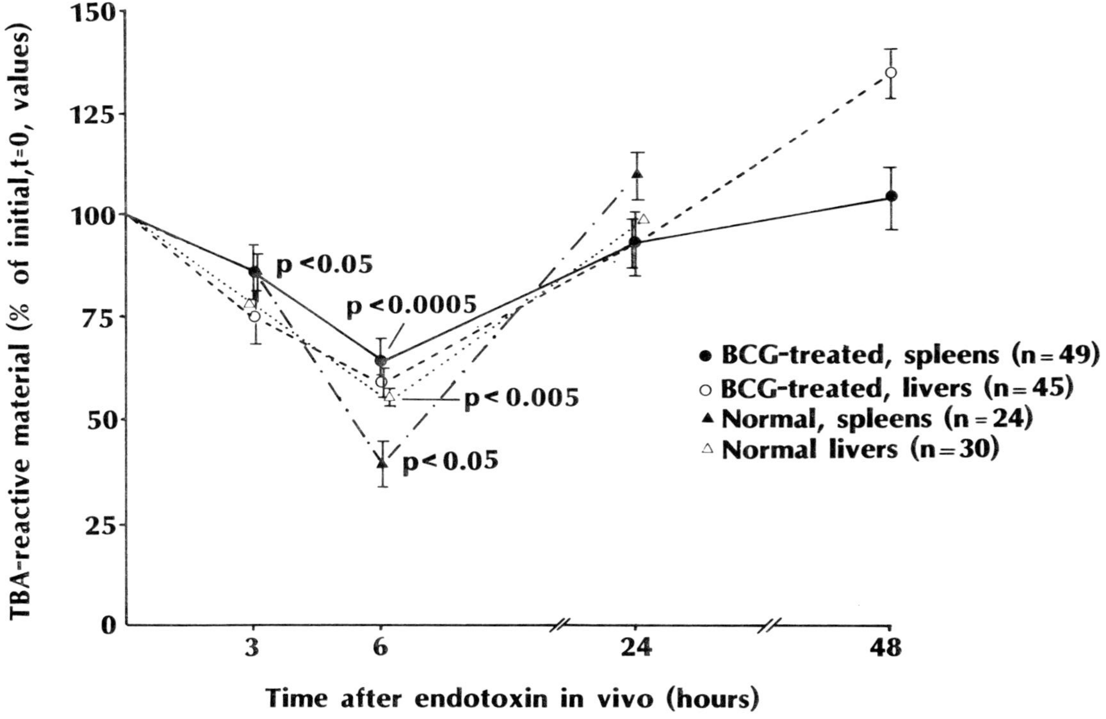

(i.e. zero time values). As can be seen, when the tissues are oxidized in vitro, the amount of TBA-reactive material in both livers and spleens decreases with time after endotoxin in vivo. This decrease in the TBA-reactive substance level reaches a minimum at about 6 hours of endotoxin in vivo and then rises again, increasing above the baseline (t=0 hours) level 24 hours after the giving of endotoxin. These patterns were essentially the same for both sensitized (BCG pre-treated) animals and non-sensitized animals, but were much more significant in the BCG-treated groups. These results might imply that the presence of endotoxin in vivo induces some resistance to oxidation of the tissues (as measured in vitro). This 'protective' effect of endotoxin is maximal after 6 hours and thereafter declines.

The amounts of all four enzymes assayed for - superoxide dismutase, catalase, glutathione peroxidase and glutathione-S-transferase, increased over the same period (Figures 2 and 3). The

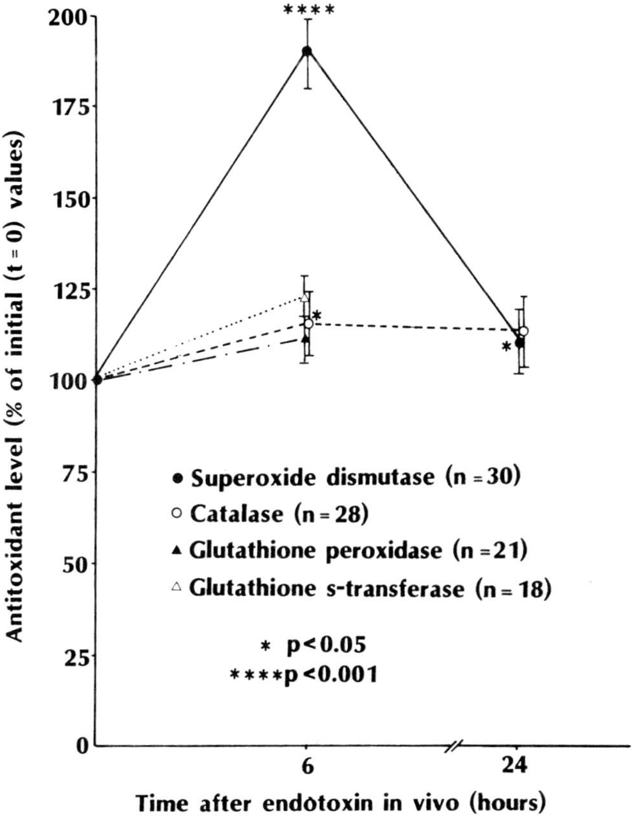

Figure 2. Antioxidant enzymes in the liver of BCG-primed mice following administration of endotoxin. Enzyme amounts are expressed as a percentage of the values obtained without endotoxin.

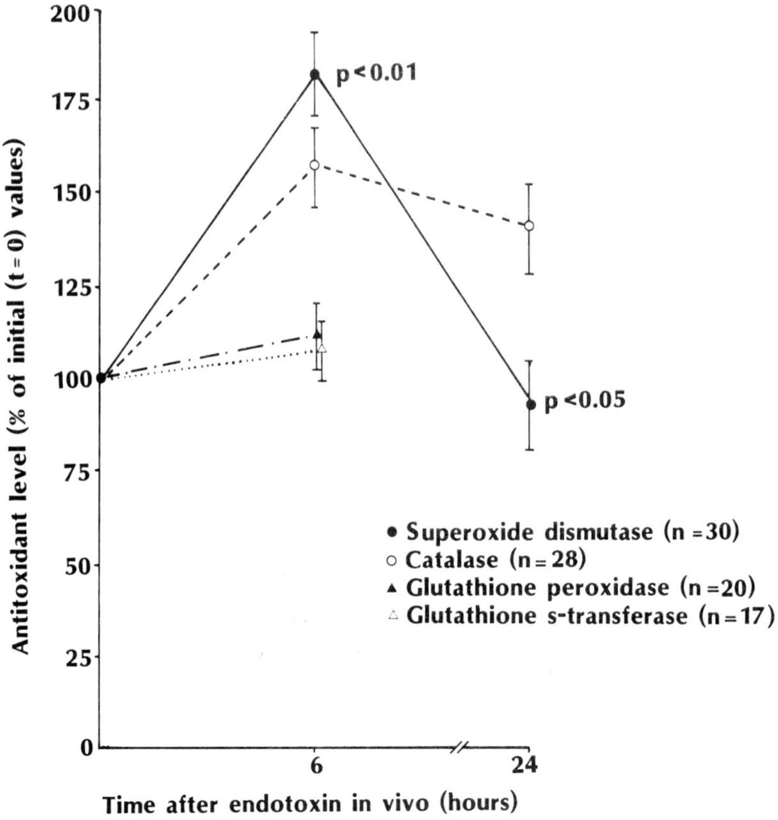

Figure 3. Antioxidant enzymes in the spleen of BCG-primed mice following administration of endotoxin. Enzyme amounts are expressed as a percentage of the values obtained without endotoxin.

increases were only significant for superoxide dismutase and catalase, however.

The amounts of metallothionein and vitamin E are shown in Fig. 4. The amounts of metallothionein present increased greatly in both spleen and liver after the giving of endotoxin. The maximal levels of metallothionein were found after 24 hours of endotoxin <u>in vivo</u> and decreased to control (t=0 hours) levels after 36 hours. No significant change in vitamin E levels could be detected during this time.

DISCUSSION

In these experiments endotoxin has not only altered the susceptibility to oxidation of both liver and spleen but has also caused a rise in the levels of superoxide dismutase, catalase and metallothionein. It would be logical to conclude that the second observation is the key to the first.

Does endotoxin really affect the oxidizability of liver and spleen or is this a consequence of

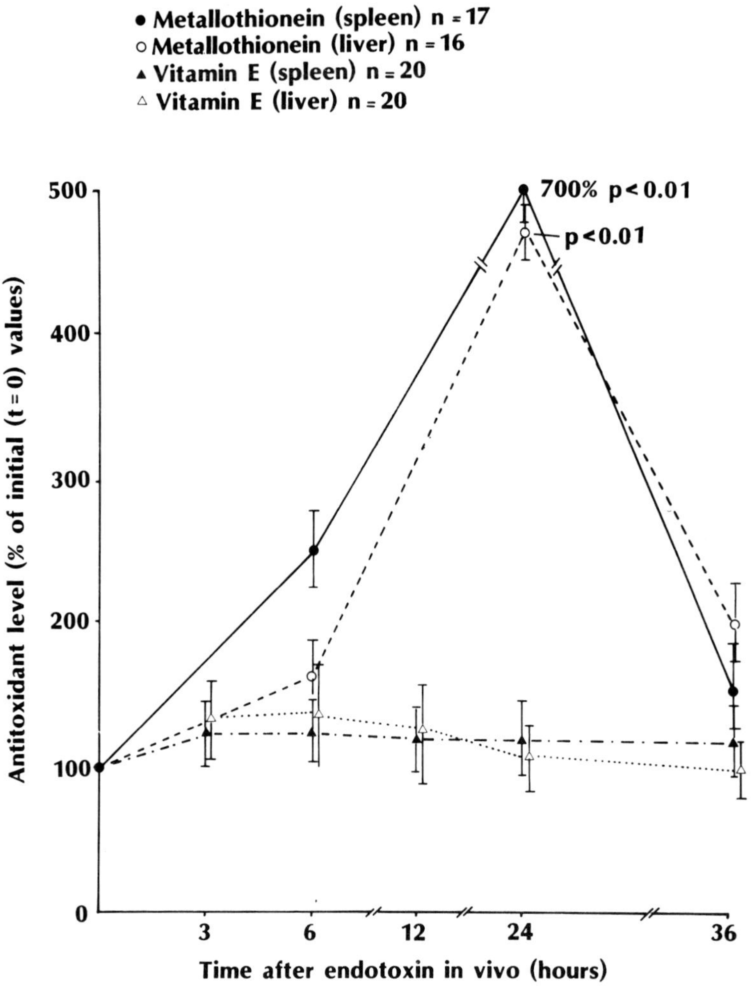

Figure 4. Relative amounts of metallothionein and Vitamin E in liver and spleens of BCG-primed mice following administration of endotoxin.

some other event induced by the endotoxic state ? It has previously been shown (14) that freeze-dried samples of spleen from endotoxic animals showed the same decrease in TBA-reactive substances. This result implies that this apparent resistance to oxidization is not due to oedematous tissue diluting out the TBA-reactive product. That the oxidizable substate is being removed is thought unlikely because increases in TBA-reactants can be measured in other tissues such as the ileum (14). Furthermore, when lipid peroxidation levels were measured in rats infused with endotoxin, while the levels of TBA reactants in serum, abdominal aortic wall and ileum were significantly increased those in kidney and liver were slightly decreased (22). These observations suggest that endotoxin is actually the cause of the TBA reactant decrease in some tissues, while increasing it in others.

There is growing evidence that endotoxin can induce the production of antioxidant enzymes. Frank et al (23) were among the first to demonstrate that endotoxin-treated rats showed increases in pulmonary superoxide dismutase, catalase and glutathione peroxidase, and suggested that these enzymes protected the animals from exposure to high oxygen tensions. They could, however, find no such increases in the lungs of mice under similar conditions. Endotoxin has been shown to increase superoxide dismutase in cultured bovine pulmonary endothelial cells (24) and human monocytes (25).

The findings of the present report suggest that endotoxin can stimulate the production of superoxide dismutase and catalase in the spleen and liver of mice sensitized with BCG.

Metallothionein was also seen to increase dramatically in the present experimental conditions. Endotoxin increases liver copper and zinc which were associated with an elevation in liver metallothionein mRNA activity (26). The time-course for this response (maximal metallothionein mRNA activity 6 hours after endotoxin administration) coincided with the decreases in TBA products measured in the present report. Moreover, peritoneal macrophages collected from mice responded to endotoxic challenge through synthesis of a sulphydryl-rich, metal-binding protein assumed to be metallothionein (27).

The time course for the protective effect of superoxide dismutase and catalase on tissue oxidation in the present report is in good agreement; the enzymes are at maximal levels when the resistance to oxidation (as measured by the decrease in TBA-reactive substances) is also maximal (Figs. 1-3). The pattern for metallothionein is not as convincing; it rises sharply at 6 hours of endotoxin, but continues to rise, reaching maximal levels at 24 hours after endotoxin, when the protection from oxidation is clearly over (Figs. 1 and 4). This would suggest that superoxide dismutase and catalase are the most important antioxidants in the liver and spleen in this form of the endotoxic crisis. When superoxide dismutase and catalase were added to oxidizing liver and spleen tissue homogenates from BCG-primed mice in an attempt to recreate the diminution in TBA-reactive material, only a slight protective effect was noted (Stark, Jackson and Parton, unpublished observations). If metallothionein was added to the mixture, the protective effect was increased still further, but the pattern of TBA-reactive material levels did not match those induced by endotoxin in the live animals. This leads to the conclusion that other antioxidants might be responsible for the resistance to oxidation induced by endotoxin. Predominant among these might be caeruloplasmin (28), and the search for these is currently in progress.

Increasing evidence suggests that endotoxic shock is an oxidative, free radical-mediated process (29,30,31). This view leads to the apparent paradox that endotoxin can protect tissues from oxidative damage while also causing such damage. The oxidative attack itself might initiate the production of antioxidant enzymes which could afford protection for the tissue until being overwhelmed by the oxidative crisis. Exposure of cultured pulmonary endothelial cells to E. coli endotoxin resulted in an increase in Mn superoxide dismutase without changing enzymes reflecting mitochondrial activity (fumarase and cytochrome-c oxidase) (24). These findings support the notion that endotoxin can induce antioxidant enzymes by causing increased intracellular production of free radicals. Mice sensitized to endotoxin do not normally begin to die until 6-8 hours after exposure to endotoxin, most succumbing to the lethal effects at times later than this (32). This fact lends weight to the hypothesis that endotoxin induces antioxidant enzymes in liver and spleen and that their protective effects become overwhelmed by the oxidative challenge after about 6 hours (in the mouse model).

In a model of endotoxin liver injury in Corynebacterium parvum treated rats (33), it was concluded that oxygen-derived free radicals were responsible for the severity and rapid onset of the lesions observed. In this model, death usually occured within 6 hours and there was a reduction in mortality and hepatic necrosis after treatment with superoxide dismutase. It was concluded that the production of superoxide radicals by hepatic macrophages and Kupffer cells was enhanced by endotoxin, and that this oxidative stress could overwhelm the natural protective mechanisms of the host cells.

Central to this hypothesis is the role of the macrophage. The macrophages, gathered in granulomatous lesions in the liver and spleen during BCG infection, are in an activated state. When challenged with endotoxin (most of which is phagocytosed within the spleen and liver) they respond by producing a range of mediators (the so-called polypeptide network (34)), as well as reactive oxygen metabolites and antioxidant enzymes. If the crisis is severe enough, the oxidative processes overcome the homeostatic defences and the host enters the decline into irreversible shock and death.

Is the infiltration of macrophages responsive to endotoxin the only circumstance favouring the development of sensitivity to endotoxin ? Recent findings in the BCG-mouse model suggest (14) that infective states can alter the percentage composition of different fatty acids in cell membrane from various tissues. In particular, the amount of arachidonic acid was increased. This might be significant in view of the important pharmacological derivatives of arachidonic acid.

Work is currently in progress to discover whether in this type of model altered fatty acid profiles can be demonstrated to influence susceptibility to oxidative changes.

ACKNOWLEDGEMENT

An outline of some findings given in this paper was presented at the International Symposium and Workshop on Verocytotoxin-producing *Escherichia coli* Infections in Toronto, July 1987.

REFERENCES

1. R.J. Hamill & D.G. Maki: Handbook of Endotoxin, Vol. 4, Ed. R.A. Proctor (Elsevier Science, 1986) p.55

2. J.L. Shenep and K.A. Mogan: J. Infect. Dis., 150, (1984) 380.

3. E.T. Rietschel, U. Schade, M. Jensen, H.W. Wollenweber, O. Lüderitz and S.G. Greisman: Scand. J.Infect. Dis. Suppl. 31, (1982) 8.

4. C. Galanos, O. Lüderitz, E.T. Rietschel and O. Westphal: Int. Rev. Biochem., 14, (1977) 239.

5. K.J. Tracey, B. Beutler, S.F. Lowry et al: Science 234, (1986) 470.

6. Z-I. Terashita, Y. Imura, K. Nishikawa and S. Sumida: Eur. J. Pharmacol. 109, (1985) 257.

7. J.T. Flynn: Handbook of Endotoxin, Vol. 2, Ed. L.B. Hinshaw (Elsevier Science, 1985) p.237.

8. A.M. Lefer: Biochem. Pharmacol. 35, (1986) 123.

9. M.A. Freudenberg, D. Keppler and C. Galanos: Infect. and Immunity, 51, (1986) 891.

10. E. Suter and E.M. Kirsanow: Immunology, 4, (1961) 354.

11. C. Adlam and M.T. Scott: J. Med. Microbiol. 6, (1973) 261.

12. A.A. Otu, R.J. Russell and R.G. White: Immunology, 32, (1977) 255.

13. J.A. Badwey and M.L. Karnovsky: Ann. Rev. Biochem., 49, (1980) 695.

14. J.M. Stark, S.K. Jackson, H.C. Ryley and B. Woodhead:Bioelectrochemistry and Bioenergetics, 18, (1987) (in press).

15. A.D. Heys and T.L. Dormandy: Clin.Sci., 60, (1981) 295.

16. M. Bradford: Anal. Biochem., 71, (1976) 248.

17. L.W. Oberly and D.R. Spitz: Handbook of Methods for Oxygen Radical Research, Ed. R.A. Greenwald (CRC Press, 1985) p.217.

18. R.F. Beers and I.W. Sizer: J. Biol. Chem., 195, (1952) 133.

19. W.A. Gunzler and L. Flohé: Handbook of Methods for Oxygen Radical Research, Ed. R.A. Greenward (CRC Press, 1985) p. 285.

20. P.C. Simons and D.L. Vander Jagt: Anal. Biochem., 82, (1977) 334.

21. S.L. Taylor, M.P. Lamden and A.L. Tappel: Lipids, 11, (1976) 530.

22. T. Yoshikawa, M. Murakami, Y. Furnkawa, H. Kato, S. Takemura and M. Kondo: Thromb. Haemost., 49, (1983) 214.

23. L. Frank, I. Summerville and D. Massaro; J. Clin. Invest., 65, (1980) 1104

24. Y. Shiki, B.O. Meyrick, K.L. Brigham and I.M. Burr: Am. J. Physiol., 252, (1987) C436.

25. K. Asayama, R.L. Janco and I.M. Burr: Am. J. Physiol., 249, (1985) C393.

26. K.R. Etzel, M.R. Swerdel, J.N. Swerdel and R.J. Cousins: J. Nutr., 112, (1982) 2363.

27. S.R. Patierno, M. Costa, V.M. Lewis and D.L. Peary: J. Immunol., 130, (1983) 1924.

28. J.M.C. Gutteridge, R. Richmond and B. Haliwell: FEBS Lett, 112, (1980) 269.

29. E.S. Lee, A.G. Greenberg, P.W. Maffuid, E.D. Melcher and T.S. Velky: J. Surg. Res., 42, (1987) 1.

30. K.A. Knox, G. Smith, P.L. Yap and H.A. Leaver: Biochem. Soc. Trans., 15, (1987) 544.

31. G.P. Novelli, P. Angiolini and R. Tani: Free Radicals in Liver Injury, Ed. T. Slater (IRL, Oxford, 1985) p.225.

32. V. Lehmann, M.A. Freudenberg and C. Galanos: J. Exp. Med., 165, (1987) 657.

33. M.J.P. Arthur, I.S. Bentley, A.R. Tanner, P.K. Saunders, G.H. Millward-Sadler and R. Wright: Gastroenterology, 89, (1985) 1114.

34. L.J. Old, Nature, 326, (1987) 330.

MEASUREMENT OF RADICAL INTERMEDIATES AND THEIR CONSEQUENCES IN BIOLOGICAL SYSTEMS

FREE RADICALS: CHEMISTRY, PATHOLOGY AND MEDICINE
edited by Catherine Rice-Evans & Thomas Dormandy, © 1988 Richelieu Press, London

Spin trapping studies on the metabolic activation of carbontetrachloride in rat liver microsomes incubated under steady-state oxygen pressures.

A.Tomasi, [i]E.Albano, [i]M.U. Dianzani, A.Iannone, [ii]T.F.Slater, [iii]H.M.Swartz, and V.Vannini.

Istituto di Patologia Generale, via Campi 287, 4110 Modena (Italy),

[i]Dipartimento di Medicina e Oncologia Sperimentale, C.so Raffaello 30, 10125 Torino (Italy).

[ii]Biochemistry Dept., Brunel University, Uxbridge Middx (U.K.)

[iii]College of Medicine at Urbana-Champaign, University of Illinois (USA).

Abstract

Spin trapping is quickly becoming a method of choice used to demonstrate the presence of short lived free radicals in biological systems. Experiments performed <u>in vivo</u>, in liver microsomes and in isolated hepatocytes give unambiguous evidence of the presence of free radical intermediates arising during the reductive bioactivation of carbontetrachloride. More recently the formation of the trichloromethyl free radical has been demonstrated in isolated mitochondria.

Formation of the trichloromethyl free radical is a key step in the process leading to carbontetrachloride toxicity; it is also well known that hypoxia favours the reductive activation of CCl_4 and potentiates its hepatotoxicity. Experimental conditions defined as hypoxic have appeared recurrently in the literature concerning the bioactivation of carbontetrachloride and similar halocompounds, though the term hypoxia usually was not well defined. Here, the use in a novel way of a gas permeable capillary tube inserted in the ESR cavity has allowed to carry out carbontetrachloride bioactivation experiments under known and well controlled partial pressures of oxygen (ranging from 0 to 14 mg/l oxygen), at 37°, using phenyl-t-butyl nitrone as a spin trap to provide evidence on free radical formation. The formation of the trichloromethyl adduct was observed in a range of oxygen concentrations between 0.2 and 8 mg/l. A decrease in the adduct formation was observed at higher oxygen concentrations. No ESR signal was detected when oxygen concentration in the sample dropped to 0.

Introduction

Direct evidence of the involvement of free radical intermediates in biochemical reactions can be obtained by the spin trapping technique which has been introduced to the biological and medical research field only a few years ago (1).

Although results obtained through the technique are not easily interpreted and great care has to be exercised to avoid erroneous deductions, nevertheless it appears that the method is very sensitive for the detection of short-lived free radical intermediates produced in a biological environment, including the possibility of *in vivo* applications.

Experiments using spin trapping have given an unambiguous answer to the old problem regarding the involvement of free radical intermediates in carbontetrachloride (CCl_4) toxicity. While CCl_4 is no longer used in clinical application (at one time it was the drug of choice for treating hookworms), it still prevails in the environment due to industrial and other uses and therefore it remains a source of concern, particularly since its possible carcinogenicity has been reported (2).

CCl_4 is the prime example of a hepatotoxic haloalkane, and the biochemical mechanisms of its toxic effects have been intensively investigated; the literature available on the subject is vast (reviews in 3 and 4).

It has been clearly established that covalent binding to proteins and lipids of the activated intermediate and the induction of lipid peroxidation are at the basis of the molecular mechanism of CCl_4 toxicity and, perhaps, carcinogenicity. CCl_4 undergoes metabolism via an aerobic pathway terminating in CO_2 production (5) and via a hypoxic path resulting in an increased yield of covalent binding (6). Spin trapping experiments have demonstrated a higher yield of trichloromethyl free radical formation under hypoxic conditions (7). It follows that a major controlling factor is the local oxygen concentration that will determine the flux through each pathway, modulating in this way the toxicity of the compound.

In vivo the liver is continuously perfused by blood in a manner that results in an oxygen gradient; higher O_2 concentrations will be present near the perilobular area, while lower ones will be established around the centrilobular area. Different metabolic pathways will be then observable in the liver depending on the area where CCl_4 metabolism is taking place.

In this work we describe, along with a brief report on the state of the art on spin trapping trichloromethyl radical, a model system where experiments can be performed with cells or cell subfractions placed in the ESR cavity in the presence of a spin trapping agent, with the oxygen partial pressure kept constant at a predetermined concentration.

SPIN TRAPPING TRICHLOROMETHYL FREE RADICAL

A number of recent reviews on spin trapping in biological systems are available (8, 9); here we will draw attention to some characteristic features of spin trapping halocarbon-derived free radicals in biological systems under normoxic or hypoxic conditions.

Formation of the PBN - trichloromethyl adduct.

The formation of an adduct between a nitrone spin trap and a free radical giving rise to a nitroxide takes advantage of the well known stability of the nitroxyl free radical function. The unpaired electron is resonance stabilized which, along with the presence of methylated carbon atoms adjacent to the nitroxide (typically a bulky tert-butyl group), prevents the formation of dimers. When the spin trap reacts with radicals, the persistent spin adduct accumulates; a ESR signal due to the nitroxide is recorded when the spectrometer detection limit is exceeded.

The accumulation of the spin-adduct depends on the concentration of the spin trapping agent, the concentration of the free radicals, the rate constant of the reaction between the spin trap and the free radical, the rate constants of competing reactions, and the rate of reactions of the spin adduct. The steady-state free radical concentration in a tissue is rather low; attempts to establish the presence of trichloromethyl radicals in liver slices incubated in the presence of CCl_4 inside the ESR cavity have demonstrated that the highest possible concentration of the free radical is below the limits of detection of the spectrometer (10).

The rate constant of the reaction between the free radical and the trapping agent are not extensively known. In the case of trapping carbon centered radicals the rate constants are rather low; for example the rate constant for PBN trapping trichloromethyl radical is lower than 10^4 M^{-1}, s^{-1} in liver microsomes (11). Hence the concentration of the spin trapping agent has to be quite high: 25 mM PBN and higher has been used in the biological system under investigation in order to trap a detectable amount of trichloromethyl radical adduct.

Little, eventually, is known about the reactivity of the spin adduct. Adducts originating from the trapping carbon centered radicals are maintained to be fairly stable, at least for various hours at room temperature.

Studies on the reactivity of $CCl_3^{\cdot}$ (12) demonstrated that the trichloromethyl radical reacts extremely rapidly with oxygen giving the much more reactive species $CCl_3O_2^{\cdot}$. This species is involved in the initiation of the lipid peroxidative process. Attempts to trap and identify the resulting trichloromethylperoxy radical generated in a biological system have been fruitless; more on this subject will be commented below.

Identification of the spin adduct.

A major problem in spin trapping is assigning the structure of the spin adduct correctly. The unpaired electron interacts with the nitrogen nucleus which has a nuclear spin I=1. The spectral feature observed is influenced as well by a β-proton ($I=\frac{1}{2}$) present in the PBN molecule. Hence the typical spectrum of PBN-carbon centered radical adducts is characterized by a triplet of doublets. In these conditions there was no possibility of an unambiguous identification of the trichloromethyl-PBN adduct. Methods have been developed to provide authentic spin adduct through reactions driven by high energy radiation, or by electrochemical reduction of halocarbons (13,14). However, minor changes in solvent polarity produce larger effects on the splitting of the nitroxyl nitrogen and the β-hydrogen than changes in the structure of the free radical trapped. Therefore the comparison between hyperfine splitting constants of PBN adducts between an authentic spin adduct spectrum obtained in chemical model systems and that obtained from experiments in biological materials is risky because the distribution of the spin adduct between lipids and water phase is not well known. A safe assignment can be done by HPLC isolation of the adduct and identification by mass spectrometry, a procedure which is possible only in the case of rather stable adducts.

Another possible approach is that of replacing a nucleus of the free radical with an isotope of different nuclear spin. In the case of CCl_4, the PBN-trichloromethyl adduct was unambiguously identified using ^{13}C-CCl_4; here a further splitting in the spectrum was observed when PBN trapped the free radical produced during ^{13}C-CCl_4 bioactivation, the new feature was definitely to be assigned to the presence of the ^{13}C isotope in β-position .

The trapping and identification of the trichloromethylperoxy radical has been attempted but has not been successful. An ESR spectrum obtained from microsomes incubated in the presence of ^{13}C-CCl_4, PBN and a high concentration of oxygen resulted in a spectrum which did not show typical ^{13}C lines. Experiments carried out with $^{17}O_2$ (I=5/2) did not show any presence of the oxygen isotope near the nitroxide radical center (7); hence there was no direct evidence of a peroxy radical adduct. A reasonable explanation of the fact was given when it was demonstrated that PBN peroxy radical adducts were unstable at room temperature and in the light (15).

Investigations in vitro and in vivo

The microsomal fraction, in particular, the Mixed Function Oxidase System (MFOS) centered on cytochrome P_{450} is the site of CCl_4 bioactivation (3). Hence most spin trapping experiments have been performed on microsomal subfractions or on isolated liver cells, systems that are particularly rich in MFOS.

Experiments also have been performed in vivo delivering the spin trap and CCl_4 to the animal; the spin adduct was then extracted in an organic solvent. PBN was able to trap specifically the $CCl_3\cdot$ along with lipid and lipid oxy radicals (16, 11); a different spin trap (trimethoxy phenyl-t-butyl nitrone) was unable to trap the CCl_4-derived radical (16). The authors interpreted the results as indicating that PBN enters the lipid domain of the endoplasmic reticulum, hence it can access the site of $CCl_3\cdot$ formation at cytochrome P_{450}.

Cytochrome P_{450} is the undisputed principal activating site of CCl_4, however the spin trapping technique has allowed the identification of a different activating site in isolated mitochondria.

This latter work has shown that purified mitochondria preparations are able to activate CCl_4 to $CCl_3\cdot$ as demonstrated by spin trapping (17). It has been known for a long time that CCl_4 administered in vivo damages mitochondria uncoupling oxidative phosphorylation and impairs Ca^{++} transport (18). More recently the possibility of mitochondrial activation was suggested by the recovery of alkylated proteins, lipids (19) and DNA (20) extracted from highly purified mitochondria incubated in the presence of CCl_4.

The data so far available suggest that the mechanism for activation in mitochondria share some analogies with the univalent reduction of oxygen occurring via the respiratory chain. CCl_4 has a reduction potential close to that of O_2 and therefore it could compete with it for reducing equivalents. Complex III and Complex IV may well participate in the activation of CCl_4 probably by a mechanism similar to that of cytochrome P_{450} mediated one. These results could lead to reconsider the role played by the damage of these organelles in the pathogenesis of CCl_4 induced toxicity.

Effect of PBN on the bioactivating system.

Another aspect to be taken in account when dealing with trapping the trichloromethyl radical is the effect of the spin trap itself. Recent investigations (21) have demonstrated that PBN 25 mM moderately affects the integrity of liver cell, decreases hepatocyte glutathione content, and competes with other substrates (hence, with CCl_4 also) for binding to cytochrome P_{450}. On the other hand, α(4-pyridyl-1-oxide-N-t-butyl nitrone (4-POBN), which has less effect on those cell parameters, is much less efficient in trapping $CCl_3\cdot$ in a biological environment.

The relative concentration of the spin trap in the phase where the bioactivation is taking place is apparently very important. Since CCl_4 is metabolically activated to a free radical by cytochrome P_{450} which is completely surrounded by the membrane, a highly lipophilic spin trap like PBN will have an advantage compared to other, less lipophilic, spin traps. The specificity in trapping a certain radical demonstrated under various experimental conditions probably depends on the structure, lipid solubility, and reaction rates of the spin trapping agent used, however such data up to now are not available.

Similarly it is still not possible to estimate the net result of the metabolic competition between PBN and carbontetrachloride, certainly it may be expected that a lower amount of the halocompound will be metabolized in the presence of the spin trap. Since CCl_4 metabolic activation is the cause of its toxicity, a protective effect of the spin trap toward the toxic effect induced by CCl_4 should be expected. And indeed it has been reported that CCl_4 induced lipid peroxidation is partially inhibited by PBN (7).

Oxygen concentration and CCl_4 bioactivation.

Trichloromethyl radical trapping in microsomes has been carried out in two different ways: the experiment was either done directly in a ESR cavity with the sample contained in a flat cell, or the microsomal suspension was incubated in a flask, the PBN adduct was extracted in chloroform methanol and the organic phase placed in the ESR cavity for study. It is obvious that oxygen concentration in a sealed flask varies during the incubation time particularly in the presence of NADPH added as cofactor in this kind of experiments. Moreover, experiments carried out in a ESR flat cell, incubating microsomes in the presence of NADPH and a halocarbon, will lead to anoxia in a very short time. It appears clear that if oxygen concentration affects the bioactivation of carbontetrachloride, these approaches have significant drawbacks.

Noll and De Groot (22) have developed an oxy-stat apparatus capable of determining a steady-state oxygen concentration over the incubation period and, using this apparatus, performed experiments using liver microsomes in the presence of NADPH and CCl_4. In this way it has been demonstrated that a maximum of CCl_4-induced lipid peroxidation was reached at a low (though not zero) oxygen concentration. The interpretation of such data leads to the conclusion that the CCl_4 bioactivation is favoured at low oxygen concentration, but not so low as to make the lipid peroxidative process insignificant. A major disadvantage of this apparatus, if applied to ESR, is due to the continuous addition of buffer, which contains a predetermined oxygen concentration, to the incubating mixture. This is the mechanism used by the authors in order to keep a steady-state oxygen concentration, however, on the other hand both cofactors and halocarbon are continuously diluted through the duration of the experiment.

Fate of the nitroxide spin adduct.

Nitroxides potentially can undergo both oxidation and reduction.

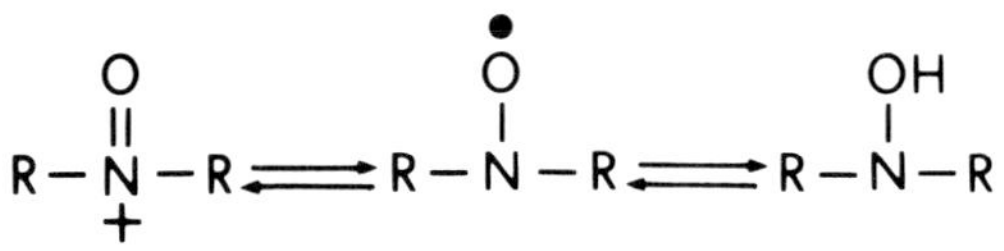

I II III

Fig.1: The nitroxonium (I) ion can be produced with strong oxidizing agents from a nitroxide (II), which can be easily reduced to hydroxylamine (III).

Ascorbic acid easily reduces many nitroxides to the corresponding hydroxylamine, which is ESR silent (23). It was known as well that nitroxides undergo reduction reactions in the presence of various biological materials (24). Recently the attention was drawn on the fact that the reduction rate of nitroxides in a biological system (cells in culture) significantly increases at low oxygen concentrations (25). Conversion of the spin adduct to diamagnetic species has been recently reported (26) for dimethyl 1-pyrroline-N-oxide trapping oxygen centered radicals. The data available on the fate of spin adduct arising from the nitrone-carbon radical reaction are scarce. However one may infer from the above mentioned data that the nitroxide adduct will be reduced in a biological system at a rate dependent on the oxygen concentration.

In the present study we sought to determine the effect of oxygen on spin trapping of CCl_4 derived radical without influencing the concentration of the compounds present in the suspension. This has been obtained by using a gas permeable tube inserted directly in the ESR cavity. Experiments have been carried out at 37°C under various oxygen partial pressures (from 0 to 14 mg/l oxygen), in liver microsomes from phenobarbitone treated animals using phenyl-t-butyl nitrone as spin trap.

Materials and Methods

α-Phenyl-tert-butylnitrone (PBN) was purchased from Aldrich, and $K_3[Fe(CN)_6]$ from Fisher Chem. Co., all other chemicals were obtained form Sigma and used without purification.

Adult male rats maintained on a standard laboratory diet, were pretreated with phenobarbitone (PB), 1g/l in drinking water for at least 5 days before use.

Liver microsomes were prepared as previously described (25). The pellets were either used the same day of preparation or subdivided in small aliquots and kept at -80°C up to the moment they were used. The microsomes incubation mixture consisted of three parts of 0.1 M phosphate buffer p 7.4, two parts of 0.15 M KCl, 10 mM $MgCl_2$, 25 mM PBN, and 0.5 mM NADPH added at the beginning of the incubation; total incubation volume was 100 ul. Some experiments were performed adding 0.5 mM $NADP^+$ instead of NADPH; CCl_4 (100 μM) was added to the mixture contained in a 1 ml Eppendorf tube and thoroughly stirred for a few seconds.

The sample was drawn into a gas permeable teflon capillary tube (Zeus Industries, Raritan N.J.) of 0.081 mm inside diameter and 0.0038 ± 0.0013 mm wall thickness about 10 cm long. The tube containing the sample was folded in two and inserted in a quartz ESR tube, open at both ends, placed in the cavity Dewar; gas of the desired composition was flowed through the dewar and around the sample. Temperature was checked by a copper-constantan thermocouple and kept at 37°c in all the experiments. Typical spectrometer settings were: modulation amplitude 2 gauss, frequency 9.4 GHz, Field modulation frequency 100 KHz, microwave power 10 mW, sweep field 100 Gauss.

Results and Discussion

Microsomes obtained from livers of not PB induced rats did not give rise to a well detectable ESR signal when incubated in the presence of CCl_4 and NADPH. Experiments were therefore carried out only on PB induced microsomal fractions, which are expected to be much more efficient in activating CCl_4.

Fig.2

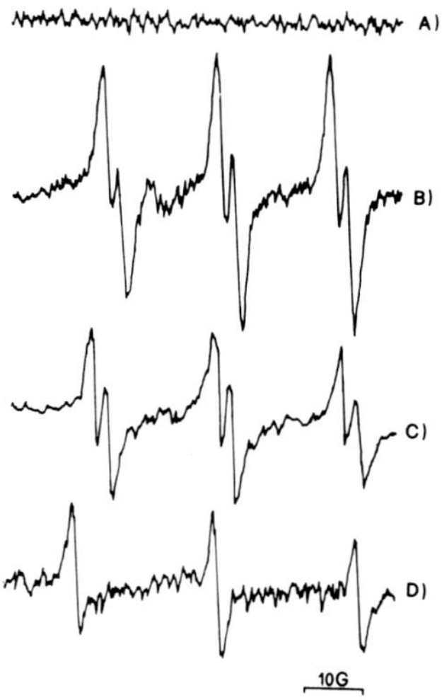

control microsomes

+ carbontetrachloride and NADPH

+ carbontetrachloride and $NADP^+$

addition of 1 mM ferricyanide following anoxic reduction of the spin adduct depicted in trace B

PB induced microsomes gave rise to a well detectable spin adduct within a minute of incubation at 37° under air flow.

Hyperfine splitting constants of the signal were ^{14}N=14 Gauss and ^{1}H=1.8, accurately fitting with the well known trichloromethyl-PBN adduct (Fig. 2B). The assignment of the spin adduct to the $CCl_3^{\cdot}$-PBN adduct, though, was performed, as discussed above, using ^{13}C-CCl_4.

The signal intensity arose very quickly within two minutes of incubation, then more slowly as shown in figure 3.

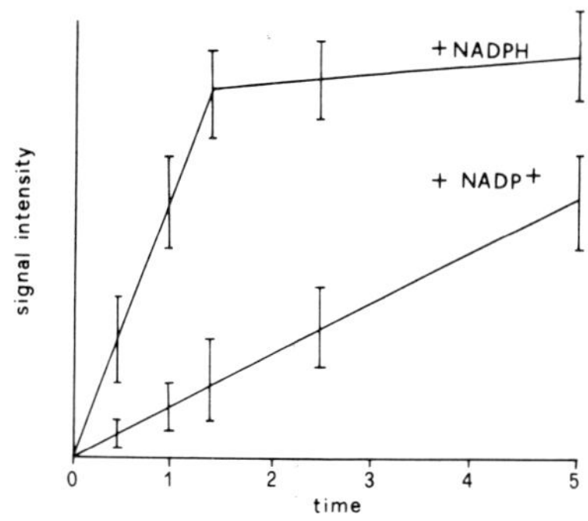

Fig. 3: The intensity of the ESR signal grew rapidly when both CCl_4 and NADPH 1mM were present in the suspension; a less intense signal was recorded even when $NADP^+$ 1 mM was added. Each point represents the mean of three different experiments ± S.E.

No significant amount of different radical species, as those described in the literature, (e.g. in 13) which were attributed to the trapping of lipid radicals, were apparently trapped during the incubation period which was extended, in some occasions, up to two hours. An unexpected result was obtained when microsomes were incubated in the presence of $NADP^+$, instead of NADPH. The experiment was performed in order to make certain that the bioactivating system involved in producing the trichloromethyl radical was indeed the system centered on cytochrome P_{450}. The ESR signal in this case arose much more slowly than when NADPH was present (Fig. 3C); the final amount of trapped radical was significantly lower. The species trapped has not been identified yet, though the hyperfine splitting constants are very close to those observed for the trichloromethyl-PBN adduct.

This is an unusual observation since no bioactivation is expected in the absence of reducing electrons. As long as the spin adduct will not be surely identified it is difficult to give a plausible explanation to this result. The effect of varying the oxygen tension is shown in figure 4. At five minutes of incubation time the amount of spin adduct observed at the indicated oxygen concentration did not show any major change in the range varying from 0.5 mg to about 9 mg O_2/ml. At both, higher and lower O_2 concentration the amount of spin adduct recorded decreased substantially. The diminution observed at high O_2 concentration has a straightforward explanation: oxygen competes with the halocarbon for reducing electrons, hence a reduction of the trichloromethyl radicals yield is to be expected at the increase of oxygen concentration. The opposite, i.e. the decrease and the disappearance of the signal under low oxygen concentrations is presumably to be assigned to a reduction process similar to that described for some nitroxides probes (25).

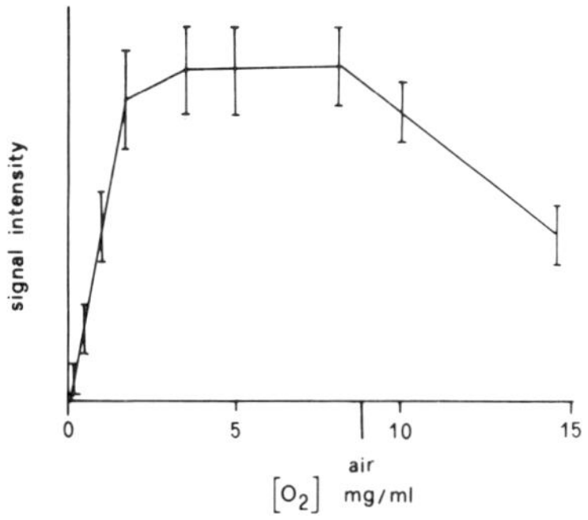

Fig.4: Intensity of the spin adduct observed after 5 min of incubation at 37°C in the presence of CCl_4 and NADPH. The incubations were carried out at different steady-state oxygen concentration. Each point represents the mean of three different experiments ± S.E.

In order to prove the existence of such reduction process samples were incubated for 30 minutes in the presence of CCl_4 and NADPH under air; then they were exposed to a different O_2 concentration for 5 minutes. Figure 5 shows the results of such an experiment and clearlyindicates that the adduct, at low oxygen tensions, is rapidly transformed in an ESR-silent species. It has been suggested that the most likely diamagnetic species to be expected under such hypoxic conditions should be the hydroxylamine (25). If this were the case a mild oxidant, like ferricyanide ($K_3[Fe(CN)_6]$, should be capable of determining a re-oxidation of the hydroxylamine back to the parent nitroxide. A new experiment was then performed: once the spin adduct was formed and recorded, as described above, oxygen free nitrogen was blown through the tubing causing a prompt disappearance of the $CCl_3^{\bullet}$-PBN adduct. At this point $K_3[Fe(CN)_6]$ was added to the suspension and the oxygen concentration was increased back to the initial one. Ferricyanide addition, as shown in figure 2 trace D, failed to restore the spin adduct suggesting that the chemical modification exceeded a mere reduction to the corresponding hydroxylamine. The ESR signal recorded has not been identified; though, it can be safely stated that the nitroxidespecies observed can be assigned to a species lacking β-proton, plausibly lost during the $K_3Fe(CN)_6$ dependent oxidation. When pure oxygen was blown through a sample prepared in the same way, no ESR signals were reappeared, corroborating the above data again indicating that the spin adduct, once reduced under hypoxic conditions, undergo further modification preventing a possible reoxidation.

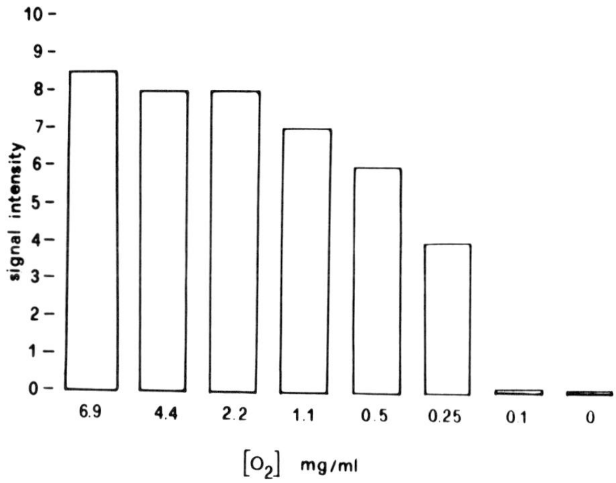

Fig.5: The trichloromethyl spin adduct, once formed under normoxic conditions, has been exposed to oxygen free nitrogen. The transformation in a diamagnetic species took place already at low, though not zero, oxygen concentration.

Conclusions

Having briefly reviewed the literature on spin trapping trichloromethyl radical and having underlined the difficulties and complications encountered in an apparently simple experiment, we have tried to indicate and, in part, carried out some experiments using a new approach. The model system under development should overcome the difficulties encountered. The model system has, in our opinion, a great potentiality and its use will be expanded to tackle problems arising in the application of the spin trapping technique to biological systems.

Acknowledgements

Research described in this paper has been supported by the Italian CNR (Progetto Oncologia contract n° 84.00837.44), and by the Ministero Pubblica Istruzione in the frame of the "Progetto Patologia da Radicali Liberi"; A.T. and A.I. are grateful to the Association for International Cancer Research (AICR, London U.K.) for financial support.

References

1) Janzen, E.G., Acc. Chem. Rev. 4 (1971), 31.
2) I.A.R.C. Monographs on the Evaluation of the Carcinogenic Risk to Chemical in Man, IARC Lyon (1979) vol 20, pp 378.
3) Slater T.F., Biochemical Mechanisms of Liver Injury, Academic Press (1978).
4) Cheeseman, K.H., Albano, E.F., Tomasi, A., Slater, T.F., Environm. Health Persp. 64 (1985), 85
5) Seawright, H., and McLean, A.E.M., Biochem. J., 105 (1967), 1055.
6) Uehleke, H., Hellmer, .H., and Tabarelli, S., Xenobiotica 3 (1073), 1.
7) Albano, E., Lott, K.A.K., Slater, T.F., Stier, A., Symons M.C.R., and Tomasi, A., Biochem. J.204 (1982), 593.
8) Janzen E.G., Spin Trapping in "Methods in Enzymology, 105 (Packer L. ed) Academic Press (1984), 188.
9) Rosen G.M., Adv. Free Rad. Biol. Med., 2 (1985), 345.
10) Calligaro,A., and Vannini, V., Pharmacol. Res. Commun. 7 (1975), 323.

11) Tomasi, A., Albano, E.Lott, K.A.K., Slater T.F., FEBS Letters, 122, (1980), 303.
12) Packer, J.E., Slater, T.F., Willson, R.L., Life Sci. 23 (1978), 2617.
13) Symons, M.C.R., Albano, E., Slater, T.F., Symons, M.C.R., Tomasi, A., J. Chem. Soc. Faraday Trans I, 78,(1982), 2205.
14) Janzen, E.G., Stronks, H.J., Dubose, C.M., Poyer, J.L., McCay, P.B., Environm. Health Persp., 64, (1985), 151.
15) Pryor W.A., Prier S.G., Church, D.F., Environm. Res. 24 (1981), 24.
16) McCAy, P.B., Lai, E.K., Poyer, J.L., Du Bose, C.M., Janzen, E.G., J. Biol. Chem., 259 (1984), 2135.
17) Tomasi, A., Albano, E., Banni, S., Botti, B., Corongiu, F.P., Dessi, A.M., Vannini, V., and Dianzani, M.U., Biochem J. 245 (1987), in press.
18) Carafoli, E., and Tiozzo, R.,Exp. Mol. Pathol. 9 (1968), 131.
19) De Castro, C.R., Bernacchi, A.S., Villaruel, M.C., Fernandez, G., Castro J.A., Agents Actions 15 (1984), 664.
20) Levy, G.N., and Brabec M.J., Toxicol. Lett. 22 (1984), 229.
21) Albano, E., Cheeseman K.H., Tomasi, A., Carini R., Dianzani M.U., Slater T.F., Biochem. Pharmacol. 35, (1986), 3955.
22) Noll, T., and De Groot H., Biochim. Biophys. Acta, 795 (1984), 356.
23) Paleos, C.M., J.C.S. Chem. Comm. (1977), 345
24) Giotta, G.J., and Wang, H.H., Biochem. Biophys. Res. Commun. 46, (1972), 1576
25) Swartz, H.M., Sentjurc, M., Morse, P.D.,II, Biochim. Biophys. Acta, 888, (1986), 82.
26) Samuni, A., Carmichael, A.J., Russo A., Mitchell, J.B., Riesz, P., Proc. Natl. Acad. Sci. (USA), 83 (1986), 7593.
27) Slater, T.F., and Sawyer, B.C., Biochem. J., 111 (1969), 317.

FREE RADICALS: CHEMISTRY, PATHOLOGY AND MEDICINE
edited by Catherine Rice-Evans & Thomas Dormandy, © 1988 Richelieu Press, London

The possible origins of human octadeca-9,11-dienoic acid

David G Wickens and Thomas L Dormandy

Department of Chemical Pathology, Whittington Hospital
Highgate Hill, London N19 5NF.

Abstract

The estimation of diene conjugation in simple lipid extracts of biological fluids by uv-spectrophotometry has been a long established but problèmatic marker of lipid peroxidation. HPLC and GC/MS studies have identified the principal diene conjugated compounds in human tissues and tissue fluids as esters of octadeca-9,11-dienoic acid (18:2(9,11)), a non-peroxide isomer of linoleic acid (18:2(9,12)). This isomer which almost certainly has the 9cis, 11trans configuration can be generated *in vitro* by the uv-irradiation of linoleic acid in the presence of protein. Simple HPLC methods have been established to measure 18:2(9,11) and 18:2(9,12) esterified as phospholipids in biological fluids and tissues. The application of these methods to the study of human pathologies has demonstrated increased concentrations of phospholipid-esterified 18:2(9,11) in the serum of chronic alcoholics, patients with primary biliary cirrhosis and pre-eclamptic toxaemia of pregnancy. More recently the phospholipid 18:2(9,11) has been shown to be elevated in exfoliated cervical cells from patients with cervical precancer compared to normals. The present paper reviews some questions relating to the origins of human 18:2(9,11).

Gilliam and colleagues in 1931 first made the observation that natural fats absorb in the ultraviolet spectrum in the 230-235nm region when left exposed to air (1). They suggested that this might reflect the molecular breakdown of polyunsaturated lipids. Two years later Edisbury *et al* showed that the absorption was due to conjugated double bonds and as such might represent a useful measure of lipid autoxidation (2). In a series of celebrated papers published between 1942 and 1946 Farmer and colleagues at the British Rubber Producers Association laid the foundations of our current understanding of lipid peroxidation (3-8). The free radical nature of lipid peroxidation was established and the primary diene conjugated products identified as hydroperoxides. These workers also proposed the generation of an intermediate hybrid radical to explain the displacement of double bonds which occurs following hydrogen abstraction. The findings of Farmer, Bergstrom (9) and others (10-12) led to the wide acceptance of the doctrine that in autoxidising systems diene conjugation represents hydroperoxides and related oxygenated products. In particular the concept became the basis of interpreting experimental findings in biological material. It should be recalled, however, that as early as 1957 Chipault *et al* suggested that the principle may not apply to all systems (13). The high-energy irradiation of methyl linoleate, for example, gave a higher diene conjugation /peroxide value than did its autoxidation, a discrepancy which could be explained in terms of non-oxygen containing diene conjugated products.

In recent years extensive HPLC studies by Cawood *et al* (14,15) have demonstrated that in fresh human serum, tissue fluids and tissues some 95% of 'total' diene conjugation is due to a single fatty acid which is not a peroxide. GC/MS analysis has identified the compound as octadeca-9,11-dienoic acid (16). Almost certainly its configuration is 9cis, 11trans (17). This fatty acid is present in esterified form in all serum lipid classes and as the free fatty acid (14,18); in abnormal states its concentration in these fractions varies independently.

18:2(9cis,11trans) can be generated, together with other 9,11 and 10,12 isomers, by exposing 18:2(9cis, 12cis) to uv-irradiation in the presence of protein (14,15,19). In the absence of protein no non-oxygenated diene-conjugated compounds are generated. It has to be assumed that the generation of 18:2(9cis,11trans) requires the interaction of a carbon-centred radical with a protein reducing group, almost certainly a thiol residue. The close proximity between protein and lipid may also be necessary to exclude oxygen. The following sequence can be proposed :-

$$18{:}2(9,12) \xrightarrow{\text{hydrogen abstraction}} 18{:}2(9,12)^{\bullet}$$

$$18{:}2(9,12)^{\bullet} \xrightarrow{\text{rearrangement}} 18{:}2(9,11)^{\bullet}$$

$$18{:}2(9,11)^{\bullet} + \text{Protein-SH} \longrightarrow 18{:}2(9,11) + \text{Protein-S}^{\bullet}$$

$$2\ \text{Protein-S}^{\bullet} \longrightarrow \text{Protein-S-S-Protein}$$

Iversen *et al* (20,21) have developed a simple HPLC

method for the estimation of phospholipid (PL) esterified 18:2(9,11) and 18:2(9,12). The method has been applied to the study of pre-eclamptic toxaemia of pregnancy (PET), paraquat poisoning, chronic alcoholics and patients with primary biliary cirrhosis (PBC).

In a preliminary report Erskine _et al_ (22) have demonstrated a significantly raised PL-18:2(9,11)/ PL-18:2(9,12) molar ratio in a PET group compared to a normotensive group at 28 weeks gestation. As the abnormality precedes the development of the symptoms of PET this method may have a predictive value.

A significantly increased concentration of PL-18:2(9,11) which could not be attributed to malnutrition or cirrhosis was demonstrated by Fink _et al_ (23-25) in chronic alcoholics. This has subsequently been confirmed by Szebeni _et al_ (26). The raised concentration of PL-18:2(9,11) rapidly reverts to normal on alcohol withdrawal. In non-addicts an acute alcohol load did not significantly alter the PL-18:2 (9,11) concentration. These findings were interpreted by the authors as pointing to a two stage abnormality : first, the slow induction of a free radical dependent detoxifying mechanism during the development of the addiction ; and second, the triggering of this mechanism by an acute alcohol load.

High concentrations of 18:2(9,11) have been reported in four cases of paraquat poisoning (27). A finding interpreted as supporting the hypothesis that paraquat toxicity is mediated by free radicals (28).

In a study of 30 epileptic patients on long-term anticonvulsant therapy the PL-18:2(9,11) concentration and the molar ratio were normal (23). The same study showed that about 20% of unselected patients with liver disease and 60% of patients with PBC had a slightly or moderately raised PL-18:2(9,11) concentration and molar ratio. The finding in PBC has been confirmed by Braganza and Day (29). The significance of these observations remains to be elucidated.

The method for the measurement of PL-18:2(9,11) and PL-18:2(9,12) has been modified to allow its application to biopsy and exfoliated cellular material (30-32). In a study of cervical biopsies from 65 women a highly significant difference in the PL-18:2(9,11) concentration and the PL-18:2(9,11)/PL-18:2(9,12) molar ratio was demonstrated between normal and precancerous samples (30). In a recently completed study of 215 women the PL-18:2(9,11)/PL-18:2(9,12) molar ratio in exfoliated cells from the cervix uteri was significantly increased in women with cervical precancer compared to normals (32).

The finding of increased concentrations of PL-18:2 (9,11) in the studies described have been interpreted as supporting the hypothesis of 'increased or modified free radical activity' in these disease states. Paradoxically, whereas diene conjugation measured as a simple uv-peak (or more often an indistinct shoulder) had been widely accepted for many years as a free-radical marker, the actual identification of the main diene conjugated compound led to the questioning of this interpretation (33,34). In particular a number of questions need to be answered.

1. How is a stereospecific isomer generated by free radical activity?; in particular why are no 10,12 isomers detectable *in vivo*?

The geometric and isomeric specificity of 18:2(9cis, 11trans) has been the principal reason for the doubts expressed about the free radical origins of this compound. Studies of the autoxidation of microsomal lipids have demonstrated that under certain conditions the initial hydroperoxides formed have predominantly the cis,trans and trans,cis configuration (i.e., for linoleic acid these would be 13-hydroperoxy-9cis,11trans -octadecadienoic acid and 9-hydroperoxy-10trans,12cis-octadecadienoic acid) (35). The generation of a 9cis, 11trans isomer is therefore in keeping with a free radical source of 18:2(9cis,11trans). The isomeric specificity of 18:2(9cis,11trans) is more perplexing. Clearly from our knowledge of the peroxidation of simple lipid systems we would expect to generate an equal or similar quantity of 18:2(10trans,12cis). Our inability to detect this compound in human fluids and tissues suggests that either 18:2(9cis,11trans) is generated by a non-radical mechanism or that our knowledge of simple systems cannot be directly applied to more complex biological structures. It is not inconceivable that in well ordered biological structures specific protein-lipid interactions could more precisely determine the outcome of free radical mediated reactions than is still widely assumed. In other words, the steric alignment of the molecules in highly organised structures, such as in membranes, could ensure the same degree of steric specificity of the products as one expects in non-structured systems as the result of enzyme catalysis. The possibilty that 18:2(9cis,11trans) is generated in human cells by an enzymic mechanism cannot, of course, be discounted. However, despite strenuous attempts to do so no such enzyme has been detected in human cells. It is also possible that the 'missing' 18:2(10trans,12cis) is in fact generated in human cells but is either metabolised more quickly than the 9cis,11trans isomer or not incorporated into the main human lipid classes.

2. Why are similar non-oxygen containing diene-conjugated products from other polyunsaturated fatty acids, e.g., arachidonate, not detected *in vivo*?

   and,

   Why have no *in vitro* free radical generating systems (with the exception of the uv-irradiation of linoleic acid in the presence of protein) so far produced the biological 18:2(9,11) isomer from linoleic acid?

Of several explanations which might theoretically provide an answer, two seem to us, on current circumstantial evidence to be the most probable. First, 18:2(9cis,11trans) may be generated in a specific site as well as by a specific mechanism. For example a linoleic acid isomer might be expected to be the main diene conjugated product of free radical reactions involving some of the cytochrome systems. Second, non-peroxide diene-conjugated products from arachidonate and other more highly unsaturated fatty acids may be more quickly metabolised than 18:2(9cis,11trans). Indeed the biological importance of 18:2(9cis,11trans) as distinct from its possible value as a free radical marker may depend on its relative stability presumably related to its esterification and incorporation in various lipid fractions. The specificity of the site and mechanism of 18:2(9cis,11trans) generation could also explain the difficulty of generating this compound *in vitro* by the exposure of a random selection of biological preparations to an equally random range of free radical generating systems.

3. Why is there a wide variation in the serum concentrations of 18:2(9cis,11trans) in different species?

   and,

   How can possible *in vivo* generation of 18:2(9cis, 11trans) be distinguished from dietary sources?

Although an enzyme capable of catalysing the conversion of 18:2(9cis,12cis) to 18:2(9cis,11trans) has not been reported in man such an enzyme has been described in certain anaerobic bacteria, the commonest of which is

Table - Animal serum phospholipid esterified 18:2(9cis, 11trans) and 18:2(9cis,12cis) and molar ratio (S.A. Iversen - unpublished data)

Source	18:2(9,12) (μmol/l)	18:2(9,11) (μmol/l)	Molar ratio
Human	749	14.6	2.0
Equine	1287	<0.5	<0.1
Rabbit	945	<0.5	<0.1
Porcine	368	6.6	1.8
Mouse	350	<0.5	<0.1
Sheep	159	20.9	13.0
Calf	135	3.3	2.4
Bovine	233	10.0	4.2
Baboon	773	3.3	0.4
Snake	652	4.4	0.7
Guinea Pig	338	0.5	0.2
Chicken	687	1.8	0.3
Dog	797	5.5	0.7
Goat	381	5.0	1.3
Rat	516	1.0	0.2
Feline	221	1.8	0.8
Hamster	716	6.0	0.8
Turkey	904	4.0	0.4

Butyrovibrio fibrosolvens (36). This micro-organism is found in the rumen of ruminants (it has not been reported in the intestinal tract of man although the presence of similar bacteria cannot be discounted) and is presumably responsible for the high concentration of phospholipid esterified 18:2(9cis,11trans) in the serum of ruminants (table). The production of 18:2(9cis,

Figure - Steps in the biohydrogenation of dietary linoleic acid in the rumen of ruminants (37)

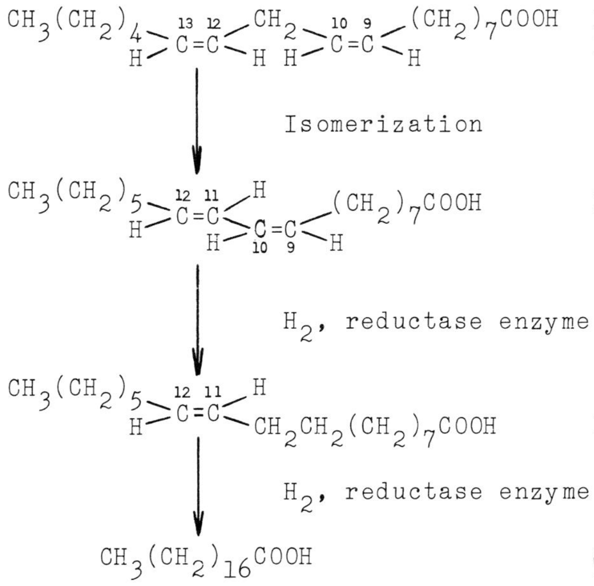

Linoleic acid 18:2(9cis,12cis)

18:2(9cis,11trans)

18:1(11trans)

Stearic acid 18:0

11trans) occurs in the pathway responsible for the hydrogenation of linoleic acid to stearic acid (Figure). Clearly if this series of reactions does not go to completion, and the finding in the sera of ruminants suggests that it does not, this would explain the high concentrations of 18:2(9cis,11trans) in cow's milk and dairy produce (unpublished observation). Another potential dietary source is partially hydrogenated fats, which, as part of our move towards a so-called healthier diet, we consume ever increasing quantities. Polyunsaturated margarines for example contain appreciable amounts of partially hydrogenated lipids in order for them to be solids at room temperatures. The introduction of trans bonds being a positive advantage in this context as it allows a solid to be achieved whilst retaining the maximum degree of unsaturation. The methods employed to partially hydrogenate polyunsaturated lipids (usually heating over nickel catalysts) would, however, be expected to generate a mixture of diene conjugated compounds of which 18:2(9cis,11trans) would be only one.

It is likely that the diet contributes significantly to the normal 'baseline' level of 18:2(9cis,11trans) in human lipids. The most compelling evidence that the diet is not the sole source comes from clinical studies. In particular the precipitous decline in PL-18:2(9cis, 11trans) in chronic alcoholics upon alcohol withdrawal and the difference between the PL-18:2(9cis,11trans) in cervical cells from women with cervical precancer and normals is difficult, if not impossible, to explain on a purely dietary basis.

References

1. Gillam, A.E., Heilbron, I.M., Hilditch, T.P. and Morton, R.A. Biochem J 25 (1931) 30-38.
2. Edisbury, J.R., Morton, R.A. and Lovern, J.A. Biochem J 27 (1933) 1451-1460.
3. Farmer, E.H. and Sundralingham, A. J Chem Soc (1942) 121-139.
4. Farmer, E.H., Bloomfield, G.F., Sundralingham, A. and Sutton, D.A. Trans Farad Soc 38 (1942) 348-356.
5. Farmer, E.H. Trans Farad Soc 38 (1942) 356-361.
6. Farmer, E.H., Koch, H.P. and Sutton, D.A. J Chem Soc (1943) 541-547.
7. Bolland, J.L. and Koch, H.P. J Chem Soc (1945) 445-447.
8. Farmer, E.H. Trans Farad Soc 42 (1946) 228-236.
9. Bergstrom, S. Arkiv For Kemi, Mineralogi Och Geologi 21A (1945) 1-18.
10. Cannon, J.A., Zilch, K.T., Burket, S.C. and Dutton, H.J. J Am Oil Chem Soc 29 (1952) 447-452.
11. Privett, O.S., Lundberg, W.O., Khan, N.A., Tolberg, W.E. and Wheeler, D.H. J Am Oil Chem Soc 30 (1953) 61-66.
12. Sephton, H.H. and Sutton, D.A. J Am Oil Chem Soc 33 (1956) 263-272.
13. Chipault, J.R., Privett, O.S., Mizuno, G.R., Nickell, E.C. and Lundberg, W.O. Ind Eng Chem 49 (1957) 1713-1720.
14. Cawood, P., Wickens, D.G., Iversen, S.A., Braganza, J.M. and Dormandy, T.L. FEBS Lett 162 (1983) 239-243.
15. Cawood, P., Iversen, S.A. and Dormandy T.L. In:

Oxygen Radicals in Chemistry and Biology (W. Bors, M. Saran and D.Tait, eds) Walter de Gruyter, Berlin (1984) 355-358.

16. Iversen, S.A., Cawood, P., Madigan M.J., Lawson, A.M. and Dormandy T.L. FEBS Lett 171 (1984) 320-324.
17. Smith, G.N., Taj, M. and Braganza, J.M. Free Rad Biol & Med (submitted for publication)
18. Harrison, K., Cawood, P., Iversen, S.A. and Dormandy, T.L. Life Chem Reports 3 (1985) 41-44.
19. Dormandy, T.L. In: Free Radicals in Liver Injury (G. Poli, K.H. Cheeseman, M.U. Dianzani and T.F. Slater, eds) IRL Press, Oxford (1985) 167-173.
20. Iversen, S.A., Cawood, P. and Dormandy, T.L. Ann Clin Biochem 22 (1985) 137-140.
21. Iversen, S.A., Cawood, P., Madigan, M.J., Lawson, A.M. and Dormandy T.L. Life Chem Reports 3 (1985) 45-48.
22. Erskine, K.J., Iversen, S.A. and Davies, R. Lancet 1 (1985) 554-555.
23. Fink, R., Clemens, M.R., Marjot, D.H., Patsalos, P., Cawood, P., Norden, A.G., Iversen, S.A. and Dormandy, T.L. Lancet 2 (1985) 291-294.
24. Fink, R., Marjot, D.H., Cawood, P., Norden, A.G. and Dormandy T.L. Life Chem Reports 3 (1985) 199-202.
25. Fink, R., Marjot, D.H., Clemens, M.R., Patsalos, P., Cawood, P., Iversen, S.A. and Dormandy T.L. In: Free Radicals, Cell Damage and Disease (C. Rice-Evans, ed) Richelieu Press, London (1986) 127-131.
26. Szebeni, J., Eskelson, C., Sampliner, R., Hartmann, B., Griffin, J., Dormandy, T. and Watson, R.R. Alcoholism: Clin & Exp Res 10 (1986) 647-650.

27. Crump, B.J., Thurnham, D.I., Situnayake, R.D. and Davis, M. Lancet 2 (1985) 955-956.
28. Bus, J.S. Aust, S.D. and Gibson, J.E. Biochem Biophys Res Commun 58 (1974) 749-755.
29. Braganza, J.M. and Day, J.P. Lancet 1 (1987) 987-988.
30. Griffin, J.F.A., Wickens, D.G., Tay, S.K., Singer, A. and Dormandy T.L. Clin Chim Acta 163 (1987) 143-148.
31. Tay, S.K., Singer, A., Griffin, J.F.A., Wickens, D.G. and Dormandy, T.L. Clin Chim Acta 163 (1987) 149-152.
32. Singer, A., Tay, S.K., Griffin, J.F.A., Wickens, D.G. and Dormandy, T.L. Lancet 1 (1987) 537-539.
33. Smith, M.T. and Thompson, S. Lancet 2 (1985) 774.
34. Thompson, S. and Smith, M.T. Chem-Biol Interact 55 (1985) 357-366.
35. Corongiu, F.P., Poli, G., Dianzani, M.U., Cheeseman, K.H. and Slater, T.F. Chem-Biol Interact 59 (1986) 147-155.
36. Hughes, P.E., Hunter, W.J. and Tove, S.B. J Biol Chem 257 (1982) 3643-3649.
37. Gurr, M.I. Int Dairy Fed Bull 166 (1983) 5-18.

FREE RADICALS: CHEMISTRY, PATHOLOGY AND MEDICINE
edited by Catherine Rice-Evans & Thomas Dormandy, © 1988 Richelieu Press, London

Ozone Toxicity Mechanisms in Plants and Animals

Horst Mehlhorn & Alan R. Wellburn

Department of Biological Sciences, University of Lancaster, Bailrigg, Lancaster LA1 4YQ, UK

Evidence accumulated over recent years concerning ozone toxicity has indicated that free radical reactions may be involved as intermediates. Recently, we have reported that ethylene reacts with ozone to yield free radicals and that the rates of ethylene formation determines ozone susceptibility in plants.

Ozone surrounds the Earth in three distinct moieties - in the upper atmosphere, in the lower troposphere and dissolved in water. Each of these ozone pools affects life in a different way. In the upper stratosphere, ozone absorbs ultraviolet solar radiation and thereby protects life from radiation damage. Lower down in the troposphere, ozone has a detrimental influence on living systems - being one of the strongest oxidants that confronts it. Ozone in the lower atmosphere will also dissolve in water where it reacts with individual water molecules to form highly reactive hydroxyl radicals. Fortunately, only small quantities of ozone dissolve in water.

Concentrations of ozone have changed both in the stratosphere and troposphere during the past century because of increasing industrial activity (1,10,25,20). Use of synthetic fertilizers (which increase microbial denitrification and cause the release of N_2O) and chlorofluorocarbons (CFCs, from aerosol spray cans and escaping refrigerants) are capable of depleting ozone in the stratosphere. This may lead to a hole being created in the protective ozone layer of the upper stratosphere. Considerable concern about possible increases of skin cancer formation has recently been voiced (20). In the lower atmosphere, where ozone is detrimental to life, ozone concentrations have almost doubled since the turn of the century (44). This has led to adverse effects on plants as shown by crop yield reductions and induced respiratory problems in sensitive humans (13,15,37). As a consequence, these observations have caused much interest about the possible biochemical mechanisms by which ozone may act.

Mechanistic studies on ozone toxicity have shown that, at high ozone concentrations, ozonolysis of cellular components such as fatty acids or proteins is likely to be responsible for tissue damage due to ozone (36). At lower ozone concentrations, such as those that normally occur in the lower troposphere (0-400 nl l^{-1}), this reaction appears unlikely because of the low solubility of ozone in aqueous solutions (30). Living systems are normally protected from direct contact with the atmosphere by a layer of extracellular water covering their cells. This may be demonstrated by the failure to produce tissue damage in algae or tissue cultures, when ozone is bubbled through the aqueous growth media at unrealistically high concentrations (8). Such levels of ozone would never occur naturally. Therefore, it seems unlikely that ozone itself is the major toxic agent involved in ozone toxicity at low ambient concentrations.

When both plants and animals are fumigated with ozone, compounds such as ascorbic acid (vitamin C), glutathione, α-tocopherol (vitamin E), or (synthetic) phenols can protect them from ozone toxicity (18,21,32,46,48,53,54,63). Much interest and research has been devoted to gaining an understanding of their chemical action inside cells. It has, for example, been established that antioxidants have to be present in the right concentrations to prevent oxidative tissue injury. If concentrations are too high then antioxidants become oxidants and promote oxidative injury rather than prevent it. This difference in behaviour has been exploited in experiments which are designed to study mechanisms of lipid peroxidation (11,14,57). These reactions are promoted by free radical linked oxidative events and can be limited by treatment with antioxidants. The active agents in lipid peroxidation are free radicals which initiate a cascade of reactions that may be terminated by antioxidants. By virtue of their stabilisation due to resonance, they are able to form radicals which not promote further chain reactions when they react with other radicals. Subsequently, they are transferred back into their original form without the formation of new radical species.

In virtually all cells, free radicals are formed *in vivo* by a number of enzymatic reactions, e.g. xanthine oxidase, glycollate oxidase etc. (14). If formed in large quantities, such radicals cannot be neutralised by the antioxidant systems of the cell and they will react with other constituents inside cells. If such damage cannot be repaired then severe tissue injury will occur. In plants, for example, this will lead to the formation of necrotic lesions on leaf surfaces (42).

Ozone toxicity has been suspected for some time to involve free radical reactions. By analysing changes in the antioxidant pools of plant and animal cells, the activities of those enzymes involved in preventing the accumulation of free radicals, protection against ozone injury after treatment with these chemicals has been detected (21). Long-term exposures of plants to ozone have shown increases of antioxidants such as glutathione, ascorbic acid or α-tocopherol (28). Possibly, these offer greater protection against free radical mediated tissue injury. In cells experiencing severe oxidative stress, an oxidation of these antioxidants can be observed (46,51,59). Simultaneously, levels of activities of enzymes involved both in regenerating the reduced forms of these antioxidants or in scavenging potentially toxic oxygen derivatives ($O_2^{\bar{\cdot}}$, H_2O_2) may increase. For example, increases in the levels of ascorbate peroxidase and glutathione reductase activities in plants or glutathione peroxidase activities in animals are known to occur during exposure to ozone (13,29,54). Similarly, changes of levels of superoxide dismutase and catalase activities have also been reported both for plants and animals (4,13,32). The simultaneous increase of both ascorbate or glutathione peroxidase and glutathione reductase in cells experiencing oxidative stress further strengthens these observations. By reducing oxidised glutathione, the enzyme glutathione reductase maintains glutathione in its reduced form. Glutathione can then reduce any oxidised ascorbic acid that is utilised by ascorbate peroxidase by eliminating toxic hydrogen peroxide (29,54). Such observations are further strengthened by the fact that treatment of animals or plants with phenolic antioxidants can protect them against ozone injury (21,53).

Free radical-mediated chain reactions have been shown to attack chemical structures which contain isolated double bonds. Studies of lipid changes in plant cells treated with ozone demonstrates one of the possible targets of these reactions (36,40). Plant membranes, which can consist up to 90% of polyunsaturated fatty acids (e.g. linolenic acid (18:3)), are more prominent targets than animal membranes. The latter have substantially less (and different) polyunsaturated fatty acids in their membranes. Thompson and coworkers have found that not only do levels of polyunsaturated fatty acids decrease after in vitro treatment with ozone but palmitic acid (16:0) also increases at the same time. NMR experiments have also shown a difference in membrane fluidity as a result of change in fatty acid components (19,40). Furthermore, they indicated changes in the distribution of lipid classes. Sterols decreased in amount relative to phospholipids. As a result, these reactions increase the amount of decomposition products such as aldehydes (9) or small hydrocarbons (6,64) which can be used to detect these cellular events. Similar results have also been found in animal systems treated with ozone (2,12,47,49,58).

As a result of treatment with ozone, other changes in cell constituents such as protein oxidation and damage to DNA might indicate an important role for ozone acting as a mutagenic and/or carcinogenic agent (17,35,38,43,45,52,60,61,66). Many of these changes resemble closely effects caused by treatments with the all-round herbicide, paraquat (13). It has been established, both in animals and plants, that this chemical exerts its toxicity via the formation of superoxide free radicals.

Leaf injury after treatment with paraquat closely resembles leaf injury after exposure to ozone. This type of leaf injury occurs both in insensitive plants, where high ozone concentrations have to be applied to induce injury, and in plants that are more susceptible. Consequently, they may be used as bioindicators displaying leaf injury even at ambient ozone concentrations. In animal systems, ozone toxicity is not so obvious. Other than alterations in breathing patterns of animals or coughing by humans it is impossible to see visible symptoms of ozone even at high concentrations (13,32,36,37).

By virtue of the difficulty of working with living tissue cultures or changes in blood chemistry to study animals, many researchers prefer to use ozone concentrations that are often more than ten times the maximum ambient ozone concentrations and these treatments may induce artefacts. Plant systems, by contrast, may be very useful systems for the study of ozone toxicity mechanisms, which may later also be applied to animal systems, because they can be studied at ambient levels of atmospheric ozone.

Fumigation studies of plants with ozone have established that large economic crop yield reductions occur every year in the U.S.A. and that a major proportion is caused by ambient ozone concentrations (15). In Europe, ozone is one of the main suspects for initiating the forest decline problems that currently affect much of Central Europe and North America (27). These suspicions have arisen from the fact that ozone concentrations today are approximately double the concentration at the beginning of the century. As a result, much interest has recently been focused on this potential toxic agent in the environment (3,5,22,41,50,65). Although growth reductions are known to be caused by ozone in the field, biochemical studies of the exact mechanisms of toxicity have been scarce. The reason is that, for a long time, ambient ozone concentrations were not considered to be particularly harmful to plants and this, in turn, was due to a lack of understanding on how to carry out meaningful fumigation studies in the laboratory. Eventually, it was recognised that ozone toxicity depends not only on the concentration of ozone but also on meteorological factors such as prevailing relative humidities and temperatures. The boundary layer resistance caused by the layer of still air above and below leaves has also been demonstrated to be an important factor limiting ozone toxicity towards plants (7,16,26,31,39).

Until recently, the exact mechanism of action of ozone has been unclear. Using plants as model systems for carrying out mechanistic studies, we have recently shown how ethylene (ethene) determines ozone toxicity in plants (30). Tingey and coworkers, as well as others, had used ethylene over the past 15 years as a general stress indicator for ozone and they were able to show that ethylene production correlated with ozone injury when studying 29 different species (55). They concluded that ethylene can be used as an indicator of ozone injury. When we studied pea responses to ozone we observed that plants continuously fumigated with quite high ozone concentrations (300 $\mu g\ m^{-3}$) did not show any injury and had depressed rates of ethylene formation. Ethylene in plants is biosynthesised from the amino acid methionine via the intermediate S-adenosyl-methionine (this also links ethylene biosynthesis to polyamine production) (33,34,62). When we inhibited ethylene biosynthesis using amino-ethoxy-vinyl-glycine (AVG) as a specific inhibitor of 1-aminocyclopropane-1-carboxylic acid (ACC) synthase, we were able to protect plants from ozone injury. This confirmed our assumption that, if ethylene was responsible for ozone toxicity, the reduction of ethylene production should result in protection of plants against ozone. Inducing ethylene production by other stresses that are known to result in increased ethylene formation rates should, and did, increase the degree of injury caused by ozone when NO or NO_2 were used to induce stress. Since then we have tried other ethylene-inducing agents like herbicides, sulphur dioxide and some pathogens to further increase ozone toxicity. Other authors who had treated plants with ozone in the presence of drought, pathogens, and wounding are able to confirm these findings (15,16,56,).

Applying these results to the older type of ozone fumigation studies has caused some re-evaluation of what levels of ozone may be toxic to plants. As fumigations with constant levels of ozone result in an adaptation of plants to ozone, by reducing their ethylene production rates which renders them less susceptible, then they are unlikely to simulate field situations where ozone occurs in intermittent peaks at intervals that may only last one or two days. In view of our observations and those of Tingey, it is interesting that conifers are affected by forest decline at high altitude in Europe and North America. Predictably, the environmental stresses in these higher locations are often more severe and ozone levels can be much higher than in the valleys. Furthermore, in correlations between ethylene emission and ozone-induced leaf injury, a conifer species (Pinus ponderosa) was the most sensitive species.

These findings may also be relevant to animal systems where ethylene may be produced in sufficient amounts as a by-product of lipid peroxidation during ozone toxicity (23,24). Lower production rates of ethylene in animals would also explain why, in general, they should be less sensitive to ozone than most plants.

Literature

1. Altshuller, A.P. (1986) The role of nitrogen oxides in nonurban ozone formation in the planetary boundary layer over N. America, W. Europe and adjacent areas of ocean. Atmos. Environ. 20, 245.
2. Arakawa, K. & M. Sagai (1986) Species differences in lipid peroxide levels in lung tissue and investigation of their determining factors. Lipids 21, 769.
3. Atkinson, R. & W.P.L. Carter (1984) Kinetics and mechanisms of the gas-phase-reactions of ozone with organic compounds under atmospheric conditions. Chem. Rev. 84, 437.
4. Bennett, E.H. Lee & H.E. Heggestad (1984) Biochemical aspects of plant tolerance to ozone and oxyradicals: superoxide dismutase. In Gaseous air pollutants and plant metabolism (M.J. Koziol & F.R. Whatley eds.) Butterworths, London, p. 413.
5. Berrang, P. & D.F. Karnosky (1986) Natural selection for ozone tolerance in Populus tremuloides. Can. J. Forest Res. 16, 1214.
6. Bucher J.B. (1984) Emissions of volatiles from plants under air pollution stress. In Gaseous air pollutants and plant metabolism (M.J. Koziol & F.R. Whatley eds.) Butterworths, London, p.399.
7. de Vos, N.E., E.J. Pell, R.R. Hill & R.H. Cole (1983) Laboratory versus field response of potato genotypes to oxidant stress. Plant Disease 67, 173.
8. Dominy, P.J. & R.L. Heath (1985) Inhibition of K^+-stimulated ATPase of the plasmalemma of Pinto bean leaves by ozone. Plant Physiol. 77, 43.

9. Esterbauer, H., A. Benedetti, J. Lang, R. Fulceri,, G. Fauler & M. Comporti (1986) Studies on the mechanism of formation of 4-hydroxynonenal during microsomal lipid peroxidation. Biochim. Biophys. Acta 876, 154.

10. Fishman, J., F.M. Vukovich & E.V. Browell (1985) The photochemistry of synoptic-scale ozone synthesis: implications for the global tropospheric ozone budget. J. Atmos. Chem. 3, 299.

11. Girotti, A.W. (1985) Mechanisms of lipid peroxidation. J. Free Rad. Biol. Med. 1, 87.

12. Goheen, S.C., L. O'Rourke & E.C. Larkin (1986) Ozone and the peroxidation of polyunsaturated fatty acids in vivo. Environ. Res. 40, 47.

13. Goldstein, B.D. (1979) The pulmonary and extrapulmonary effects of ozone. In Oxygen free radicals and tissue damage (Ciba Foundation Series No. 65, new series) Excerpta Medica, Amsterdam, p. 295.

14. Halliwell B. & J.M.C. Gutteridge (1985) The importance of free radicals and catalytic metal ions in human diseases. Molec. Aspects Med. 8, 89.

15. Heck, W.W. (1968) Factors influencing expression of oxidant damage to plants. A. Rev. Phytopathol. 6, 165.

16. Heggestad, H.E., T.J. Gish, E.H. Lee, J.H. Bennett & L.W. Douglass (1985) Interaction of soil moisture stress and ambient ozone on growth and yields of soybeans. Phytopathol. 75, 472.

17. Ito, O., K. Okano & T. Totsuka (1986) Effects of NO_2 and O_3 exposure alone or in combination on kidney bean plants: Amino acid content and composition. Soil Sci. Plant Nutr. 32, 351.

18. Konings, A.W.T. (1984) Mechanisms of ozone toxicity in cultured cells. I. Reduced clonogenic ability of polyunsaturated fatty acid-supplemented fibroblasts. Effect of vitamin E. Environ. Res. 33, 17.

19. Ladbroke, B.D. & D. Chapman (1969) Thermal analysis of lipids, proteins and biological membranes. Chem. Phys. Lipids 3, 304.

20. Larsen, S.H.H. & G. Volden (1986) Atmospheric ozone: hypothesized trend and a comment with respect to increased incidence of skin cancer. Photobiochem. Photobiophys. 11, 217.

21. Lee, T.T. (1966) Chemical regulation of ozone susceptibility in Nicotiana tabacum. Can J. Bot. 44, 487.

22. Lefohn, A.S. & V.A. Mohnen (1986) The characterisation of ozone and sulfur dioxide, and nitrogen dioxide for selected monitoring sites in the Federal Republic of Germany. J. Air Pollut. Contr. Assoc. 36, 1329.

23. Lieberman, M. & L.W. Mapson (1964) Genesis and biogenesis of ethylene. Nature 204, 343.

24. Lieberman, M. & L.W. Mapson (1966) Ethylene formation in rat liver microsomes. Science 152, 213.

25. Liu, S.C., F.C. Fehsenfeld, D.D. Parrish, E.J. Williams, D.W. Fahey, G. Huebler, P.C. Murphy (1987) Ozone production in the rural troposphere and the implications for regional and global ozone distributions. J. Geophys. Res. 92, 4191.

26. Losch, R., J.D. Tenhunen (1981) Stomatal responses to humidity - phenomenon and mechanism. In Stomatal Physiology (P.G. Jarvis, T.A. Mansfield eds.) Society for Experimental Biology, Seminar Series, No. 8, Cambridge University Press, Cambridge, pp. 137.

27. McLaughlin, S.B. (1985) Effects of air pollution on forests. J. Air Pollut. Contr. Assoc. 35, 512.

28. Mehlhorn, H., G. Seufert, A. Schmidt & K.J. Kunert (1986) Effect of SO_2 and O_3 on production of antioxidants in conifers. Plant Physiol. 82, 336.

29. Mehlhorn, H., D.A. Cottam, P.W. Lucas & A.R. Wellburn (1987) Induction of ascorbate peroxidase and glutathione reductase activities by interactions of mixtures of air pollutants. Free. Rad. Res. Comms. 3, 193.

30. Mehlhorn, H. & A.R. Wellburn (1987) Stress ethylene formation determines plant sensitivity to ozone. Nature 327, 417.

31. Mehlhorn, H. & A.R. Wellburn (1987) Environmental modulation of stomatal opening modifies ozone toxicity. (submitted to Plant Physiol).

32. Menzel, D.B. (1984) Ozone: an overview of its toxicity in man and animals. J. Toxicol. Environ. Health 13, 183.

33. Meyer, A., P. Mueller & G.Sembdner (1987) Air Pollution and Plant Hormones. Biochem. Physiol. Pflanzen 182, 1.

34. Miyazaki, J.H. & S.F. Yang (1987) The methionine salvage pathway in relation to ethylene and polyamine biosynthesis. Physiol. Plantarum 69, 366.

35. Montgomery, M.R., P. Raska-Emery & J.U. Balis (1987) Age-related difference in pulmonary response to ozone. Biochim. Biophys. Acta 890, 271.

36. Mudd, J.B. & B.A. Freeman (1977) Reaction of ozone with biological membranes. In Biochemical effects of environmental pollutants (S.D. Lee ed.) Ann Arbor Science, Ann Arbor, p. 97.

37. Mustafa, M.G. & D.F. Tierney (1978) Biochemical and metabolic changes in the lung with oxygen, ozone, and nitrogen dioxide toxicity. American Rev. Respir. Disease 118, 1061.

38. Mustafa, G.M., N.M. Elsayed, J.J. Ospital & A.D. Hacker (1985) Influence of age on the biochemical response of rat lung to ozone exposure. Toxicol. Environ. Health 1, 29.

39. Otto, H.W. & R.H. Daines (1969) Plant injury by air pollutants: influence of humidity on stomatal apertures and plant response to ozone. Science 163, 1209.

40. Pauls, K.P. & J.E. Thompson (1981) Effects of in vitro treatment with ozone on the physical and chemical properties of membranes. Physiol. Plantarum 53, 255.

41. Pell, E.J. & N.S. Pearson (1984) Ozone-induced reduction in quantity and quality of two potato cultivars. Environ. Pollut. A 35, 345.

42. Pell, E.J. & W.C. Weissenberger (1976) Histopathological characterisation of ozone injury to soybean foliage. Phytopathol. 66, 856.

43. Pell, E.J. & E. Brennan (1973) Changes in respiration, photosynthesis, adenosine 5'-Triphophate, and total adenylate content of ozonated Pinto Bean foliage as they relate to symptom expression. Plant Physiol. 51, 378.

44. Penkett, S.A. (1984) Ozone increases in ground level European air. Nature 311, 14.

45. Rietjens, I.M.C.M., M.C.M. Poelen, R.A. Hempenius, M.J.J. Gijbels, G,.M. Alink (1986) Toxicity of ozone and nitrogen dioxide to alveolar macrophages. J. Toxicol. Environ. Health 19, 555.

46. Rietjens, I.M.C.M., H.H. Lemmink, G.M. Alink & P.J. van Bladeren (1987) The role of glutathione and glutathione S-transferase in fatty acid ozonide detoxification. Chem. Biol. Interactions 62, 3.

47. Roehm, J.N., J.G. Hadley & D.B. Menzel (1971) Oxidation of unsaturated fatty acids by ozone and nitrogen dioxide. Arch. Environ. Health 23, 142.

48. Roehm, J.N., J.G. Hadley & D.B. Menzel (1971) Antioxidants vs lung disease. Arch. Intern. Med. 128, 88.

49. Roeth, E., B. Toeroek, T. Zsoldos & B. Matkovics (1985) Lipid peroxidation and scavenger mechanism in experimentally induced heart infarcts. Bas. Res. Cardiol. 80, 530.

50. Rombout, P.J.A., P.J. Lioy & B.D. Goldstein (1986) Rationale for an eight-hour ozone standard. J. Air Pollut. Contr. Assoc. 36, 913.

51. Schmidt, A. & K.J. Kunert (1986) Lipid peroxidation in higher plants: the role of glutathione reductase. Plant Physiol. 82, 700.

52. Smith, C.E., M.S. Stack & D.A. Johnson (1987) Ozone effects on inhibitors of human neutrophil proteinases. Arch. Biochem. Biophys. 253, 146.

53. Smith, G., B. Greenhalgh, E. Brennan & J. Justin (1987) Soybean yield in New Jersey relative to ozone pollution and antioxidant application. Plant Disease 71, 121.

54. Tanaka, K., Y. Suda, N. Kondo & K. Sugahara (1985) O_3 tolerance and the ascorbate-dependent H_2O_2 decomposing system in chloroplasts. Plant Cell Physiol. 26, 1425.

55. Tingey, D.T., C. Standley & R.W. Field (1976) Stress ethylene evolution: a measure of ozone effects of plants. Atmos. Environ. 10, 969.

56. Thompson, J.E., Legge, R.L. & R.F. Barber (1987) The role of free radicals in senescence and wounding. New Phytol. 105, 317.

57. Upham, B.L. & L.S. Jahnke (1986) Photooxidative reactions in chloroplast thylakoids. Evidence for Fenton-type reaction promoted by superoxide or ascorbate. Photosynthesis Res. 8, 235.

58. van der Zee, J., K. Tijssen-Christianse, T.M.A.R. Dubbelman & J. van Steveninck (1987) The influence of ozone on human red blood cells. Biochim. Biophys. Acta 924, 111.

59. van der Zee, J., T.M.A.R. Dubbelman, T.K. Rap & J. van Steveninck (1987) Toxic effects of ozone on murine L929 fibroblasts: enzyme inactivation and glutathione depletion. Biochem. J. 242, 707.

60. Veninga, T.S. & P. Evelyn (1986) Activity changes of pulmonary macrophages after in vivo exposure to ozone as demonstrated by cell adherence. J. Toxicol. Environ. Health 18, 483.

61. Veninga, T.S. & V. Fidler (1986) Ozone-induced elevation of creatinine kinase activity in blood plasma of rats. Environ. Res. 41, 168.

62. Wang, C.Y. (1987) Changes of polyamines and ethylene in cucumber seedlings in response to chilling stress. Physiol Plantarum 69, 253.

63. Wayner, D.D.M., G.W. Burton, K.U. Ingold, L.R.C. Barclay & S.J. Locke (1987) The relative contributions of vitamin E, urate, ascorbate and proteins to the total peroxyl radical-trapping antioxidant activity of human blood plasma. Biochim. Biophys. Acta 924, 408.

64. Wellburn, A.R. (1987) Biochemical mechanism of combined action of atmospheric pollutants upon plants. In Methods for assessing the effect of mixtures of chemicals (V.B. Vouk, G.C. Butler, A.C. Upton, D.V. Park & S.C. Asher eds.), Scope, p. 813.

65. Wolff, G.T., P.L. Lioy & R.S. Taylor (1987) The diurnal variations of ozone at different altitudes on a rural mountain in the Eastern United States. J. Air Pollut. Contr. Assoc. 37, 45.

66. Wright, E.S., D. M. White, A.N. Brady, L.C. Li, J.B. D'Arcy & K.L. Smiler (1987) DNA synthesis in pulmonary alveolar macrophages and Type II cells: effects of ozone exposure and treatment with -difluoromethylornithine. J. Toxicol. Environ. Health 21, 15.

FREE RADICALS: CHEMISTRY, PATHOLOGY AND MEDICINE
edited by Catherine Rice-Evans & Thomas Dormandy, © 1988 Richelieu Press, London

γ-Irradiation of Spices

I. Davidson and A.R. Forrester

Chemistry Department, University of Aberdeen
Meston Walk, Old Aberdeen AB9 2UE, Scotland.

The concentration of free radicals in γ-irradiated spices depends on the dose, the nature and origin of the spice and the time interval between γ-irradiation and esr measurement. The radicals present in spices may be trapped using nitrosodurene and evidence is adduced that the radicals trapped arise from the major components of the oleoresins of the spices. Hydrogen abstraction from the components of the oleoresins during γ-irradiation is the most likely origin of the radicals present in the spices.

Preservation of food by γ-irradiation is a topical and somewhat controversial subject in the U.K. at present. The recent report[1] published by the Advisory Committee on Irradiated and Novel Foods (ACINF) concluded that ionising radiation upto an overall average dose of 10 k Gy would not prejudice the safety and wholesomeness of the food. However, debate on and responses to this conclusion are encouraged before legislation is drafted.

Our contribution to the subject has been centred on γ-irradiation of spices. Spices were chosen because they have a particularly high microbial content[2] and there have been several outbreaks of food poisoning due to contaminated spiced meats. Accordingly, the highest doses recommended in the ACINF report were for spices (3-10 k Gy). The chemical changes which such doses cause are more likely to be detectable by chemical and spectroscopic means than are those which arise in low-dosed foods. In particular γ-irradiation generates free radicals in the spices and these are easily detectable by esr spectroscopy.

Measurement of Radical Concentrations

Initially we measured the free radical content of a variety of γ-irradiated spices, some of which had been given doses at five different levels. It was immediately obvious from the first measurements made after γ-irradiation that the free radical content was approximately proportional to the dose although a five-fold increase in dose produced only a two-fold increase in radical content. Hence, the technique appeared to have potential as a dosimetric method since at present there is no established method of measuring the extent to which food has been irradiated. Clearly if legal limits are to be imposed as recommended in the ACINF report[1] dosimetric methods must be rapidly developed otherwise the law will be unenforceable. However, it soon became clear (Fig. 1 and 2) that the free radical content steadily decreased after irradiation. The rate of decrease was greatest during the first 20 days and after 40 days there was little change. Significantly the free radical content never reached the

Fig. 1

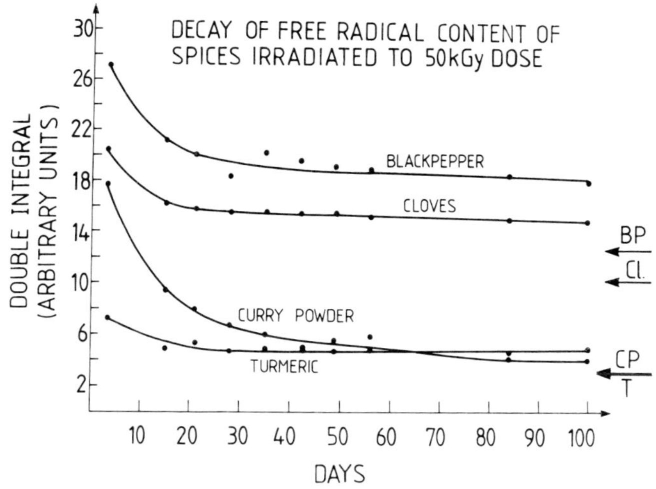
DECAY OF FREE RADICAL CONTENT OF SPICES IRRADIATED TO 50kGy DOSE
DOUBLE INTEGRAL (ARBITRARY UNITS)
30
26
22
18
14
10
6
2
BLACKPEPPER
CLOVES
CURRY POWDER
TURMERIC
BP
Cl.
CP
T
10 20 30 40 50 60 70 80 90 100
DAYS

Fig. 2

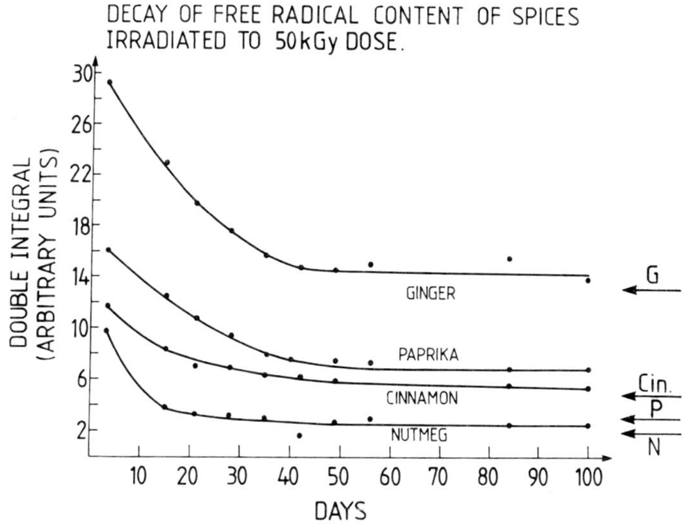
DECAY OF FREE RADICAL CONTENT OF SPICES IRRADIATED TO 50kGy DOSE.
DOUBLE INTEGRAL (ARBITRARY UNITS)
30
26
22
18
14
10
6
2
GINGER
PAPRIKA
CINNAMON
NUTMEG
G
Cin.
P
N
10 20 30 40 50 60 70 80 90 100
DAYS

level present before γ-irradiation irrespective of the dose (Fig. 3 and 4). The pre-irradiation free-radical content of the spices varies substantially which is presumably due to the different processing of the spice before marketing. Grinding, exposure to u.v. irradiation (sunlight) and heat drying will all generate free radicals in the spice. Highest levels detected were for ginger and black pepper and lowest for nutmeg and turmeric which were only about 30% of the former pair. In Figs. 5 and 6 it is shown how the large difference in radical content of differently dosed samples of paprika and curry powder measured after 3 days had all but disappeared after 30 days. This has obvious implications for free radical dosimetry. Not knowing when the spices had been irradiated would make it difficult to assess dose on the basis of free radical content. A further limitation came to light when a number of samples of the same spice were purchased from different sources (Table 1) and their esr spectra measured. This base line variation in radical content introduces a further error in the correlation of dose against radical content.

Fig.3

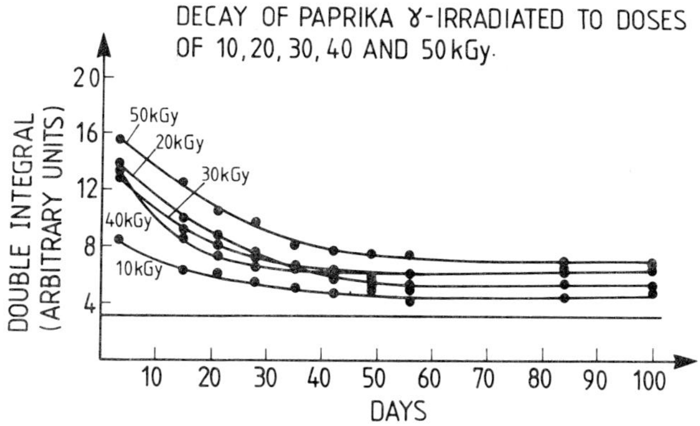

DECAY OF PAPRIKA γ-IRRADIATED TO DOSES OF 10, 20, 30, 40 AND 50 kGy.
DOUBLE INTEGRAL (ARBITRARY UNITS)
20
16
12
8
4
50kGy
20kGy
30kGy
40kGy
10kGy
10
20
30
40
50
60
70
80
90
100
DAYS

Fig.4

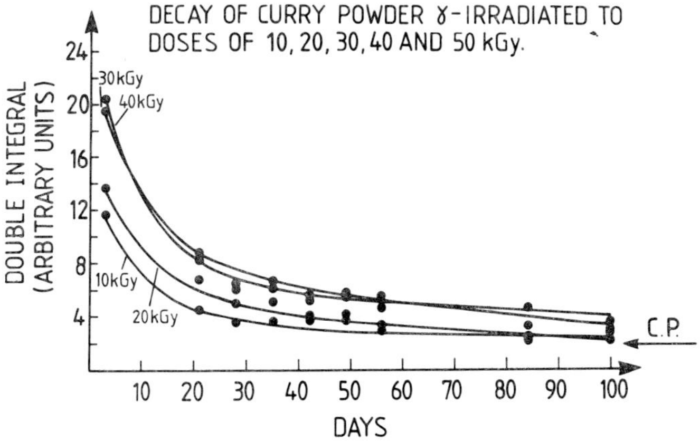

DECAY OF CURRY POWDER γ-IRRADIATED TO DOSES OF 10, 20, 30, 40 AND 50 kGy.
DOUBLE INTEGRAL (ARBITRARY UNITS)
24
20
16
12
8
4
30kGy
40kGy
10kGy
20kGy
C.P.
10
20
30
40
50
60
70
80
90
100
DAYS

Fig.5 COMPARISON OF RADICAL CONCENTRATIONS IN SPICES IRRADIATED TO DIFFERENT DOSES

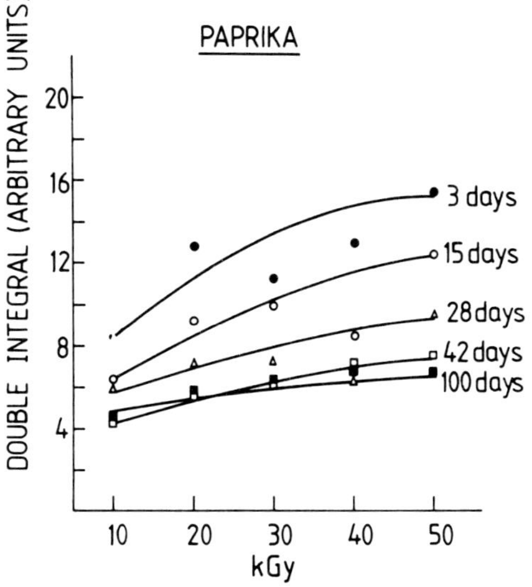

Fig.6

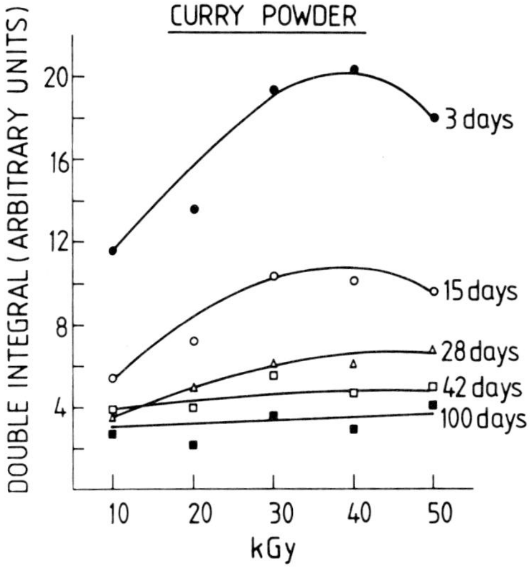

Table 1

Free Radical Content of Unirradiated Spices

Spice	Origin of Spice		Radical Content (Arb. Units)
Curry powder	Norwich Market 1		5.6
	Norwich Market 2		4.7
	T.R. Suterwalla		3.1
	Sharwoods		6.8
		Av.	5.1
White pepper	Norwich Market 1		5.7
	Norwich Market 2		3.5
	Schwartz		2.9
		Av.	4.1
Ginger	Aberdeen Market		18.4
	Rajah		16.6
	Schwartz		12.3
		Av.	15.8

Identification of Radicals

The measurements just described provided little or no information about the types of radical being formed in the spice. The signals showed no fine structure and were centred around g = 2.0045(5). The observation that the radical signals decayed but did not fall to pre-irradiation levels imples that most of the radicals produced are relatively small and are not held rigidly in the matrix of the spice. Those which persist must occupy sites in the matrix which are held rigidly apart and the number of these increases with increasing dose. It is highly desirable that as many as possible of the radicals formed be identified for two main reasons. First, if any of the radicals present is uniquely formed by γ-irradiation then this would provide at least a qualitative measure of dose. In this respect it should be remembered that γ-irradiation can lead to the breaking of relatively strong bonds not easily broken by heating or u.v. irradiation. Secondly, it is important to establish that none of the radicals formed represents a health hazard, e.g. is a known proximate or ultimate carcinogen.

Spin trapping[3] offered the most likely means of radical identification and accordingly samples of the irradiated spices were added to solutions of the spin trap nitrosodurene in chloroform. In almost all cases the resulting solution gave rise to relatively intense esr spectra immediately. A typical spectrum is shown in Fig. 7. When the solutions were remeasured after 24 h many had developed additional signals. Similar additions of unirradiated spice to solutions of nitrosodurene gave rise to very weak or no spectra although on standing for 24 h signals did develop in most cases. Use of 2-methyl-2-nitrosopropane as trap also gave well resolved spectra but offered no advantage over nitrosodurene. Indeed the actual or potential formation of di-t-butyl nitroxide from 2-methyl-2-nitrosopropane favoured the use of nitrosodurene. Useful though these spectra were they could not

SPECTRUM OF SPIN ADDUCTS DERIVED FROM γ-IRRADIATED BLACK PEPPER AND NITROSODURENE.

Fig.7

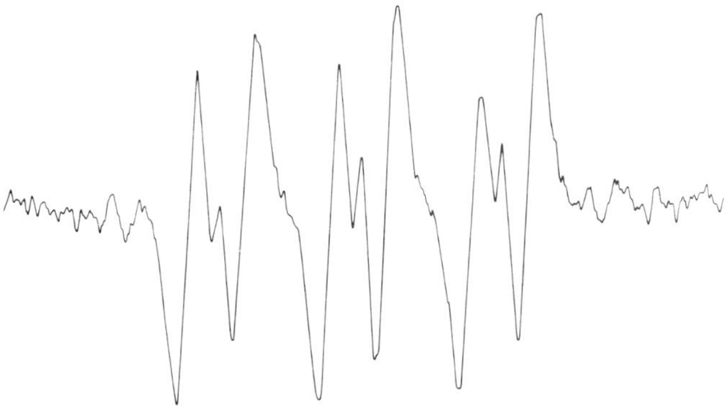

BLACK PEPPER (WHOLE SPICE)

SPECTRUM OF SPIN ADDUCTS DERIVED FROM γ-IRRADIATED BLACK PEPPER OLEORESIN AND NITROSODURENE.

Fig.8

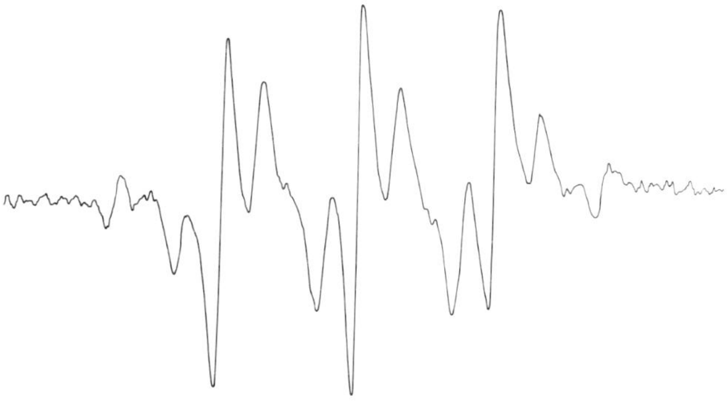

BLACK PEPPER (OLEORESIN)

by themselves provide the identity of the radical species in the spices.

Spices are a complex mixture of inter alia polysaccharide, protein and natural oils but the significance of the spin trapping results is that most of the radicals present in the spice are soluble in organic solvents and so are likely to be derived from the natural oils. The extractable essential oils or oleoresins constitute about 5-10% by weight of the spice depending on the spice and its origin. The oleoresins from five spices, turmeric, black pepper, nutmeg, cinnamon and ginger were extracted using methylene chloride and the dried waxy solids γ-irradiated. When these irradiated oleoresins were subjected to spin trapping with nitrosodurene the spectra observed were in all cases similar to (Fig. 8) but not identical with those produced from the corresponding spice. Coupling contents of the spin adducts observed are given in Table 2.

Identification of the components of the volatile oils and oleoresins of spices and measurement of their percentage composition mainly by gas chromatography is well-reported. For example, the oleoresin from ginger[4] has over 30 components. Many of these are present in trace amounts and in Table 3 the main components of ginger, cinnamon, nutmeg, black pepper and cloves are given.

Table 2

ESR Parameters derived from Spectra of
γ-Irradiated Spices and Oleoresins

Spice	Irradiated Spice	Irradiated Oleoresin
Turmeric	(a) a_N=14.1	(a) a_N=14.0
	(b) a_N=14.1, a_H=8.9(CH_2)	(b) a_N=14.0, a_H=9.0(CH_2)
Nutmeg	(a) a_N=13.9	(a) a_N=14.1
	(b) a_N=14.0, a_H=7.6(CH)	(b) a_N=13.8, a_H=8.5(CH)
	(c) a_N=14.0, a_H=11.0(CH_2)	(c) a_N=14.1, a_H=8.1(CH_2)
Cinnamon	No signal	(a) a_N=13.8
		(b) a_N=14.0, a_H=6.6(CH)
		(c) a_N=13.8, a_H=11.2(CH_2)
Cloves	(a) a_N=14.2, a_H=5.4(CH)	(a) a_N=14.2, a_H=5.4(CH)
Black pepper		(a) a_N=13.8
	(b) a_N=14.1, a_H=5.6(CH)	(b) a_N=14.0, a_H=6.0(CH)
		(c) a_N=11.4, a_H=11.0(CH_2)
Ginger	(a) a_N=14.0	(a) a_N=14.0
	(b) a_N=14.0, a_H=6.1(CH)	(b) a_N=14.0, a_H=7.2(CH)
	(c) a_N=14.0, a_H=11.0(CH_2)	(c) a_N=14.2, a_H=10.6(CH_2)
	(d) a_N=a_H=14.4	
Curry powder	(a) N = 13.6	
Paprika	(a) No signal	

Table 3

Main Components of the Oleoresins of Certain Spices

Spice	Components
Ginger[4]	α- and β-zingiberene 20-35%, ar-curcumene 18%, 6-gingerol 18-36%.
Cinnamon[5]	cinnamaldehyde 65%, eugenol 15%.
Nutmeg[6]	α- and β-pinenes 25-52%, sabinene 30-40%, R-limonene 10%.
Black pepper[7]	α- and β-pinene 37%, R-limonene 25%, β-caryophyllene 18%, β-phillandrene 4%, piperine 35-40%.
Cloves[8]	eugenol 78-98%, acetyleugenol 10-17%.

Eugenol (1), α-pinene (3), R-limonene (4), β-caryophyllene (5), and cinnamaldehyde (6) all major components of the oleoresins of the above species were adsorbed onto silica gel (0.1 g oil onto 3 g SiO_2) and the "dry solids" γ-irradiated to a dose of 50 k Gy. Acetyl eugenol (2), which is a solid, was irradiated directly to the same dose. The radicals so produced were trapped by reaction with nitrosodurene dissolved in chloroform. Interpretation of the resulting spectra of the spin adducts is tabulated in Table 4. Only with cinnamaldehyde was no spectrum observed. In a series of control experiments the oils similarly adsorbed on silica gel (and neat acetyleugenol) but not γ-irradiated were added to solutions of nitrosodurene in chloroform. β-Caryophyllene alone

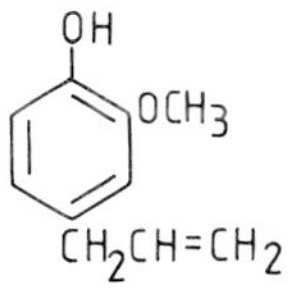

(1)

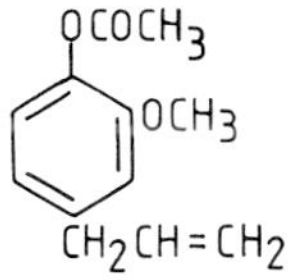

(2)

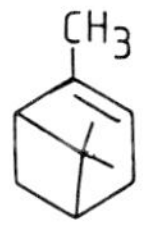

(3)

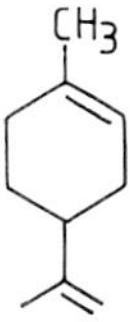

(4)

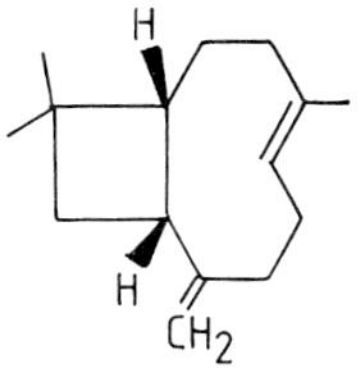

(5)

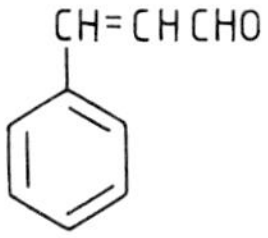

(6)

Table 4

Spice Component	Radical Generation: γ-Irradiated	Radical Generation: (Bu^tO_2CO)
Eugenol	a_N=14.8, a_H=8.4(CH_2)	
Acetyleugenol	(a) a_N=14.2, a_H=8.4(CH_2)	(a) a_N=13.6, a_H=6.1(CH_2)
	(b) a_N=14.0, a_H=4.2(CH)	(b) a_N=13.8, a_H=2.6(CH)
		(c) a_N=13.7, a_H=9.5(CH_2)
α-Pinene	(a) a_N=14.0	(a) a_N=14.2
	(b) a_N=14.0, a_H=11.0(CH_2)	(b) a_N=13.9, a_H=11.4(CH_2)
	other weak signals	(c) a_N=14.0, a_H=4.6(CH)
		(d) a_N=13.9, a_H=7.9(CH_2)
R-Limonene	(a) a_N=13.8	(a) a_N=14.1
	(b) a_N=13.8, a_H=11.6(CH_2)	other weak signals
	(c) a_N=13.9, a_H=8.5(CH_2)	
	(d) a_N=13.8, a_H=10.3(CH)	
	(e) a_N=13.5, a_H=7.4(CH)	(e) a_N=11.0, a_H=6.2

gave rise to a spectrum immediately; the others developed weak spectra over a period of two days. The spin adduct produced from caryophyllene is the secondary alkyl duryl nitroxide (7) [a_N=14.2, a_H=5.3G(1H)] and by analogy with earlier work[9] must arise by an "ene reaction" followed by oxidation (by nitrosodurene) of the ensuing hydroxylamine as indicated below. The reaction is regiospecific otherwise a tertiary alkyl duryl nitroxide would have been formed by reaction at C-4 or a primary alkyl one by

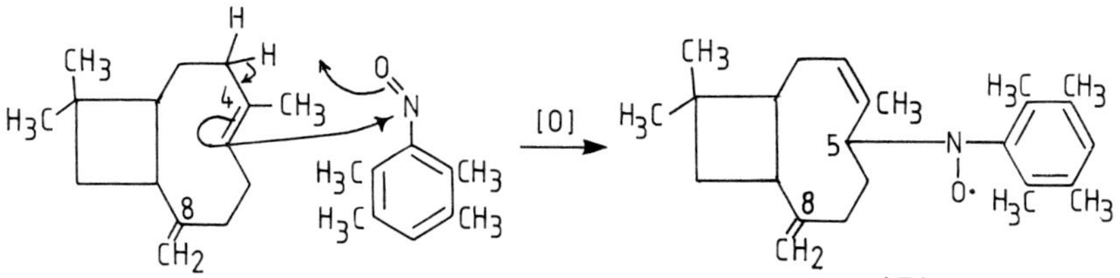

(5) (7)

reaction at the exocyclic methylene group. Presumably steric factors prevent the former but it is not clear why reaction is not favoured at the exposed methylene group. In principle, "ene reactions" could also occur with the alkenes (1-4) and nitrosodurene but do not or if they do only very slowly.

Radicals derived from eugenol and acetyleugenol

Radicals formed by overall loss of hydrogen atoms (H^++e^-) are commonly formed on γ-irradiation of organic substrates and the spin adducts derived therefrom frequently show which hydrogen atoms have been abstracted. Thus, the spin adducts derived from eugenol and acetyleugenol both give rise to a triplet of triplets.

[a_N=14.2, a_H=8.4G(CH_2)] as the main component in their esr spectra. These are almost certainly due to the spin adducts formed from (8b; R=H and $COCH_3$). The alternative spin adduct derived from

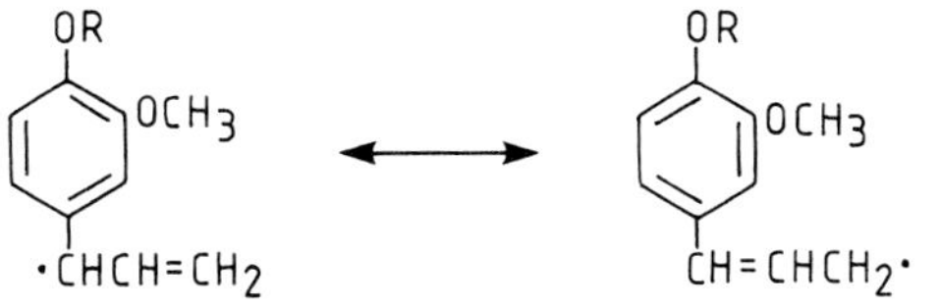

(8a) (8b)

(8a) [a_N=14.0, a_H=4.2G (triplet of doublets)] is clearly present for acetyleugenol but is barely discernible for eugenol.

Radicals derived from α-pinene

An intense triplet a_N=14.0G accompanied by a much weaker triplet of triplets a_N=14.0, a_H=11.0(2H)G characterised the spectrum. The latter spectrum we attribute to the spin adduct of (9a), the parameters quoted being similar to those previously

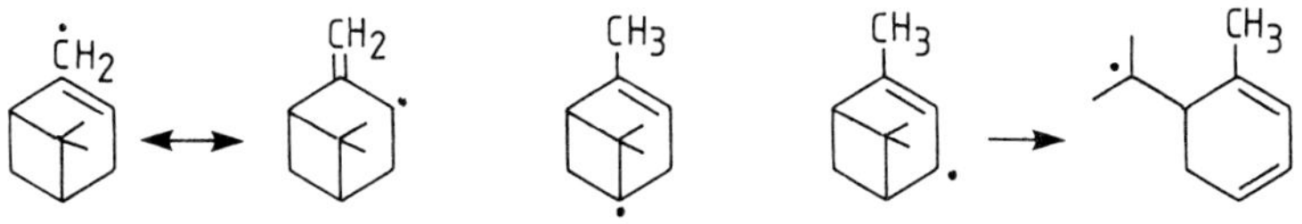

(9a) (9b) (10) (11a) (11b)

reported.[10] Removal of a bridgehead hydrogen to give the tertiary alkyl radical (10) would be relatively difficult and hence the intense triplet is more likely to be due to the spin adduct of the rearranged radical (11b) formed by transformation of the allylic radical (11a) (cf. rearrangement of radicals derived from β-pinene[11]).

Radicals derived from R-Limonene

Not surprisingly the spectrum which arises from the spin adductsderived from R-limonene was much more complex than the others. Allylic hydrogen removal from limonene can give a wide variety of primary, secondary and tertiary alkyl radicals (12-16) as shown below. The intense triplet (a_N=13.8G) must be due to the adduct of either (14b) or (15b) or both. The triplets of

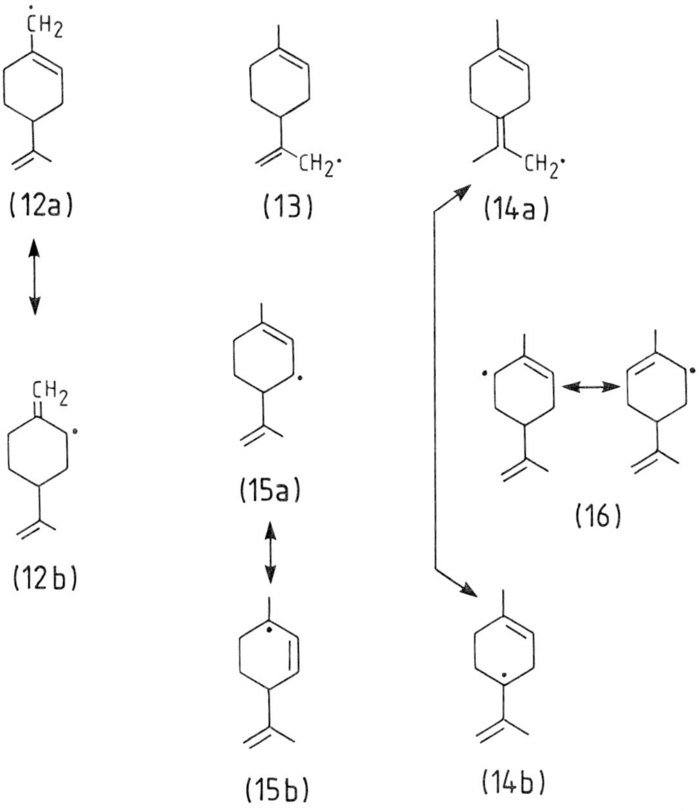

doublets a_N=13.4, a_H=7.4 and a_N=13.4, a_H=10.3G arise from two of (12b), (15a) or (16) but cannot be specifically assigned at this stage. Similarly two weak triplets of triplets a_N=13.8, a_H=11.6 (2H)G and a_N=13.8, a_H=8.5(2H)G are due to adducts two from (12a) (13) and (14a).

We have assumed that the radicals produced from these oils arise by hydrogen abstraction. Confirmation of this is clearly desirable and accordingly we sought to generate such radicals in an unequivocal way. t-Butoxyl radicals generated (a) at room temperature using di-t-butylperoxyoxalate and (b) at -40°C by u.v. irradiation of di-t-butyl peroxide were used as the hydrogen abstracting species.

With α-pinene, R-limonene and acetyleugenol using di-t-butyl peroxyoxalate as the source many of the spin adducts had esr parameters very similar to if not identical with those produced from the corresponding γ-irradiated samples (see Fig. 9). However, in general the spectra were more complex and showed additional lines only some of which have been assigned. Only with eugenol was the spectrum (triplet of doublets a_N=14.3, a_H=3.6G) fundamentally different from that obtained from the γ-irradiated samples (triplet of triplets a_N=14.8, a_H=8.4G). The differences may be attributable simply to the longevity of the several radicals produced during γ-irradiation since circumstances determined that the spin trapping experiments on the γ-irradiated samples were

Fig. 9 COMPARISON OF SPECTRA OF SPIN ADDUCT DERIVED FROM LIMONENE BY (a) γ-IRRADIATION (b) HYDROGEN ABSTRACTION BY t-BUTOXYL

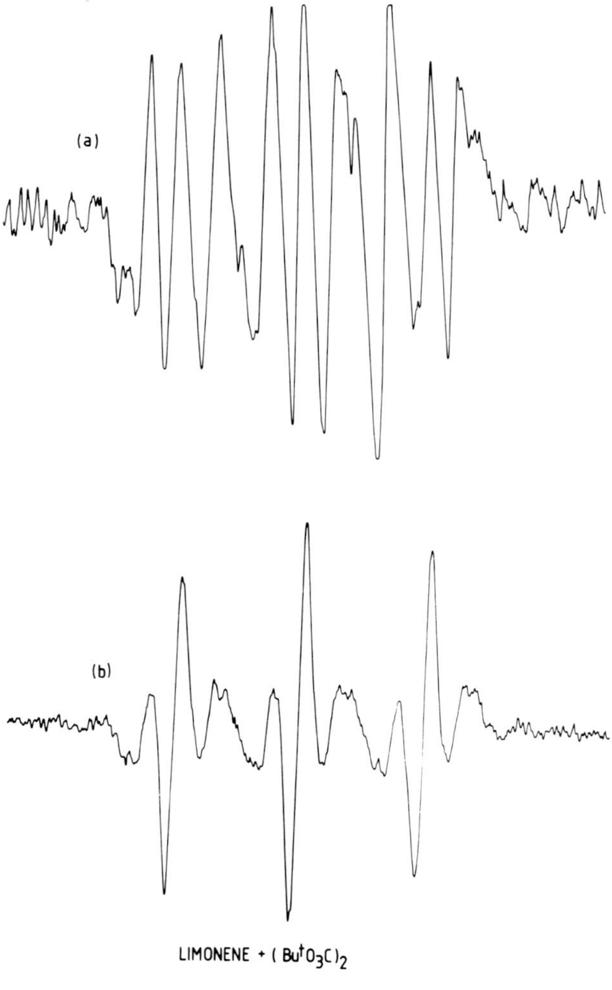

done 48 h after γ-irradiation. There was reasonably good correspondence between the spectra observed after hydrogen abstraction by t-butoxyl radicals generated (a) at room temperature from di-t-butylperoxyoxalate and (b) at -40°C by photolysis of di-t-butyl peroxide.

Further progress in the identification of the radicals formed in γ-irradiated spices requires that the spectra which arise from mixtures of spin adducts be fully interpreted. To aid this an HPLC coupled ESR technique is being used which allows separation of the nitrosodurene spin adduct and subsequent measurement of their spectra. Details of this will be given in a subsequent paper.

Experimental

ESR measurements were made on an E104A Varian spectrometer. Spin concentrations were measured for samples sieved to regulate particle size before being packed into tubes of identical diameter. Duplicate tubes were prepared for each spice and the values given are the average of measurements on two tubes.

Spin Trapping: General Procedure

(a) Spices. The whole spice (100 mg) was added to a 0.05 M solution of nitrosodurene in chloroform (1 ml). The residue was collected and the filtrate degassed for 30 s by nitrogen bubbling before esr measurement.

(b) Oleoresins. The oleoresin (50 mg) was added to a 0.05 M solution of nitrosodurene in chloroform (1 ml). The solution was degassed for 30 s by nitrogen bubbling before esr measurement.

(c) Spice components. The spice component (33 mg) adsorbed onto silica gel (chromatographic) (1 g) after γ-irradiation was added to a solution of nitrosodurene (16 mg) in chloroform (5 ml). The silica was removed by filtration, and the filtrate was reduced to 1 ml *in vacuo*. The concentrate was degassed by nitrogen bubbling for 30 s and its esr spectrum measured.

Hydrogen Abstraction using (a) Di-t-butylperoxyoxolate. The substrate (100 mg) was added to a 0.05 M solution of nitrosodurene in chloroform. The solution was degassed as before and then di-t-butylperoxycarboxylate (1-2 mg) was added. The esr spectrum was measured immediately.

(b) di-t-butyl peroxide. The substrate (100 mg) and di-t-butyl peroxide (2-5 drops) were dissolved in a solution of chloroform which was 0.05 M with respect to nitrosodurene. After degassing, the solution was briefly irradiated with a 1 kw u/v. lamp in the cavity of an esr spectrum cooled to -30°C.

Extraction of Oleoresins from Spices

The spice (1 g) was treated with chloroform and the mixture

stirred for 24 h before the undissolved solid was collected and the filtrate was evaporated. In this way 1 g of each of the following spices gave the amount of oleoresin indicated; cinnamon (0.018 g), cloves (0.106 g), ginger (0.074 g), nutmeg (0.537 g), black pepper (0.095 g), turmeric (0.102 g).

References

1. "Report on the Safety and Wholesomeness of Irradiated Foods" by the Advisory Committee on Irradiated and Novel Foods, HMSO (1986), ISBN 0113210590.
2. J. Farakas in "Preservation of Food by Ionising Radiation", ed. E.S. Josephsen and M.S. Petersen, C.R.C. Press (1983), Vol. 3, Chapter 4, p.110.
3. M.J. Perkins in "Advances in Physical Organic Chemistry", Academic Press (1980), Vol. 17, p.1.
4. V.S. Govindarajan, C.R.C. Critical Reviews in Food Science and Nutrition (1982), Vol. 17, p.1.
5. R.O.B. Wrjesekera, C.R.C. Critical Reviews in Food Science and Nutrition (1978), Vol. 10, p.1.
6. J. Baldry, J. Dougan, W.S. Mathews, J. Nabney, G.R. Pickering and F.V. Robinson, Int. Flavours (1976), 28.
7. V.S. Govindarajan, C.R.C. Critical Reviews in Food Science and Nutrition (1977), Vol. 9, p.115.

8. Thorpe's Dictionary of Applied Chemistry, 4th edit. (1947), Vol. 8, p.657.

9. A.B. Sullivan, J. Org. Chem. 3 (1966) 2811; G.T. Knight and B. Pepper, Tetrahedron 27 (1971) 6201.

10. M.D. Crozet and J. Muzart, J. Organometal. Chem. 244 (1983) 191.

11. J.A. Berson, C.J. Olson and J.S. Walia, J. Am. Chem. Soc. 82 (1960) 500.

Acknowledgements. We thank MAFF Food Science Division for financial support and the Scottish Universities Reactor Centre at East Kilbride for γ-irradiating samples of spices, oleoresins and individual spice components for us.

FREE RADICALS: CHEMISTRY, PATHOLOGY AND MEDICINE
edited by Catherine Rice-Evans & Thomas Dormandy, © 1988 Richelieu Press, London

Production of Reduced Forms of Oxygen by Adrenochrome in the presence of Ascorbate, Microsomes and Submitochondrial Particles.

A. Bindoli, M.P. Rigobello and L. Galzigna
Centro Studio Fisiologia Mitocondriale, CNR and Istituto di Chimica Biologica, Università di Padova (Italy).

Abstract

Adrenochrome, an oxidation product of adrenaline with psychotomimetic properties, can be reduced by ascorbate in a reaction that shows a rapid oxygen consumption. Superoxide dismutase and catalase strongly inhibit this oxygen uptake. Similarly, the addition of adrenochrome to rat liver microsomes or beef heart submitochondrial particles supplemented with reducing substrates determines a large consumption of oxygen. The reported results are compatible with a redox cycle where reducing equivalents supplied by ascorbate or the electron transport chains of microsomes or mitochondria are able to reduce adrenochrome which is subsequently reoxidized by oxygen with the production of superoxide anion and hydrogen peroxide.

Introduction

Catecholamines can easily undergo an oxidative cyclization to indole-5,6-quinones, generally referred to as aminochromes. In particular adrenochrome is the oxidation product of adrenaline and the adrenaline-adrenochrome transformation can also be induced by the superoxide anion (1).

Adrenochrome has been reported to be involved in the development of some forms of mental illness (2) possibly after formation of a complex with acetylcholine (3). Some toxic properties of adrenochrome appear to depend on the formation of adrenolutin, while its reduction to 5,6-dihydroxy N-methyl indole is thought to eliminate its toxicity. According to an old hypothesis (4), schizophrenia might in fact result from an alteration of the balance between the two pathways. Although the speculation concerning the pathogenetic role of adrenochrome in the malfunctioning of the central nervous system hitherto escaped any attempt of an experimental verification, the physiological role of adrenochrome and other aminochromes in the biosynthesis of melanins is well established (5). Despite this, the mechanism of redox changes of the aminochrome molecule is still rather poorly understood.

In the present communication we wish to describe the interaction of adrenochrome with ascorbate and the involvement of the microsomal and mitochondrial electron transport chains in its reduction. This occurs in the presence of reducing equivalents and is accompanied by a marked oxygen uptake.

Materials and Methods

Rat liver microsomes were prepared essentially as described by Ernster and Nordenbrand (6) and beef heart submitochondrial particles according to the procedure "3" of Smith (7). Protein content was estimated by the biuret method (8) and oxygen uptake was followed with a platinum electrode assembly of the Clark type (9).

Results and Discussion

Several years ago it has been reported that ascorbate is able to reduce adrenochrome (10, 11). In addition, concomitantly with this reduction, a rapid oxygen uptake occurs as observed by Galzigna (12). The oxygen uptake (Table 1) increases by increasing pH and does not depend on an autoxidation of ascorbate catalyzed by contaminant metals since, in our conditions, the oxidation of ascorbate in the absence of adrenochrome is negligible (not shown).

Table 1. Effect of superoxide dismutase, catalase and antioxidants on ascorbate-induced stimulation of oxygen uptake by adrenochrome.

Additions	Oxygen uptake (nmol O_2/min·mg prot)	% Inhibition
None	120	-
Superoxide dismutase	95	21
Catalase	79	34
Superoxide dismutase + Catalase	39	67
Butylated hydroxytoluene	110	8
Chlorpromazine	108	10
Adrenochrome semicarbazone	6	-

Reactions were carried out in 50 mM K-phosphate buffer (pH 7.4) at 25°C, in the presence of 8 mM ascorbate and 40 μM adrenochrome. When indicated were also added: 130 μg/ml superoxide dismutase, 70 μg/ml catalase, 0.04 mM butylated hydroxytoluene and 0.8 mM chlorpromazine. When adrenochrome semicarbazone (40 μM) was added, adrenochrome was omitted.

Consequently, the observed oxygen uptake appears to arise from an autoxidation of a partially reduced adrenochrome free-radical intermediate. The latter form may autoxidize with the production of superoxide anion which, secondarily, after dismutation, originates hydrogen peroxide. In fact either superoxide dismutase or catalase (or the combination of the two) strongly inhibit oxygen consumption (Table 1) and this can be explained on the basis of a continuous restoration of the oxygen in the reaction vessel, due to the dismutation of superoxide or to the catalase activity. On the other hand, typical antioxidants such as butylated hydroxytoluene (BHT) or chlorpromazine appear to be scarcely effective since their inhibitory effect does not exceed 10% (Table 1). When adrenochrome semicarbazone (where an imino group substitutes a quinone moiety) is used instead of adrenochrome, almost no oxygen uptake occurs in the presence of ascorbate, indicating that the o-quinone structure is essential for the phenomenon.

The redox cycling, which we postulate as an explanation of such phenomenon, is not restricted to the chemical system ascorbate-adrenochrome but also occurs when adrenochrome is added to rat liver microsomes in the presence of NADH and even more in the presence of NADPH (Table 2). In the latter case, the microsomal NADPH-cytochrome P-450 reductase evidently delivers electrons that are able to reduce adrenochrome to its semiquinonic form. The consumption of oxygen, in excess if compared to the amount of the adrenochrome present, is once again suggestive of a redox cycling.

Also beef heart submitochondrial particles supplemented with NADH and rotenone display an oxygen consumption due to the presence of adrenochrome (Table 3); catalase, added to the medium after most of the oxygen uptake has taken place, determines a return of oxygen into the incubation medium, indicating that hydrogen peroxide was formed (not shown). Nevertheless, when succinate was used as a substrate in the presence of antimycin, no stimulation of the

Table 2. Effect of adrenochrome on oxygen uptake in rat liver microsomes.

Additions	Oxygen uptake	
	Initial rate	Total O_2
	(nmol O_2/min·mg prot)	(nmol O_2/10min·mg prot)
NADH	5.3	22.8
NADH + Adrenochrome	20.4	72.2
NADPH	7.6	29.7
NADPH + Adrenochrome	47.1	83.6
NADH + NADPH	7.6	36.5
NADH + NADPH + Adrenochrome	52.2	147.7

Microsomes (1 mg/ml) were incubated in 50 mM K-phosphate buffer (pH 7.4) at 25°C. When indicated were also added: 1.6 mM NADH, 1.6 mM NADPH and 80 μM adrenochrome.

Table 3. Effect of adrenochrome on oxygen uptake in beef heart submitochondrial particles.

Additions	Initial rate of O_2 uptake (nmol O_2/min·mg prot)
NADH	151.4
NADH + Rotenone	3.1
NADH + Rotenone + Adrenochrome	41.3
Succinate	23.7
Succinate + Antimycin	4.5
Succinate + Antimycin + Adrenochrome	4.5

Beef heart submitochondrial particles (2.5 mg/ml) were incubated in 50 mM K-phosphate buffer (pH 7.4) at 25°C. When indicated were also added: 0.5 mM NADH, 2μg/mg prot rotenone, 0.2 mM adrenochrome, 5 mM succinate and 2μg/mg prot antimycin.

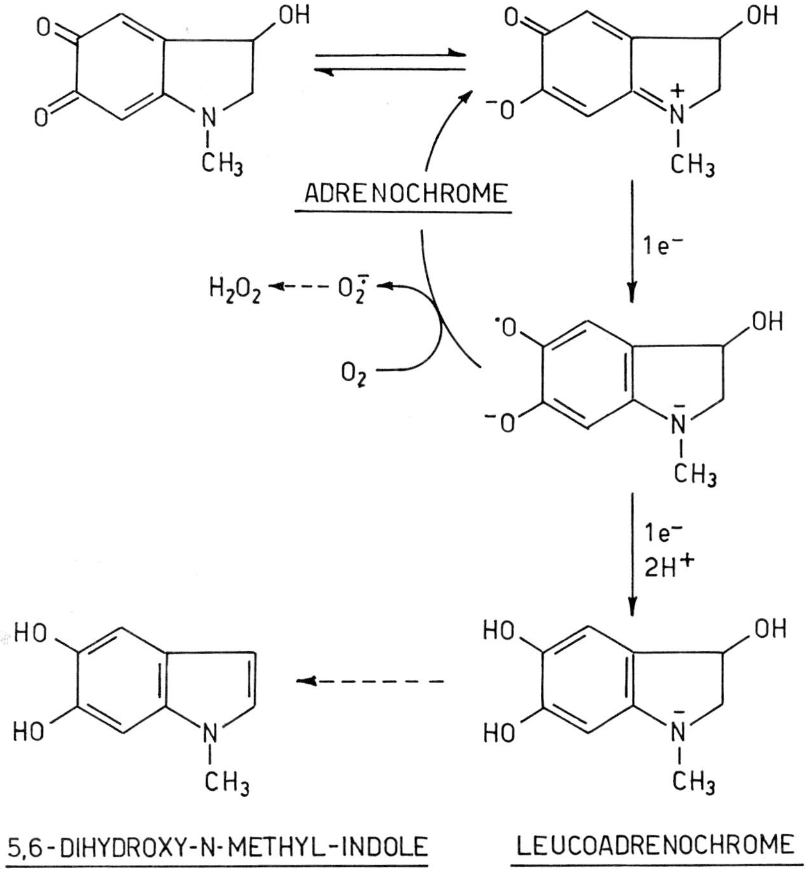

OH
O
O
N
CH3
ADRENOCHROME
-O
1e-
H2O2
O2
2H+
HO
5,6-DIHYDROXY-N-METHYL-INDOLE
LEUCOADRENOCHROME

oxygen uptake was observed (Table 3) suggesting that the NADH dehydrogenase flavin of complex I is the mitochondrial site of adrenochrome reduction.

The reported results are all compatible with the hypothesis of a redox cycling (Figure 1) in which the reducing equivalents supplied by ascorbate, or the electron transport chains of microsomes and mitochondria, are able to reduce adrenochrome which is subsequently reoxidized by oxygen with the production of superoxide anion and hydrogen peroxide.

References

1. H.P. Misra and I. Fridovich. J. Biol. Chem. 247 (1972) 3170.
2. R.A. Heacock. Bull. Chimie Ther. N°4 (1971) 300.
3. L. Galzigna. Nature 225 (1970) 1058.
4. A. Hoffer, H. Osmond and J. Smythies. J. Mental Sci. 100 (1954) 29.
5. R.C. Sealy, C.C. Felix, J.S. Hyde and H.M. Swartz. Free Radicals in Biology , Vol. IV (Academic Press 1980) pp. 209-259.
6. L. Ernster and K. Nordenbrand. Methods Enzymology 10 (1967) 574.
7. A.L. Smith. Methods in Enzymology 10 (1967) 81.
8. A.G. Gornall, C.J. Bardawill and M.M. David. J. Biol. Chem. 177 (1949) 751.
9. R.W. Estabrook. Methods in Enzymology 10 (1967) 41.
10. S. Roston. Nature 194 (1962) 1079.
11. G.L. Mattok. J. Chem. Soc. (1965) 4728.
12. L. Galzigna. Biochem. Pharm. 21 (1972) 203.

FREE RADICAL SPECIES AND THEIR PATHOLOGICAL IMPLICATIONS

FREE RADICALS: CHEMISTRY, PATHOLOGY AND MEDICINE
edited by Catherine Rice-Evans & Thomas Dormandy, © 1988 Richelieu Press, London

Detection of free radical production during myocardial reperfusion injury using electron spin resonance spin trapping.

Michael J. Davies, Pamela B. Garlick*, Trevor F. Slater and David J. Hearse*.

Biochemistry Department, Brunel The University of West London, Uxbridge, Middlesex UB8 3PH and *Cardiovascular Research, The Rayne Institute, St. Thomas' Hospital, London SE1, U.K.

ABSTRACT

Reperfusion of isolated rat hearts which have previously been deprived of oxygen for a short period of time (ischemia) can precipitate lethal ventricular arrhythmias and accelerate tissue necrosis. The triggers responsible for these events remain controversial although recently attention has been directed towards the involvement of oxygen-derived free radicals. Indirect evidence for the mediation of such species in the reperfused rat myocardium is reviewed and recent experimental data from electron spin resonance spin trapping experiments presented. These studies provide direct evidence for a short, intense burst of radical production during the first few minutes of aerobic myocardial reperfusion.

INTRODUCTION

Myocardial ischemia involving an interruption of coronary blood flow initiates a series of cellular events which, if left unchecked, culminate in cell death. Reperfusion of tissue after a short period of ischemia during the relief of coronary artery spasm or spontaneous or induced thrombolysis is, however, not without danger and can lead to potentially lethal ventricular arrhythmias. In cases of severe injury this can accelerate tissue necrosis and hemorrhage {1,2}.

The electrophysiological and metabolic mechanisms that underly the genesis of ischemia and reperfusion-induced arrhythmias are controversial {3,4,5}. A large number of different triggers such as the stimulation of alpha or beta adrenoceptors {6}, the elevation of cyclic AMP content {7}, the release of lysophosphatides {8}, the disturbance of ionic homeostatis (particularly Na^{+}, K^{+}, and Ca^{2+}; {9}), the disruption of fatty acid metabolism {10} and the heterogeneity of tissue injury and recovery {3}, have been suggested as contributory factors in establishing the observed electrical perturbations. Data relating to such triggers are conflicting and suggest either that multiple mechanisms may be involved or that some further trigger mechanism acts as the precipitating factor. Attention has recently been focussed on the possibility that free radical production may contribute to the observed tissue injury {11,12}.

FREE RADICALS AS ARRHYTHMIC TRIGGERS?

The role of free radicals in tissue injury has been reviewed {13-16}. It has been concluded that oxygen-derived species such as superoxide ($O_2^{-\cdot}$) and hydroxyl ($HO^{\cdot}$) radicals as well as hydrogen peroxide are important mediators of cellular injury. Many systems have been postulated as free radical sources in the reperfused myocardium; these include the xanthine oxidase pathway {17}, the arachidonic acid pathway {18}, invading leukocytes {19}, the degradation of catecholamines {20}, and electron leakage from the mitochondrial electron transport chain {21}. Whatever their source, there is undisputed evidence as to the cytotoxicity of these species by, for example, destruction of cellular membranes or alteration of critical enzyme systems {22}. The phospholipid component of cellular membranes is a highly vulnerable target due to its susceptibility to peroxidation. This can lead to changes in its permeability characteristics and its ability to maintain transmembrane ionic gradients {13-16, 23-25}. Free radical-mediated oxidation of important sulphydryl groups of membrane-bound ion pumps such as the sodium-potassium ATPase, can also lead to ionic imbalance due, for example, to the loss of intracellular potassium to the extracellular space and to the build-up of intracellular Na^+ and Ca^{2+} {25-27}. This will result in profound electrophysiological disturbances which may predispose the tissue to electrical instability and arrhythmias. Although the cell possesses several endogenous protective mechanisms such as the antioxidants alpha-tocopherol, ascorbate, urate and glutathione and the protective enzymes superoxide dismutase (SOD), catalase and glutathione peroxidase, which effectively prevent free radical-induced injury under normal circumstances, it has become obvious that under certain conditions (such as ischemia and reperfusion) these protective mechanisms can be overloaded and severe radical-mediated damage occurs {28}. McCord {28} and Hess and colleagues {14} have developed the theory that ischemia can "prime" the myocardium for free radical-induced injury. Thus ischemia can promote the calcium-dependent proteolytic conversion of xanthine dehydrogenase to xanthine oxidase {14,28}; the former utilizes NAD^+ as an electron acceptor whereas the latter uses oxygen and yields superoxide radicals. This conversion occurs at a time when both hypoxanthine and xanthine, which are substrates for this enzyme, are accumulating in myocardial cells due to the hypoxic-induced breakdown of ATP. Moreover, ischemia can activate the arachidonic acid pathway, provoke the dissociation of the mitochondrial electron transport chain, and cause the release of catecholamines which are subsequently degraded. The former events can result in the production of superoxide radicals which have been suggested to activate a latent chemotactic factor present in extracellular fluids, thereby causing the invasion of activated leukocytes {29}. These factors create a scenario for free radical production and tissue injury during the reperfusion phase and have led several groups to study the effects of various protective agents {11,12,30-32}.

INDIRECT EVIDENCE FOR RADICAL GENERATION

Studies using the isolated rat heart with transient (10 minute) regional ischemia (produced by ligation of the left anterior descending coronary artery; see reference {11,30}) and reperfusion have produced the results in the Table.

Although it should be noted that alternative modes of action can be proposed for almost all of the interventions studied, the only known common denominator of all these compounds is their ability to interfere with the flux of free radicals. This would imply a link between the extent of free radical generation and the incidence of arrhythmias. Further evidence to reinforce this circumstantial link has been provided by the discovery that free radical generating systems greatly increased the vulnerability of heart tissue to arrhythmias {30-32}. Hypoxanthine (10 μM) and xanthine oxidase (1000 IU/dm^3) were shown to increase the incidence of ventricular fibrillation in vitro from 20% in controls to 100% in treated groups, and a $FeCl_3$/ADP (0.1 μM:1.0 μM) system increased fibrillation from 20% in controls to 93% in treated systems {31-32}.

Intervention	Dose	Condition	Ref.	Incidence of ventricular fibrillation (%)[a] Control	Drug
Allopurinol[b]	20mg/kg	in vivo	11	67 (18)	11 (18)
SOD	$1x10^5$ IU/dm^3	in vitro	30	80 (15)	0 (15)
Catalase	$5x10^5$ IU/dm^3	in vitro	30	80 (15)	13 (15)
Mannitol	50 mM	in vitro	30	80 (15)	7 (15)
Glutathione	10 mM	in vitro	30	80 (15)	20 (15)
Desferroxamine[c]	150 uM	in vitro	30	80 (15)	7 (15)
Folic Acid Solution[d]	10mg/kg	in vivo	32	67 (18)	17 (12)
Soybean Trypsin Inhibitor[e]	4mg/kg	in vivo	32	75 (12)	42 (12)
Ascorbic Acid	0.5 mM	in vivo	12	91 (21)	22 (9)

a) Numbers in parenthesis indicate number of animals in group

b) Inhibitor of xanthine oxidase

c) Extremely good chelator of Fe^{3+} (binding constant 10^{31}).

d) Active component is pterinaldehyde, a potent inhibitor of xanthine oxidase

e) Inhibits conversion of xanthine dehydrogenase to xanthine oxidase.

These results suggest that free radical spin traps which also act essentially as free radical scavengers by converting highly reactive radicals into relatively unreactive spin adducts would also possess anti-arrhythmic properties. Studies by Hearse and Tosaki using the spin traps α-phenyl-N-tert-butylnitrone (PBN) and 5,5-dimethyl-1-pyrroline-N-oxide (DMPO) have shown that these compounds do indeed exert a dose-dependent effect on reperfusion-induced ventricular fibrillation in the isolated rat heart {33,34}.

DIRECT EVIDENCE FOR RADICAL GENERATION

This association between the production of free radicals and the genesis of reperfusion-induced arrhythmias, though strong, is indirect and may, therefore, be circumstantial. In order to prove a direct relationship between these two factors, it is necessary to detect and quantify free radical production under identical conditions to those used in the arrhythmia studies mentioned above. Unfortunately radicals such as $HO^{\bullet}$ and $O_2^{\bullet -}$ (and lipid-derived species which may arise from further reactions of these species) are very difficult to detect directly using electron spin resonance spectroscopy (which is specific for species with unpaired electrons, i.e. free radicals and some metal ions, and is the only available technique for directly detecting radicals in biological systems). This may be due to technical reasons (e.g. the highly anisotropic spectra of species such as $HO^{\bullet}$ which result in extremely broad undetectable lines), or the low steady state concentrations and high reactivity of such radicals. However, the technique of spin trapping {35,36} enables most of these problems to be overcome; the resultant adducts tend to have well defined spectra with sharp lines and are, in many cases, stable for considerable periods of time. We have, therefore, employed e.s.r. in conjunction with the spin trap PBN to try and identify and quantify radical production in isolated rat hearts under similar conditions to those used in the complementary arrhythmia studies {37}.

Isolated rat hearts were perfused at 37°C in the Langendorff mode with oxygenated Krebs-Henseleit {38} buffer containing PBN (3 mM). After 35 minutes of aerobic stabilization and 15 minutes of total, global ischemia, the hearts were reperfused with oxygenated buffer for 30 minutes. Sequential one minute aliquots of coronary effluent were collected and the PBN-spin adducts present concentrated by extraction into toluene. E.s.r. spectra were recorded from

a) oxygenated buffer before passage through the heart, and

b) at the time points indicated in Figure 1. Typical spectra are shown in Figure 2. Spectrum C is characteristic {35,36} of a nitroxyl radical formed by the trapping, by PBN, of a reactive radical generated during the reperfusion of the myocardium. Since the intensity of the spin-adduct signal obtained (i.e. signal

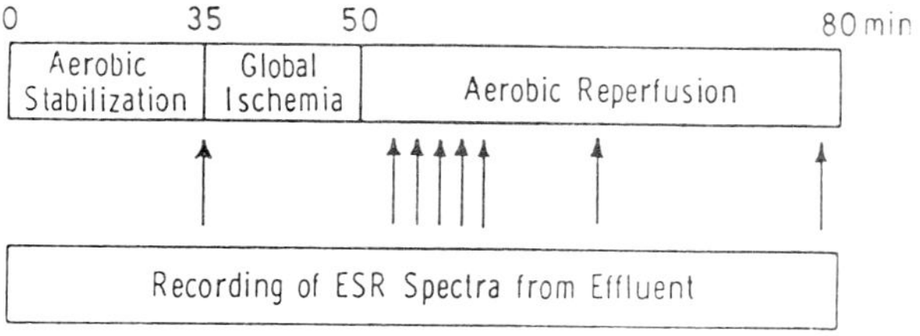

Figure 1: Perfusion protocol showing times at which e.s.r spectra were recorded.

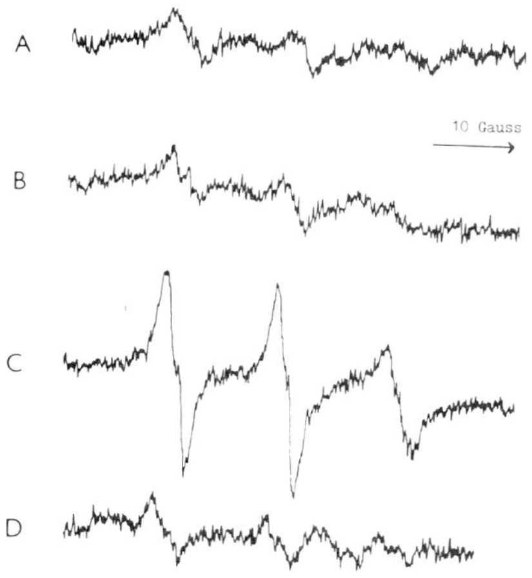

Figure 2: E.S.r spectra of spin adducts after extraction from A) oxygenated buffer before passage through the heart, b) the perfusate from the heart after 35 minutes stabilization, C) the perfusate during the 2nd minute of aerobic reperfusion and, D) the perfusate during the fifteenth minute of aerobic reperfusion. All spectra were recorded under identical spectrometer conditions.

height) is directly proportional to the radical concentration present in the coronary effluent a time course of spin-adduct formation can be constructed (Figure 3).

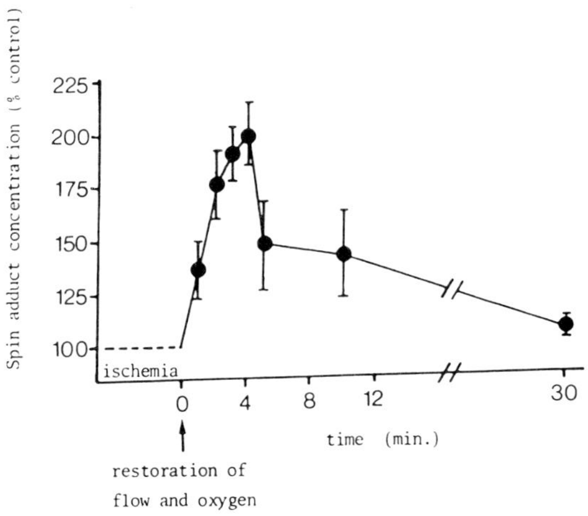

Figure 3: Time-course of PBN spin adduct formation during aerobic reperfusion. The bars represent the standard error of the means.

These experiments demonstrate that there is a large, sudden increase in spin adduct production during the early moments of reperfusion with oxygenated buffer and strongly suggest that there is a sudden burst of free radical production in the myocardium immediately preceding this.

This burst of spin adduct formation could be also due either to radicals produced and trapped during the ischemic phase, or to spin adducts generated by metabolism of the spin trap. These possibilities were tested in experiments similar to those above except that the first 10 minutes of reperfusion were carried out using anoxic buffer. Perfusate samples for e.s.r. analysis were collected at the time points indicated in Figure 4. The absence of a sudden increase in spin adduct

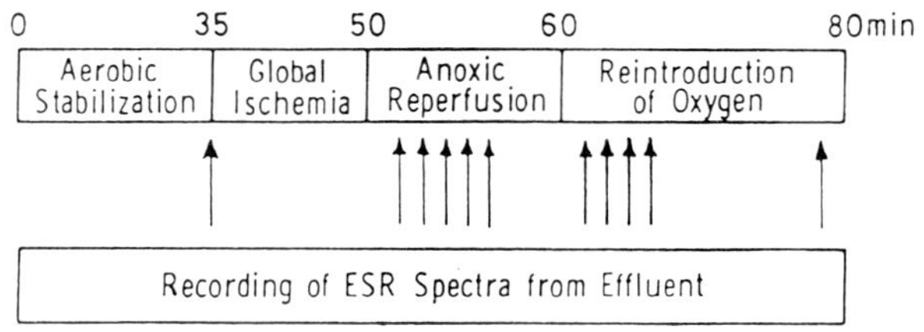

Figure 4: Perfusion protocol showing times at which e.s.r. spectra were recorded.

concentration during the 10 minutes of reperfusion with anoxic buffer excludes the above as possible sources of spin adduct formation in this system. Subsequent re-introduction of oxygen after the first 10 minutes of (anoxic) reperfusion, resulted in an increase in spin adduct concentration similar to that observed in the wholly oxygenated system (Figure 5). The spin adduct signals observed in both sets of experiments were identical in nature and differed only in intensity. The hyperfine coupling constants a_N 13.60 a_H 1.63 G of this spin adduct (which can be observed at room temperature for many hours) were calculated from spectra obtained from degassed samples under high resolution conditions (Figure 6). These values are consistent with the spin trapping of an alkoxyl radical (c.f. values of a_N 14.22, a_H 2.10 G and a_N 14.11, a_H 1.83 G for the trapping of methyl linoleate and t-butyl alkoxyl radicals respectively in benzene {39}) or perhaps a carbon-centred species with electron-withdrawing groups attached to the formal radical centre. The broadening of the high field lines of the observed spin adduct, which is apparent even at room temperature, indicates that this species in undergoing relatively slow rotational motion; this is indicative of the trapping of a relatively large molecule such as a free fatty acid or phospholipid molecule. The observed adduct probably arises via attack on membrane components by initially produced oxygen-dependent radicals

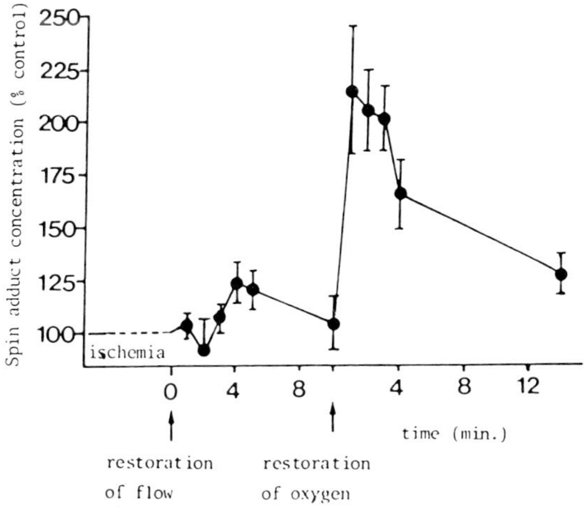

Figure 5: Time-course of PBN spin adduct formation during experiments where reperfusion was initially carried out using anoxic buffer before switching to oxygenated buffer. The bars represent the standard errors of the means.

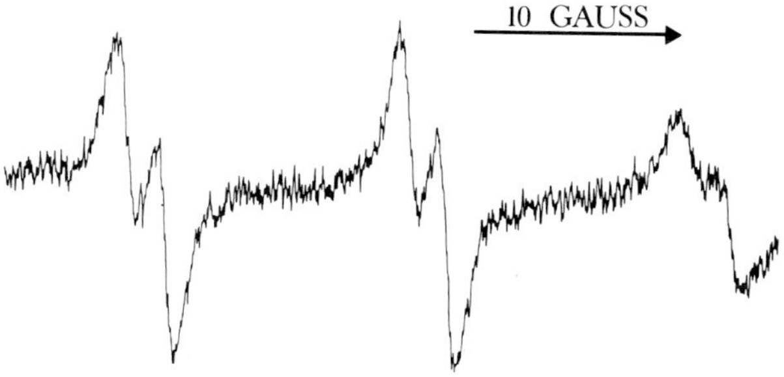

Figure 6: High resolution spectrum of spin adduct, after extraction and degassing, from 2nd minute of a erobic reperfusion.

(such as $HO^{\cdot}$); the adducts formed by reaction of $O_2^{\cdot -}$ and $HO^{\cdot}$ with PBN are unstable and both have a larger a_H value (2.01 and 2.25 G for the $HO^{\cdot}$ and $O_2^{\cdot -}$ adducts respectively; {40,41}) than that of the observed species.

CONCLUDING REMARKS

Using the isolated rat heart, evidence has been obtained to indicate that a diverse collection of interventions, which are all known to modulate free radical production and reactivity, are able to improve the resistance of the myocardium to potentially lethal arrhythmias. This provides indirect evidence that free radicals act as arrhythmic triggers. Further studies using electron spin resonance spectroscopy and a spin trap have provided direct evidence for a sudden burst of oxygen-dependent free radical production during the first few minutes of reperfusion of the previously ischemic myocardium; the peak of radical generation occurs at approximately the same time as the maximum vulnerability to arrhythmias.

ACKNOWLEDGMENTS

We are grateful to Sarah Shuter for assistance in the preparation of this manuscript and to the British Heart Foundation, the St Thomas' Hospital Research Endowments Fund and the National Foundation for Cancer Research for financial support.

REFERENCES

1. Hearse, D.J., J. Mol. Cell. Cardiol. 9 (1977) 605.

2. Hearse, D.J., Clin. Res. Reviews 4 (1984) 58.

3. Manning, A.S. & Hearse, D.J., J. Mol. Cell. Cardiol. 16 (1984) 497.

4. Hoffman, B.E. & Rosen, M.R., Circ. Res. 49 (1981) 1.

5. Greenberg, H.M., Kulbertis, H.E., Moss, A.J. & Schwartz, P., Ann. N.Y. Acad. Sci. 427 (1984) 1.

6. Sheridan, D.J., Penkoske, P.A., Stoll, B.E. & Corr, P.B., J. Clin. Invest. 65 (1980) 81.

7. Podzuweit, T., Dalby, A.J. Cherry, G.W. & Opie, L.H., J. Mol. Cell. Cardiol. 10 (1978) 81.

8. Corr, P.B. & Sobel, B.E. in Early Arrhythmias Resulting from Myocardial Ischemia (Ed. J.R. Parratt), MacMillan Press, London (1982) 199.

9. Hirche, H. Friedrich, R., Kebbel, L., McDonald, F. & Zyeka V. in Early Arrhythmias Resulting from Mycardial Ischemia (Ed. J.R. Parratt), MacMillan Press, London (1982) 139.

10. Most, A.S., Capone, R.J. & Mastrofrancesco, P.A., Cardiovasc. Res. 11 (1977) 198.

11. Manning, A.S., Coltart, D.J. & Hearse, D.J., Circ. Res. 55 (1984) 545.

12. Woodward, B. & Zakaria, M., J. Mol. Cell. Cardiol. 17 (1985) 485.

13. McCord, J.M., New Eng. J. Med. 312 (1985) 159.

14. Hess, M.L. & Manson, N.H., J. Mol. Cell. Cardiol. 16 (1984) 969.

15. Bulkley, G.B., Surgery 94 (1983) 407.

16. Slater, T.F., Biochem. J. 222 (1984) 1.

17. Chambers, D.E., Parks, D.A., Patterson, G., Roy, R., McCord, J.M., Yoshida, S., Parmley, L.F. & Downey, J.M., J. Mol. Cell. Cardiol. 17 (1985) 145.

18. Romson, J., Bush, L., Jolly, S. & Lucchesi, B.R. J. Cardiovasc. Pharmacol. 4 (1982) 187.

19. Romson, J., Hook, B., Kunkel, S., Abrams, G., Schork, A. & Lucchesi, B.E., Circulation 67 (1983) 1016.

20. Singal, P.K., Beamish, R.E. & Dhalla, N.S., Adv.Exp.Med.Biol. 161 (1983) 391.

21. Turners, J.F. & Bovaris, A., Biochem. J. 191 (1980) 421.

22. Hess, M.L., Okabe, E. & Kontos, H.A., J. Mol. Cell. Cardiol. 13 (1981) 767.

23. Halliwell, B. & Gutteridge, J.M.C., Free Radicals in Biology and Medicine, Clarendon Press, Oxford (1985).

24. DelMaestro, R.F., Thaw, H.H., Bjork, J., Planter, M. & Arfors, K.E., Acta Physiol. Scand. Suppl. 492 (1982) 465.

25. Kramer, J.H., Mak, I.T. & Weglicki, W.B., Circ. Res. 55 (1984) 120.

26. Noronha-Dutra, A.A. & Steen, A., Lab. Invest. 47 (1982) 346.

27. Harris, W.E. & Stahl, W.L., Biochem. J. 218 (1984) 331.

28. McCord, J.M., in Therapeutic Approaches to Myocardial Infarct Size Limitation (Eds. D.J. Hearse and D.M. Yellon), Raven Press, New York (1984) 209.

29. Petrone, W.F., English, D.K., Wong, K. and McCord, J.M., Proc.Natl.Acad.Sci. USA 77 (1980) 1159.

30. Bernier, M., Hearse, D.J. & Manning, A.S., Circ.Res. 58 (1986) 331.

31. Bernier, M., Hearse, D.J. & Manning, A.S., J.Mol.Cell. Cardiol. 18 (Suppl. 1) (1986) 100.

32. Manning, A.S., J.Mol.Cell.Cardiol. 18 (Suppl. 1) (1986) 197.

33. Hearse, D.J. & Tosaki, A., Circ.Res. (1987) in press.

34. Hearse, D.J. & Tosaki, A., Cardiovascular Pharmacology (1987) in press.

35. Janzen, E.G. Acc. Chem. Res. 4 (1971) 31.

36. Perkins, M.J., Adv. Phys. Org. Chem. 17 (1980) 1.

37. Garlick, P.B., Davies, M.J., Hearse, D.J. & Slater, T.F., Circ. Res. (1987) in press.

38. Krebs, H.A. & Henseleit, K., Hoppe Seylers Z. Physiol. Chem. 210 (1932) 33.

39. Yamada, T., Niki, E., Yokoi, S., Tsuchiya, J., Yamamoto, Y. & Kamiya, Y., Chem. Phys. Lipids 36 (1984) 189.

40. Janzen, E.G., Nutter, D.E., Davis, E.R., Blackburn, B.J., Poyer, J.L. & McCay, P.B., Can. J. Chem. 56 (1978) 2237.

41. Ohto, N., Niki, E. & Kamiya, Y., J.Chem.Soc. Perkin Trans. 2 (1977) 1770.

FREE RADICALS: CHEMISTRY, PATHOLOGY AND MEDICINE
edited by Catherine Rice-Evans & Thomas Dormandy, © 1988 Richelieu Press, London

Changes in Redox Status of Blood in Patients with Acute Myocardial Infarction

J.J.F. Belch*, M. Chopra*
A. Tweddel, I. Hutton, S. Cobbe, W.E. Smith°

Department of Medicine* and Cardiology,
Glasgow Royal Infirmary,
10 Alexandra Parade,
Glasgow G31 2ER,
Scotland,
U.K.

Department of Pure & Applied Chemistry°,
University of Strathclyde,
Glasgow,
Scotland,
U.K.

Abstract

We have investigated changes in redox status of blood following reperfusion in patients with myocardial infarction. Extracellular parameters followed were plasma thiol concentration and caeruloplasmin activity and intracellular parameters were residual thiol concentration in lysate after haemoglobin precipitation and superoxide dismutase activity. The thiols give one a measure of redox status, whereas the enzymes give an indication of oxidative ability and superoxide scavenging ability respectively. Results show that patients have a more oxidised extra-cellular and more reduced intracellular environment creating a greater redox stress across the membrane. Following reperfusion further oxidation of plasma thiol and increase of caeruloplasmin levels was observed. There was however no significant change in intracellular thiol levels.

Introduction

Oxygen derived free radicals have been implicated in many disease processes such as inflammation[1,2,3], digestive tract diseases[4], nervous disorders[5] and ischaemia[6]. Recently there has been increasing evidence for free radical damage in reperfusion injury of the heart[7,8]. During ischaemia a cytosolic protease is activated which attacks the cytosolic enzyme xanthine dehydrogenese converting it by limited proteolysis to a new enzyme activity, xanthine oxidase. Hypoxanthine then provides the substrate for xanthine oxidase to generate the superoxide radical ($O_2^{\cdot -}$) via univalent reduction of O_2 [7]. Once this conversion has been triggered, the availability of molecular oxygen will result in the production of free radicals. Other sources of free radicals in ischaemia are mitochondrial NADH dehydrogenase[9] and neutrophils[10].

Reperfusion is thus associated with extension of the injury initially produced by ischaemia. Experimental studies have shown that free radical scavengers and agents that prevent free radical production can reduce myocardial infarct size in animals[11]. It has been suggested recently that there are points of interplay between reperfusion injury and inflammation[12] whereby an ischaemia/ reperfusion injury would lead to subsequent inflammation and conversely whereby inflammation could lead to impairment of the circulation and hence to ischaemic injury.

There have been several reports suggesting an alteration in redox status of blood in inflammatory states. It has been reported that plasma thiol levels are low[13,14] and caeruloplasmin levels are significantly higher in these patients[14,15,16]. Recently, we have found that in vascular diseases the plasma is more oxidised and the intracellular environment is more reduced. Furthermore, levels of caeruloplasmin (Cp) and superoxide dismutase (SOD) were found to be abnormal in these patients[17]. As we believe that redox status of blood as reflected by these parameters alters in the presence of free radicals we have further investigated changes in these following reperfusion in patients with acute myocardial infarction.

Patients and Methods

Patients with a history of chest pain of at least 30 minutes duration due to acute myocardial infarction (AMI) were enrolled in the study. AMI was subsequently confirmed by ECG changes and elevation in cardiac enzymes. Effect of treatment with a fibrinolytic agent was assessed and compared with placebo treatment. Both groups of patients received heparin and ancillary therapy (Warfarin and Timolol). Ethical permission for the study had been received from the Regional Ethical Committee.

Results of 13 patients are reported here. These results were also compared with 12 age and sex matched controls.

Methods

10 mls of heparinised blood was centrifuged at 3000 rpm for 10 minutes and plasma was removed by suction. Plasma was kept at 4°C for thiol and caeruloplasmin (Cp) measurements. The red blood cells were washed twice with saline and lysed for 2 hours in distilled water (1:1 v/v) at 4°C. After approximately 2 hours 2 mls of hemolysate was removed and to it was added 0.8 ml of a mixture of chloroform and ethanol (3:5 v/v) at 0°C to precipitate haemoglobin, followed by 0.30 ml of distilled water. The resulting mixture was then centrifuged (3000 rpm, 10 minutes) and pale yellow clear supernatant was removed by suction. This lysate was used to determine superoxide dismutase activity (SOD) and intracellular thiol concentration (LSH).

Superoxide dismutase (SOD)activity of the lysate was measured by the method of Misra and Fidovich[18]. Thiol concentration of plasma and lysate was measured on same day by the technique described by Ellman[19]. The caeruloplasmin oxidase activity (Cp) was measured by a modification of the method of Menden[20].

Results and Discussion

Results of this study suggest that redox status of blood is altered in patients with acute myocardial infarction. All the four parameters, PSH, LSH, Cp and SOD are different from normal values.

Plasma thiol levels are most significantly decreased indicating that plasma is more oxidised (Tables 1,2 and 3).

Table 1: Median, I.Q.R. and Range Values for normal subjects

	PSH μml^{-1}	Cp mg dl^{-1}	LSH μml^{-1}	SOD μg ml^{-1}
(n)	(12)	(12)	(12)	(12)
Median	511	43.2	159	70.2
I.Q.R.	468-531	34.4-45.6	134-203	61.2-75.8
Range	441-595	24.0-52.5	110-358	51.1-97.1

Table 2: Median, I.Q.R. and Range Values for Active Group (Significance of difference from Normal Values is also given [Mann-Whitney]).

(a) Prior to Reperfusion

	PSH μml^{-1}	Cp mg dl^{-1}	LSH μml^{-1}	SOD μg ml^{-1}
(n)	(6)	(6)	(6)	(6)
Median	410	51.2	243	64.9
I.Q.R.	397-418	43.5-58.6	125-359	55.1-78.3
Range	389-420	38.0-63.0	71-388	41.3-95.1
Significance of diff. from normal	$p<.001$	$p<.06$	N.S.	N.S.

(b) 7 Days After Reperfusion

	PSH μml^{-1}	Cp mg dl^{-1}	LSH μml^{-1}	SOD μg ml^{-1}
(n)	(6)	(6)	(6)	(6)
Median	320	72.5	257	57.0
I.Q.R.	292-349	65.6-83.6	246-313	55.7-62.1
Range	247-388	62.5-90.0	167-353	49.1-78.2

Table 3: Median, I.Q.R. and Range Values for Placebo Group

(Significance of difference from Normal Values is also given [Mann-Whitney]).

(a) Prior to Reperfusion

	PSH μml^{-1}	Cp $mg\ dl^{-1}$	LSH μml^{-1}	SOD $ug\ ml^{-1}$
(n)	(7)	(7)	(7)	(7)
Median	428	50.3	287	74.8
I.Q.R.	390-452	48.7-50.7	201-310	54.3-75.8
Range	354-497	47.0-64.2	133-443	35.5-77.4
Signif. of diff. from Normal	$p<.001$	$p<.001$	N.S.	N.S.

(b) 7 Days After Reperfusion

	PSH μml^{-1}	Cp $mg\ dl^{-1}$	LSH μml^{-1}	SOD $ug\ ml^{-1}$
(n)	(5)	(5)	(5)	(5)
Median	360	65.0	319	48.3
I.Q.R.	355-404	47-78	309-380	42.4-48.4
Range	242-416	36-85	240-387	40.3-53.7

Plasma thiol levels can be affected by the radicals in many ways. Free radicals induced tissue damage in ischaemic patients will lead to the release of low molecular weight thiol species in blood. Caeruloplasmin (Cp) has been shown to affect circulating low molecular weight thiol molecules[21,22] but it cannot affect the albumin thiol (major contributor of plasma thiol) directly due to steric hindrance. However, low molecular weight disulphides produced by Cp can readily exchange with albumin thiol leading to a decrease in plasma thiol levels. It has been reported that the superoxide radical can activate a latent chemotactic factor present in extracellular fluids, thereby enabling a single activated neutrophil (which is liberating O_2 from its surface into the extracellular fluid) to signal its location to other nearby resting neutrophils. This superoxide dependent chemoattractant consists of a lipid component noncovalently bound to serum albumin. It is thus possible that superoxide ion interaction with this chemoattractant exposes previously hidden thiol group of albumin, making it readily available for oxidation and/or thiol disulphide interchange reaction.

Figure I (PLACEBO GROUP)

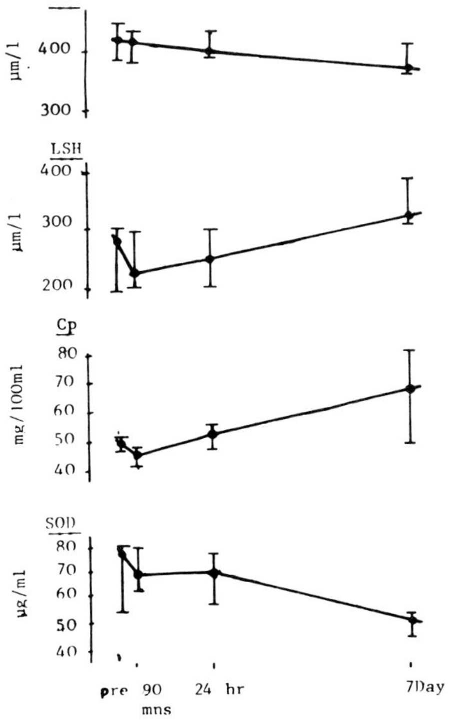

Figure II (ACTIVE GROUP)

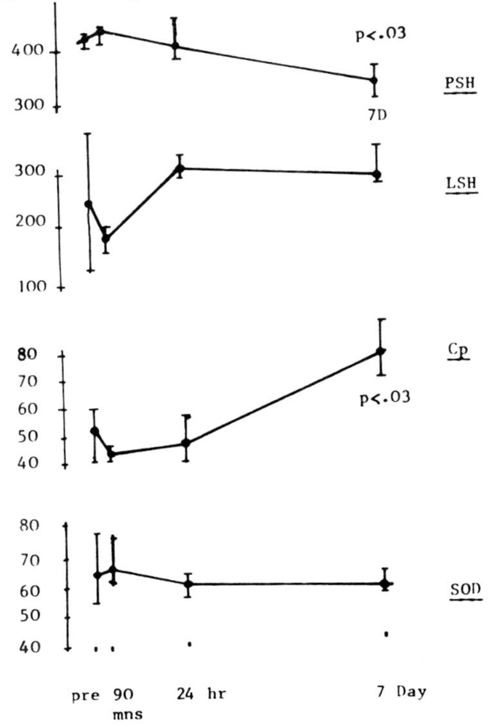

Figure I and II show the median values and I.Q.R. for PSH, LSH, Cp and SOD prior to and 90 minutes, 24 hours and 7 days after reperfusion.

Significance of the changes are also given (Wilcoxon-Rank Sum).

There was a slight improvement in all the four parameters in the active group (Figure II), 90 minutes after reperfusion. These patients along with placebo group patients also received heparin 6 hours after dosing and Timolol and Warfarin 24 hours after inclusion in the study. Both groups of patients show a similar trend in deterioration 24 hours after inclusion in the study. We suggest that deterioration is due to free radical production following dissolution of intravascular thrombus. These four parameters have been previously reported to be abnormal in rheumatoid arthritis (a disease in which there are several reports of free radical pathology). Also, it has been reported that treatment of rheumatoid patients with anti-inflammatory agents penicillamine, gold and sulphasalazine (SASP) leads to improvement in these parameters[23,24,25,26]. Thus, although the enzymes responsible for producing free radicals in inflammation and ischaemia may be different, there seems to be similar effect of free radicals on these parameters and particularly on PSH and Cp levels.

There was a decrease in PSH and increase in Cp levels in both groups 7 days after reperfusion. This change was more significant in active group (Figure II). Active group patients received an additional fibrinolytic agent. It appears although initial effect of the drug is beneficial as shown by improvement in all the four parameters 90 minutes after administration, drug does seem to have similar effect on free radical pathology as other agents such as heparin, Timolol and Warfarin. The only difference between the placebo group and the active group was that reperfusion with the drug appears to prevent the 7 day fall in RBC SOD levels.

Although these are preliminary results and a larger sample population is required before any definite statement can be made, results of this study do show a trend of deterioration in PSH and Cp levels following reperfusion and we suggest that this is due to free radical production.

References

1. Babior, B.M., Kipnes, R.S. and Curnutte, J.T.
J. Clin. Invest., 52, (1973), 741.

2. McCord, J.M.
Science, 185, (1974), 529.

3. Root, R.K., Metcalf, J., Oshino, N. and Chance, B.
J. Clin. Invest., 55 (1975), 945.

4. Parks, D.A., Bulkley, G.B. and Granger, D.N.
Surgery, 94 (1983), 415.

5. Demopoulous, H.B., Flamm, E.S., Pietronigro, D.D., and Seligman, M.L.
Acta Physiol. Scand., 492 (1980), 91.

6. McCord, J.M. and Roy, R.S.
Can. J. Physiol. Pharmacol., 60 (1982), 1346.

7. McCord, J.M.
N. Engl. J. Med., 312 (3), (1985), 159.

8. Rowe, G.T., Eaton, L.R. and Hess, M.L.
J. Mol. Cell Cardiol., 16 (1984), 1075.

9. Guarnieri, C., Muscari, C., Ventura, C. and Mavelli, I.
Free Rad. Res. Comm., 1, No.2, (1985), 123.

10. Engler, R.
Fed. Proc., 46, No.7, (1987), 2395.

11. Jolly, S.R., Kane, W.J., Bailie, M.B., Abrams, G.D., Lucchesi, B.R.
Circ. Res., 54 (1984), 277.

12. McCord, J.M.
Fed. Proc., 46, No.7 (1987), 2402.

13. Hall, N.D., Maslen, C.L. and Blake, D.R.
Rheumatol. Int., 4, (1984), 35.

14. Banford, J.C., Brown, D.H., Hazelton, R.A., McNeil, C.J., Sturrock, R.D. and Smith, W.E.
Ann. Rheum. Dis., 41, (1982), 458.

15. Brown, D.H., Buchanan, W.W., El-Ghobarey, A.F., Smith, W.E. and Teape, J.
Ann. Rheum. Dis., 38 (1979), 174.

16. Velo, G.P., Franco, L., Conforti, A. and Milamino, P.
In: Inflammatory disease and copper (1982). J.R.J. Sorensen (ed.), New Jersey, Humana Press, p.329.

17. Chopra, M., Belch, J.J.F., Hutchison, S., Lorimer, R., Sturrock, R.D., Forbes, C.D. and Smith, W.E. (1987).
Accepted for Free Radicals in Biology and Medicine.

18. Misra, H.P. and Fridovich, I.
Arch. Biochem. Biophys., 181 (1977), 308.

19. Ellman, G.E.,
Arch. Biochem. Biophys., 82, (1959), 70.

20. Menden, E.E., Boiano, J.M., Murthy, L. and Petering, H.G.
Anal. Lett., 10 (1977), 197.

21. Albergoni, V. and Cassini, A.
Febs. Letts., 55, (1974), 261.

22. Albergoni, V. and Cassini, A.,
Bull. Mol. Biol. Med., 3 (1978), 79.

23. Banford, J., Brown, D.H. Hazelton, R.A., McNeil, C.J., Smith, W.E. and Sturrock, R.D.
Rheumatol. Int. 2 (1982), 107.

24. Rae, K., McKay, C.N.N., Brown, D.H., Smith, W.E., Lewis, D. and Capel, A.A.
Ann. Rheum. Dis., 45 (1986), 839.

25. Hall, N., Blake, D.R. and Bacon, P.A.
J. Rheumatol., 9 (1982), 593.

26. Bird, H.A., Dixon, J.S. and Pickup, M.E.
J. Rheumatol., 9 (1982), 36.

Acknowledgement

This work was funded by a British Heart Foundation Grant 86/72 (2D31).

FREE RADICALS: CHEMISTRY, PATHOLOGY AND MEDICINE
edited by Catherine Rice-Evans & Thomas Dormandy, © 1988 Richelieu Press, London

Generation of Reactive Oxygen Species by Human Spermatozoa:

R.J. Aitken and J.S. Clarkson, MRC Reproductive Biology Unit,

37 Chalmers Street, Edinburgh EH3 9EW

Studies of the cellular mechanisms regulating human sperm function indicate the presence of a calcium-dependent mechanism for the generation of reactive oxygen species, which can be activated by divalent cation ionophores and stimulators of protein kinase C. In purified membrane preparations or triton permeabilized cells a burst of reactive oxygen species production can be induced with NADH or NADPH, the latter being the preferred substrate. The production of reactive oxygen species in either intact or permeabilized cells is not disrupted by the mitochondrial inhibitors antimycin A, oligomycin or rotenone. Scavengers of singlet oxygen or hydroxyl radicals are also without effect although cytochrome C is active, suggesting the presence of superoxide anion.

The clinical significance of these findings is indicated by the fact that infertile men, whose spermatozoa are unable to fuse with oocytes, are characterised by extremely high (40x normal) basal levels of reactive oxygen species production which are elevated still further upon stimulation with A23187.

Defective sperm function is now recognised as the most common, defined, cause of human infertility, being exhibited by at least a third of the clinic population (Hull, 1986; Aitken and Clarkson, 1987). Despite its prevalence, this condition presents severe problems of patient management, since very little is known about the exact nature of the defects present in the spermatozoa of such men or the aetiology of their condition. Although some empirical treatments exist, this lack of knowledge concerning the cause and origin of defective sperm function has led to a general lack of any effective therapies; the only possible exceptions to this rule being immunosuppressive therapy for cases of idiopathic autoimmunity against sperm antigens (Hendry et al., 1981; Mathur et al., 1981) and spermatic vein embolisation in instances of varicocoele.

In order to address the problems encountered in the diagnosis and treatment of male infertility we have undertaken a programme of research into the fundamental cell biology of the human spermatozoon to elucidate the mechanisms involved in the regulation of sperm function. Once these processes are better understood we should be able to turn to the infertile population and determine, in precise, biochemical terms, the nature of the defects responsible for disrupting the fertilizing capacity of the spermatozoa.

Our recent studies have focused on the role of calcium as an intracellular mediator of sperm function. The resting cytoplasmic level of free ionic calcium in human spermatozoa is maintained at about 70-100 nM, largely through the activity of ATP-dependent Ca^{2+} pumps and Na^{+}/Ca^{2+} antiporters in the sperm plasma membrane and mitochondria (Breitbart et al., 1983; Breitbart and Rubinstein, 1983; Rufo et al., 1984; Irvine and Aitken, 1986). However, if cytoplasmic calcium levels are elevated through the addition of a calcium ionophore, such as A23187, then a chain of events is set in motion resulting in a secretory event, the acrosome reaction, and acquisition of the spermatozoon's ability to recognise and fuse with the vitelline membrane of the ovum (Aitken et al., 1984). However, this ability to respond to a calcium signal by undergoing sperm-oocyte fusion is not shown by the spermatozoa of all men. Within the infertile population, a high proportion of patients produce spermatozoa which are refractory to a calcium stimulus and will not exhibit any biological responses when A23187 is added to the culture medium (Aitken et al., 1984).

In order to determine the reason for this refractoriness we have undertaken a detailed examination of the changes induced in human spermatozoa in response to a sudden influx of calcium. In the course of these studies we have uncovered evidence for a calcium-sensitive superoxide generating system within the human spermatozoon, which exhibits the characteristics of an NADPH oxidase. This system is membrane bound, is not affected by rotenone, antimycin or oligomycin but responds rapidly to activators of protein kinase C, such as the 12-myristate, 13-acetate phorbol ester, or membrane permeant diacylglycerol analogues. The clinical significance of these observations was suggested when cases of defective sperm function, exhibiting refractiness to A23187 in biological tests, were discovered to exhibit highly elevated levels of reactive oxygen species production, 40x that observed in normal functional cells, even in the absence of ionophore. These findings suggest a relationship between defective sperm function and oxidative damage induced in the sperm plasma membrane through the generation of excess reactive oxygen species. These are the first studies to describe a biochemical lesion in the spermatozoa of infertile men and they have clear implications for the diagnosis and treatment of this condition.

Materials and Methods

Unless otherwise stated, samples of human semen were obtained from normal healthy donors with normal semen profiles (more than 20×10^6 spermatozoa/ml, with more than 40% normal morphology and progressive motility: Aitken et al. 1982). The semen samples were diluted with 8 ml of medium BWW (Biggers et al., 1971) and the spermatozoa pelleted by centrifugation at 500g for 5 min. The sperm pellets were then taken through 2 further cycles of centrifugation and resuspension and were finally adjusted to a concentration of 20 x 10^6 cells/ml.

Sperm-oocyte fusion The capacity of human spermatozoa to exhibit sperm-oocyte fusion in response to the calcium signal generated by A23187 was examined according to Aitken et al., (1984). In essence, this assay system is based upon the unique susceptibility of hamster oocytes to fusion with acrosome reacted human spermatozoa, once the block to interspecies fertilization, the zona pellucida, has been removed by trypsin treatment (Yanagimachi et al., 1976). Zona-free hamster eggs, prepared in medium BWW, were incubated with human spermatozoa which had been previously incubated for 3 hours with the divalent cation ionophore, A23187, in order to induce the acrosome reaction - a prerequisite for sperm-oocyte fusion. The human spermatozoa and zona-free hamster eggs were then incubated together for a further 3 hours, at the end of which the oocytes were recovered and examined for the presence of human spermatozoa which had fused with the vitelline membrane of the ovum and become incorporated into the ooplasm. Such fused spermatozoa were identified by the presence of decondensing sperm heads with an attached or closely associated tail (Yanagimachi et al., 1976; Aitken et al., 1984; Aitken 1986).

Reactive oxygen species

Human spermatozoa were washed free of seminal plasma, as previously described, and resuspended at a concentration of 20 x 10^6 cells/ml. 1ul of luminol (5-amino-2, 3-dihydro-1, 4-phthalazinedione) stored as a 100 mM stock solution in dimethyl sulfoxide, was added to 500ul of spermatozoa followed by the addition of either 500 ul medium BWW (control) or an equal volume of BWW containing an aqueous suspension of the Ca^{2+}, Mg^{2+} salt of A23187 (0.05 mg/ml). Luminescence was subsequently recorded on a Berthold, Biolumat LB 9500 T luminometer in the integration mode.

Results

Generation of reactive oxygen species by intact cells

Suspensions of normal, functional human spermatozoa produced low levels of reactive oxygen species in their resting state (Fig. 1). However, upon addition of A23187, a sudden burst of reactive oxygen species production was observed which plateaued after 3-5 min. This response to A23187 was recorded with spermatozoa which had been prepared by the conventional centrifugation technique, or isolated using either a 'swim-up' procedure (Berger et al, 1985) or discontinuous Percoll

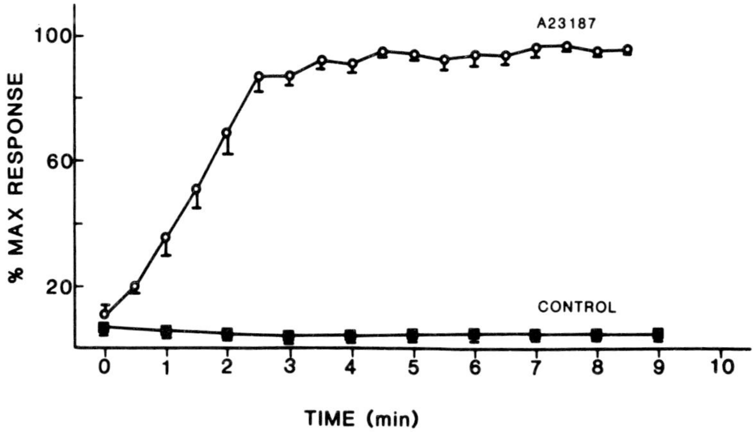

Fig. 1 Production of reactive oxygen species by human spermatozoa in response to A23187.

gradients; the activity recorded must, therefore, have originated from the spermatozoa and not from any other cell type (eg. leucocytes) contaminating the seminal plasma.

This response of intact cells to A23187 was calcium dependent, since it was not observed if the cells were suspended in calcium-depleted medium, containing no added calcium (medium BWW normally contains 1.7 mM $CaCl_2$) and supplemented with 100 uM EGTA. Mediation by calmodulin was also indicated since the response to A23187 in the presence of exogenous calcium, could be disrupted by pre-exposing the spermatozoa for 1 hour to the calmodulin antagonists calmidazolium (0.5 uM) and trifluoperazine (30 uM), (Fig. 2).

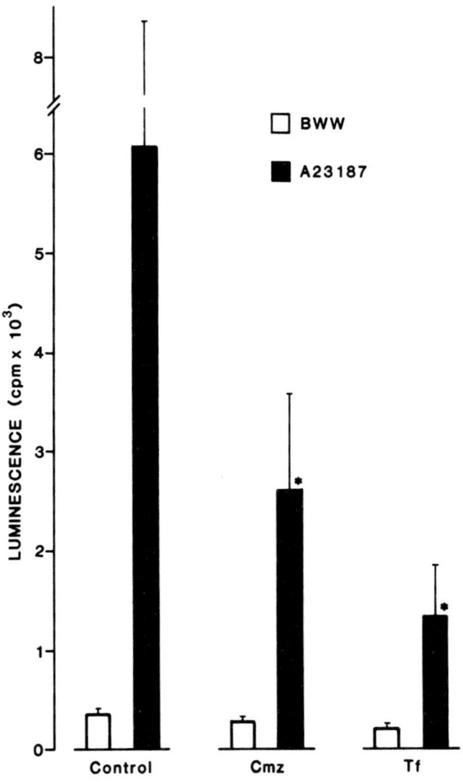

Fig. 2 Production of reactive oxygen species by human spermatozoa in the presence of calmodulin antagonists.

Several features of this reactive oxygen species generating system also suggested regulation by protein Kinase C. Hence, treatment of intact cells with activators of this kinase system, such as phorbol 12-myristate, 13-acetate ester and 1-oleoyl, 2-acetyl glycerol (Figs 3 and 4) resulted in an immediate and profound stimulation of activity. The specificity of these

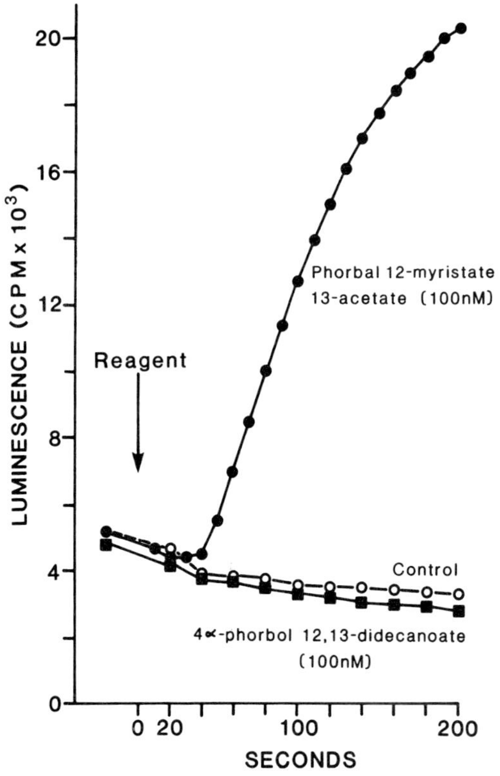

Fig. 3 Influence of phorbol esters on production of reactive oxygen species by human spermatozoa.

responses was indicated by the fact that related compounds which do not influence this kinase system, including the 4α-phorbol 12,13-didecanoate or 1,2-dioleoyl glycerol, exhibited a corresponding inability to influence reactive oxygen species production (Figs 3 & 4).

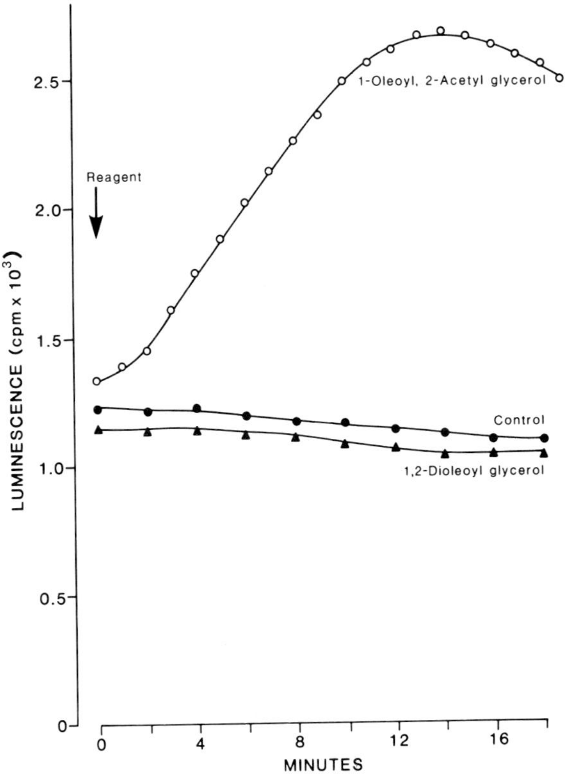

Fig. 4 Influence of diacylglycerol analogues on production of reactive oxygen species by human spermatozoa.

The generation of reactive oxygen by human spermatozoa was dependent upon the presence of a glucose in the external medium. If the glucose was removed or replaced with an equivalent concentration (5.56 mM) of 2-deoxyglucose (Fig. 5) both the resting level of reactive oxygen species production and the response to A23187 were both considerably reduced, even though

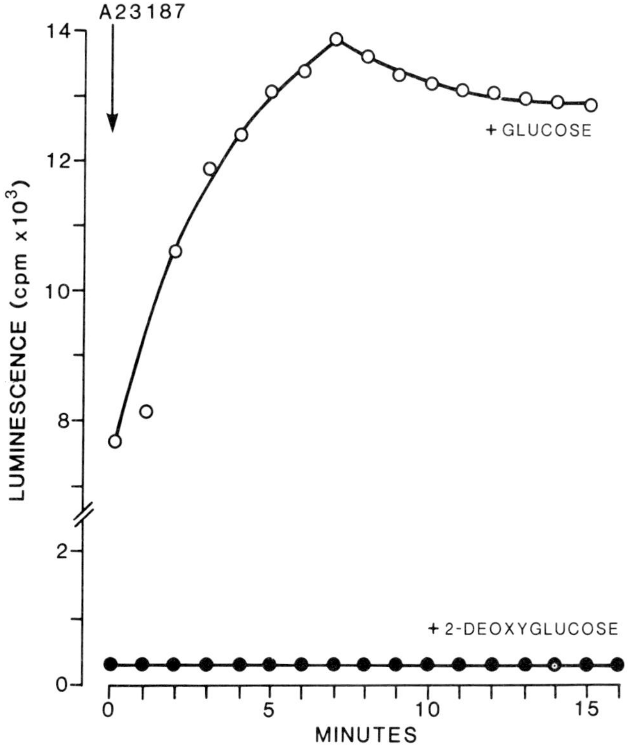

Fig. 5 Influence of glucose and 2-deoxyglucose on production of reactive oxygen species in response to A23187.

the medium contained alternative substrates (pyruvate and lactate) and sperm motility was unimpaired. Direct comparison of glucose with a variety of other sugars, for their ability to support the generation of reactive oxygen species, has emphasised the key role played by glucose. Even substrates which will readily enter the glycolytic pathway, such as fructose and mannose, are significantly less effective than glucose in the support of reactive oxygen species production.

The genesis of reactive oxygen species by intact human spermatozoa did not appear to involve the mediation of the sperm mitochondria. Hence, when the mitochondrial inhibitors oligomycin, antimycin A and rotenone were added to suspensions of human spermatozoa prior to A23187, no interference in the production of reactive oxygen species was observed (Fig. 6).

Identity of reactive oxygen species An indication as to the identity of the reactive oxygen species generated by intact human spermatozoa was obtained through the selective use of appropriate scavengers. Reagents capable of interacting with singlet oxygen (dimethyl furan, bilirubin and histidine HCl) or the hydroxyl radical (histidine HCl, sodium benzoate and mannitol) did not influence the response to A23187, while cytochrome C, a scavenger of superoxide anion, consistently reduced this activity by about 50%. Preliminary attempts to monitor superoxide anion production

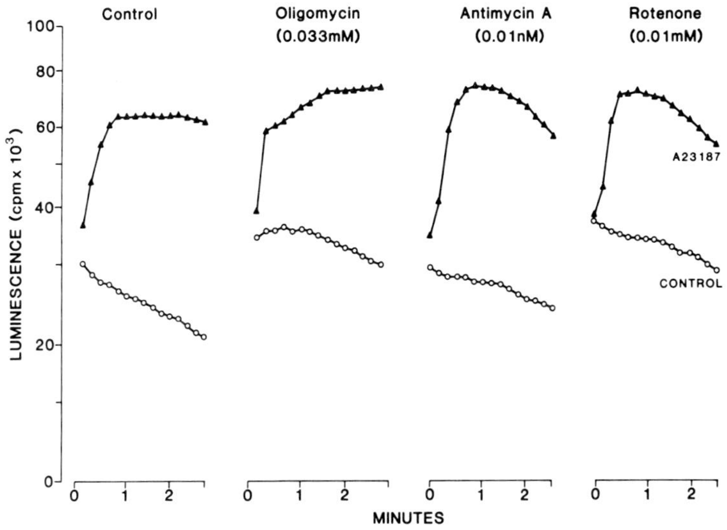

Fig. 6 Influence of mitochondrial inhibitors on the production of reactive oxygen species in response to A23187.

directly through the conventional ferricytochrome C assay, have proved unsuccessful, possibly because this system is less sensitive that luminol and/or a high proportion of the reactive oxygen species produced may not be released to the outside of the cell.

Reactive oxygen production by permeabilized cells Since the generation of reactive oxygen species by neutrophils and macrophages is known to depend upon the activity of a membrane bound NADPH oxidase, experiments were conducted to determine whether this pyridine nucleotide might be involved in the case of human spermatozoa. Neither NADH nor NADPH influenced the generation of reactive oxygen species by intact spermatozoa. However, in broken cell preparations, 100,000g particulate fractions and purified membrane vesicles (Aitken et al., 1987) both of these compounds were effective in stimulating an immediate burst of activity. Permeabilization of the cells with triton-x also rendered human spermatozoa responsive to NADH and NADPH; the latter being the preferred substrate yielding a km value of 3.8 uM compared with 12 uM for NADH (Fig. 7).

In such permeabilized cells, alternative sources for the generation of reactive oxygen species including xanthine, ascorbate, or the amino acids phenylalanine, tyrosine and tryptophan (Tosic and Walton, 1950) were found to be ineffective. Furthermore the responses to NAD(P)H were not impaired by the addition of the mitochondrial inhibitors oligomycin (0.033 mM) antimycin (0.01M) and rotenone (0.01mM). Scavangers of hydroxyl radicals (ethanol, mannitol, sodium benzoate and histidine HCl) and singlet oxygen (histidine HCl, diazabicyclo-octane, bilirubin and dimethyl furan) were also inactive in modifying the luminol response to NAD(P)H, whilst cytochrome C was completely effective

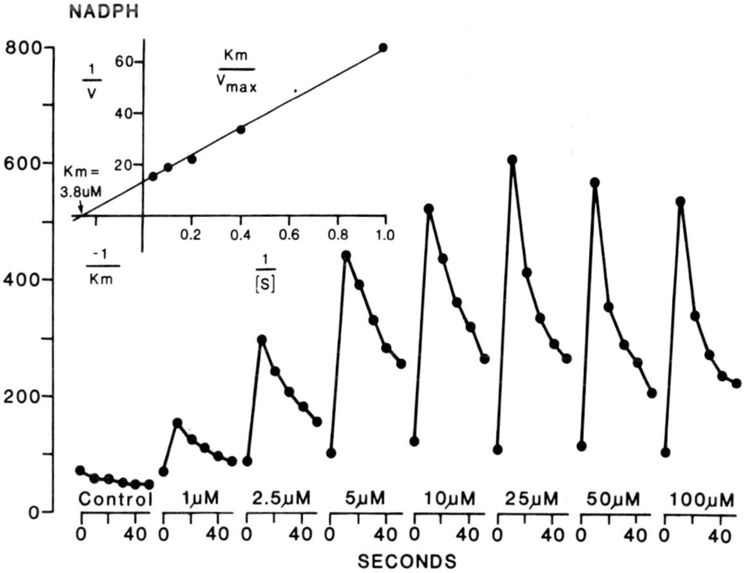

Fig.7 Influence of NADPH on the generation of reactive oxygen by triton-permeabilized cells.

in this regard.

Pathological significance One of the features of the response of intact human spermotozoa to A23187 was the considerable variation between individuals in the magnitude of the activity recorded (Fig. 8). In order to determine whether such variation bore any relationship to the differences between samples in their capacity for fertilization, a study was conducted in which simultaneous

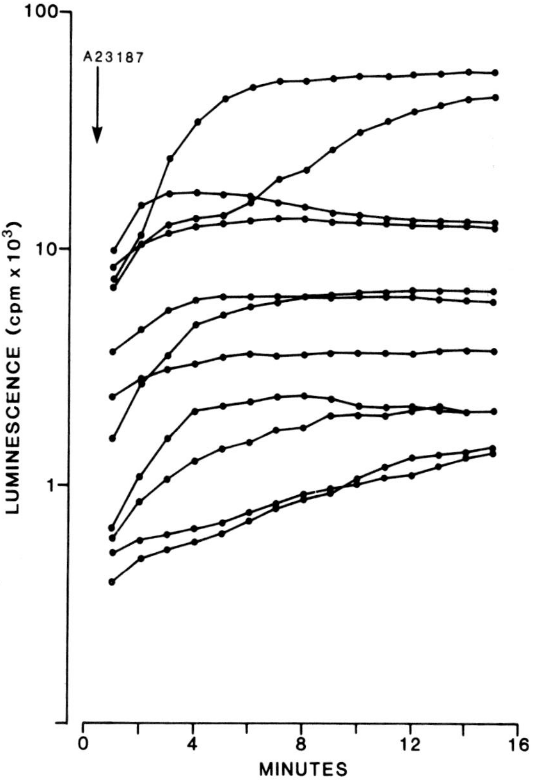

Fig. 8 Interindividual variation in the responses of the spermatozoa to stimulation with A23187.

measurements were made of both the generation of reactive oxygen species and the capacity of the same cells for sperm-oocyte fusion when stimulated by A23187. The results of this analysis are presented in Fig. 9. This figure reveals that samples exhibiting a capacity for sperm-oocyte fusion within the normal fertile range (25-100%) were characterized by a low level of reactive oxygen species production which showed a small, discrete rise on addition of A23187. In contrast, samples exhibiting a subnormal capacity for fertilization in response to A23187, particularly those in which no biological response to the ionophore was observed (0% fertilization) were characterized by extremely high levels of reactive oxygen species production. The mean, basal level of activity exhibited by such defective samples was 40 fold higher than the corresponding value recorded with normal fertile men and was increased still further upon addition of the ionophore.

Discussion

The results obtained in these studies indicate that the human spermatozoon is capable of generating reactive oxygen species. Although hydrogen peroxide and superoxide anion production by mammalian spermatozoa have previously been recorded, these reactive oxygen species were thought to arise from the deamination of certain amino acids (Tosic and Walton, 1950) or as a by-product of mitochondrial activity (Holland et al., 1982; Chapman et al., 1985). Neither of these sources of reactive oxygen species appears to apply to the human spermatozoon. In the latter the production of reactive oxygen exhibits a number of parallels with superoxide generation in neutrophils and

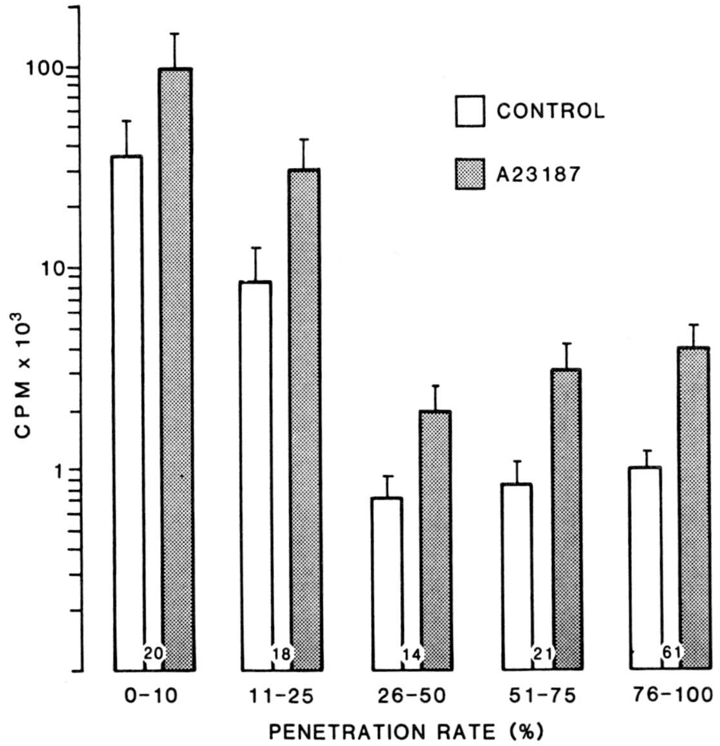

Fig. 9 Relationship between the capacity of human spermatozoa for sperm oocyte fusion (% oocytes penetrated) and production of reactive oxygen species. Open bars = basal levels; shaded bars = response recorded 3 min after A23187 addition. Numbers at base of each column = number of samples analysed.

macrophages in that the system is calcium (and possibly calmodulin) dependent, responds to activators of protein kinase C and utilizes NADPH as a source of electrons. Recent studies monitoring the production of ^{14}C labelled CO_2 from glucose labelled in positions 1 and 6, further indicates that the induction of reactive oxygen species generation by human spermatozoa is associated with accelerated glucose oxidation through the hexose monophosphate shunt, in order to provide the putative oxidase system with NADPH, in a similar manner to that described for leucocytes (R.J. Aitken, unpublished observations).

The precise nature of the reactive oxygen species produced by human spermatozoa requires further elucidation. The studies employing scavengers of singlet oxygen, hydroxyl radicals and superoxide anion indicated that the latter is probably the primary product of the NAD(P)H oxidase system, since cytochrome C was effective in suppressing or completely eliminating the luminol responses obtained with intact and permeabilized cells respectively. The complete effectiveness of cytochrome C with triton-permeabilized, as opposed to intact cells, may indicate that only a proportion of the superoxide generated by the cell is released to the exterior. This might also explain why a membrane permeant probe, such as luminol, is a much more sensitive indicator of reactive oxygen species production by human spermatozoa than ferricytochrome C itself. Since superoxide dismutase is present in both human spermatozoa and seminal plasma (Mennella & Jones, 1980) any superoxide produced by these cells should rapidly dismutate to hydrogen peroxide.

The latter must then be removed from the environment of the cells if it is not to cause peroxidative damage. Since mammalian spermatozoa do not possess catalase activity, this end is probably achieved by glutathione peroxidase, although direct evidence for this enzyme system in human spermatozoa has not yet been obtained.

The lack of catalase activity, the presence of large quantities of unsaturated fatty acids in the sperm plasma membrane (Jones et al., 1979) and the existence of this superoxide generating system, suggest that the human spermatozoon is highly susceptible to oxidative stress (Jones & Mann, 1976). The artificial induction of lipid peroxidation in human spermatozoa with ferrous ion and ascorbate has been shown to induce a loss of membrane function, inhibiting the potential of the spermatozoa for sperm-oocyte fusion and, ultimately, their motility (R.J. Aitken, unpublished observations; Jones et al., 1979). In view of this susceptibility to peroxidative damage, it is possible that the high levels of reactive oxygen species production recorded in cases of male infertility, play a causative role in the disruption of sperm function. In normal, functional, spermatozoa the mechanisms responsible for the production of reactive oxygen species are held under suppressive control. When such cells are stimulated by a calcium stimulus to undergo the acrosome reaction, a controlled, discrete rise in the genesis of reactive oxygen species is observed and may play a role in the attainment of the acrosome reaction itself or in the interaction between the spermatozoon and the ovum. In cases of defective sperm function this suppressive control is, for some reason, ineffective and the spermatozoa exhibit a high basal level of reactive oxygen species production without the need for a calcium stimulus. Such defective spermatozoa are also highly perturbable however, and if they are subjected to A23187 treatment, the generation of reactive oxygen species shows a corresponding exaggeration.

The primary defect in the spermatozoa may therefore be a disruption of the control mechanisms regulating the activity of the putative NADPH oxidase which then, through the excessive generation of reactive oxygen species, induces peroxidative damage in the sperm plasma membrane and a resultant loss of sperm function. Future studies on the pathological significance of oxidative stress in human spermatozoa will now focus on such factors as the relationship between the genesis of excited oxygen species and lipid peroxidation, the availability of iron in the seminal plasma, the possible therapeutic value of antioxidants and the fundamental mechanisms regulating the activity of this superoxide generating system.

References

Aitken, R.J. Int. J. Androl. Suppl 6 (1986) 1.

Aitken, R.J. and Clarkson, J.S. J. Reprod. Fert. (1987) In Press

Aitken, R.J., Best, F.S.M., Richardson, D.W., Djahanbakhch, O., Mortimer, D., Templeton, A.A. and Lees, M.M. Fertil. Steril. 38 (1982) 212.

Aitken, R.J., Ross, A., Hargreave, T., Richardson, D.W. & Best, F.S.M. J. Androl. 5 (1984) 321.

Aitken, R.J., Hulme, M.J., Henderson, C.J., Hargreave, T.B. and Ross, A. J. Reprod Fert (1987) In press.

Berger, T., Mars, R.P. and Moyer, D.L. Fertil. Steril. 43 (1985) 268.

Biggers, J.D., Whitten, W.K. and Whittingham, D.G. Methods in Mammalian Embryology., Freeman (1971) 86-116.

Breitbart, H., Stern, B. and Rubinstein, S. Biochem Biophys Acta 728 (1983) 349.

Breitbart, H. and Rubinstein, S. Biochem. Biophys Acta 732 (1983) 464.

Chapman, D.A., Killian, G.J., Gelerinter, E. and Jarrett, M.T. Biol Reprod 32 (1985) 884.

Holland, M.K., Alvarez, J.G. and Storey, B.T. Biol Reprod 27 (1982) 1109.

Hendry, W.F., Stedronska, J., Parslow, J and Hughes, L. Fertil Steril 36 (1981) 351.

Hull, M.G.R. Human Embryo Research: Yes or No Ciba Foundation (1986) 24-38.

Irvine, D.S. and Aitken, R.J. Gamete Res.15 (1986) 57.

Jones, R. and Mann, T. Proc. R. Soc Lond B. Biol Sci 193 (1976) 317.

Jones, R., Mann, T. and Sherins, R. Fertil Steril 31 (1979) 531.

Mathur, S., Baker, E.R., Williamson, H.O. Derrick, F.C., Teague, K.J. and Fudenberg, H.H. Fertil Steril 36 (1981) 486.

Mennella, M.R.F. and Jones, R. Biochem J. 191 (1980) 289.

Rufo, G.A., Schoff, P.K. and Lardy, H.A. J. Biol. Chem 259 (1984) 2547.

Tosic, J. and Walton, A. Biochem J 47 (1950) 199.

Yanagimachi, R., Yanagimachi, H. and Rogers, B.J. Biol. Reprod. 15 (1976) 471.

FREE RADICALS: CHEMISTRY, PATHOLOGY AND MEDICINE
edited by Catherine Rice-Evans & Thomas Dormandy, © 1988 Richelieu Press, Lond

Free Radicals and Pancreatitis

Joan M Braganza, Royal Infirmary, Oxford Road, Manchester M13 9WL

ABSTRACT

It is proposed that oxidant stress initiates both acute and chronic pancreatitis and, therefore, that antioxidants could revolutionize the management of these diseases.

The overlap in so many aspects of acute pancreatitis(AP) and chronic pancreatitis(CP) suggests a pathogenetic link between them(1). The notion that oxidant stress initiates each condition offers scope for treatment and prevention.

ACUTE PANCREATITIS

Protection against autodigestion

The exocrine pancreas is arguably the foremost protein factory in the body. It achieves this feat through a perfected conveyor-belt system that commences with the uptake of amino acids at the base of the acinar cell and culminates at its apical pole with the discharge, upon demand, of zymogen granules that contain the digestive hydrolases. Perhaps as a safeguard against any impedence to the congested traffic lanes, there is an alternative exit route - probably at the basolateral membranes - whereby secretory products are extruded directly into the interstitium and bloodstream(2-5). Apart from amylase and lipase, the digestive hydrolases are secreted as inactive precursors that are activated within the duodenum, in cascade fashion, only when enterokinase lops off an octapeptide from trypsinogen: this catalytic step is facilitated by bile salts. There are several other shields against pancreatic autodigestion(6), not least that trypsin inhibitor is cosecreted in an amount that can trap up to 20% of potential tryptic activity. Within the bloodstream(5-7) trypsinogen (? and also chymotrypsinogen) exists as the free zymogen but proelastase is complexed with α_1 protease inhibitor(α_1PI). This acute phase reactant avidly binds to, and inactivates, elastase from polymorphonuclear leucocytes(PMNL) as well as any pancreatic proteases that may have been prematurely activated. It acts as a transporter, slowly off-loading to α_2 macroglobulin(α_2M) where some proteolytic activity is regained against small peptides: this threat is reduced by the rapid removal of α_2M- complexes by the reticulo-endothelial system (half-life $<$ 10 min). There is no known physiological inhibitor of phospholipase A_2.

The Disease

Although AP occurs in diverse settings(Table I), the clinical and pathological features are the same - a stereotyped response suggesting a common detonator at intracellular level. Any concept must rationalise four sets of observations.

TABLE I MAIN INITIATING FACTORS/POSTULATED ATTACK ROUTES

HUMAN	EXPERIMENTAL
<u>Ductal</u>	
migrating gall stone	impeded outflow+secretin
pancreas divisum/cancer of ampulla	retrograde bile/trypsin
duodenal diverticulum	prograde infusion after
afferent loop syndrome	oral aspirin/ethanol
alcoholic debauch	
<u>Intra-acinar</u>	
drugs/inhaled chemicals/insecticide	puromycin/DL-ethionine
fasting-surfeit/anorexia nervosa/TPN	choline deficiency
chronic renal failure/post-transplant	hyperstimulation
infections/scorpion venom	cholinesterase inhibitor
trauma	
Reye's/haemolytic uraemic syndrome	
idiopathic	
<u>Arterial</u>	
hypothermia/cardio-pulmonary bypass	ischaemia-reperfusion
hyperchylomicronaemia	free fatty acids
hypercalcaemia	trypsin

1) The inciting agent can attack by three different routes (1,8). 2) The earliest disturbance in AP is a loss of secretory polarity, so that an abnormally large fraction of secretions - zymogens, not activated enzymes(5,9-11) - is discharged by the endocrine route: some of the material may represent escape through damaged ductal epithelium(12). This disproportionation - similar to that which occurs in hepatocytes treated with microtubular poisons(13) - is well demonstrated in animal experiments that simulate the "road-block" phase of

human AP, as caused by a migrating gall stone (without reflux), pancreatic ischaemia, hyperchylomicronaemia, or hyperstimulation(14-17). 3) The human disease is usually self-limiting. 4) The pathological progression from pancreatic oedema to interlobular/periglandular fat necrosis, haemorrhage and coagulative necrosis(HPN) is accompanied by the appearance in the gland of active elastase and phospholipase A_2. The entry of these enzymes into the bloodstream, along with products of activated cascades - coagulation, complement, kinin, renin, etc(18) - depletes α_2M stores. Death from multiple organ failure, especially the shock lung syndrome, is the usual outcome.

Guilt by Association

No hypothesis satisfies each of these requirements but it is taken for granted that in AP the gland is cannabalized by its own prematurely activated enzymes. The proposed "triggers" for trypsinogen activation include(1): enterokinase in refluxed duodenal juice or abnormal bile; plasmin and thrombin that seep into the pancreatic bed; enhanced autoactivation; and lysosomal cathepsin B which is an activator at PH $<$4. In fact, the evidence does not favour this sequence of events. For example, in the earliest phase of human AP the serum contains zymogens but not active trypsin(9). Furthermore, the injection of trypsin between pancreatic lobules does not cause HPN (18);

whilst the presence of active enzymes, whether with or without bile, in the pancreatic duct is innocuous so long as the ductal permeability barrier is preserved(12). Therefore, trypsinogen should be regarded as "not guilty until proved otherwise". Yet, there is general agreement that phospholipase A_2 and elastase put the nails in the coffin. Where might these enzymes come from, if not from their pancreatic precursors? Postmortem histology of the pancreas(Figure 1) from patients who died within a few days of admission(8) shows legions of PMNL at the scene of the crime, but pancreatologists generally dismiss these cells as innocent bystanders.

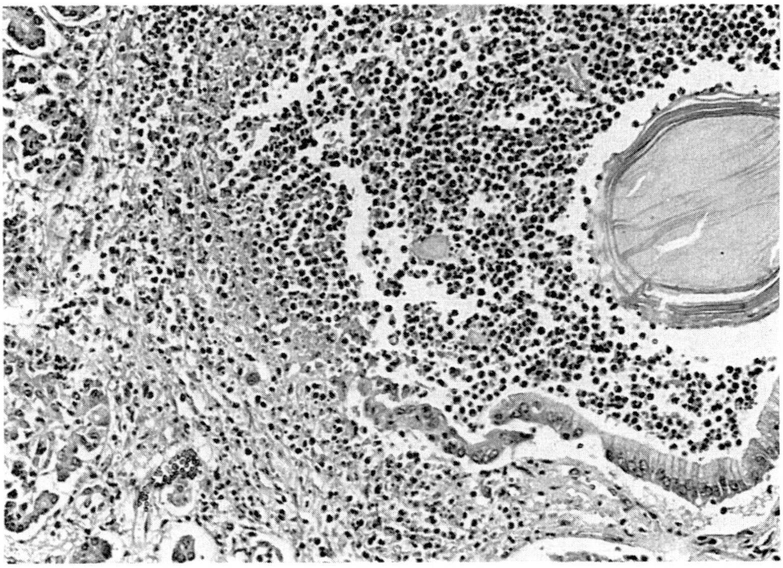

Fig 1 An interlobular duct (right) is filled with PMNL, and contains proteinaceous material. The cells have invaded the parenchyma through a necrotic segment of the duct. H&E x 170 (courtesy of Dr EW Benbow, Manchester University).

Observations in two experimental models, in each of which death from HPN occurs with fearful speed and regularity, lead one to question this interpretation. Firstly, young mice that are fed a choline deficient/DL-ethionine supplemented diet develop a "road-block" within the acinar cells by 24 h, so that crinophagic vacuoles seek egress via the basolateral membrane. PMNL invade the interstitium by 48-72 h, at which time 70-90% of glandular elastolytic activity, but $<$ 5% of tryptic or chymotryptic activity is in the form of the free protease(19-21). Secondly, depletion of PMNL prevents the pulmonary disturbances(22) that are otherwise inevitable when AP is initiated by injecting a trypsin/bile salt mixture into the pancreas.

A "Radical" Hypothesis

Very recent observations incriminate free radicals as the common detonator in AP, and suggest that these species - along with elastase, phospholipase A_2 and other substances that are discharged into the pancreatic interstitium by activated leucocytes(23,24) - may effect the transformation to HPN. The debacle could represent the effects of one or more of the following agents: the primary metabolites of oxygen, the secondary oxidant HOCl, organic free radicals, free radical oxidation products, and secretions from leucocytes. These novel concepts represent the translocation to the pancreas of notions that are gaining ground in the spheres of sepsis-induced shock lung syndrome and cigarette smoke-induced pulmonary emphysema - in both of which entities the suscept-bility of key methionine residues (in α_1PI and the repair enzyme methionine sulfoxide reductase) to inactivation by $OH^{\bullet}$ and HOCl, and, consequently, unmitigated elastase activity, is seen as the Achilles heel(25) - but with a prior step that involves free radical-induced dysfunction of microtubular transport systems(13,26-28).

Experiments in the isolated perfused ex-vivo pancreas, by Cameron's group(29-32), show that concurrent treatment with superoxide dismutase and catalase (virtually) aborts AP that otherwise ensues if the gland is stimulated to secrete against resistance, or if it is rendered ischaemic for 2 h and then reperfused, or if oleic acid is infused into the arterial line. The addition of albumin - a sacrificial antioxidant(33) - to the perfusate protects in the second variant, whilst the addition of a calcium channel-blocking agent protects in the third. The antioxidant enzymes also ameliorate pancreatic injury caused by supramaximal stimulation(34). HPN is not a feature in these two models which mimic only the earliest (? self-limiting) stage of the corresponding human condition. The findings indicate that oxygen free radicals detonate AP by imposing a functional "road-block" within acinar cells and, since this occurs even when the initial attack is likely to be directed against ductal epithelium or capillary endothelium, it appears that the damage may somehow be exerted via the pancreatic interstitium which quickly becomes oedematous. Extracellular free radicals are known to injure plasmalemmal membranes and, although their effect on intracellular function is less clear, it is recognised that damage to lysosomal membranes and alterations in intracellular proteins may follow (23). In the cited models of AP, xanthine oxidase - derived from the dehydrogenase by the action of a calcium/calmodulin-dependent protein - seems to be the source of oxygen free radicals. It has been suggested that chymotrypsin may be responsible for the conversion and this enzyme has been shown to exacerbate oxidant injury of the isolated perfused lung(35).

TABLE 2 INCOMPLETE LIST OF SUSPECTED CHEMICALS IN AP &/OR(*) CP

azathioprine*	tetracycline	food cyanogens*
sodium valproate*	L-asparaginase	ethanol*
rifampicin	phenformin, etc	tobacco smoke*
frusemide etc	acetaminophen, NSAID	corn oil*
steroids, OCP*	methylene chloride	(hypercalcaemia)*
nitrofurantoin	pentachlorophenol	(hyperlipidaemia)*
metronidazole	diethylformamide	
sulphasalazine, etc	trichloroethylene*, etc	
methyldopa	paint solvents*	
diphenoxylate	burnt diesel/petrol*	

The list of chemicals associated with AP is long(eg ref 36, 37) and ever-growing(Table II). A few of these (eg L-asparaginase) may initiate AP by interfering with protein synthetic pathways - in effect causing a functional "road-block". However, the obvious link between several of the chemicals is their influence on the cytochrome P450 complex. Many are known to be powerful enzyme inducers, and/or to deplete glutathione, and/or to yield a reactive intermediate: it is recognised that concurrent enzyme induction - due to chronic alcoholism, tobacco smoke, anticonvulsant drugs, etc - increases the yield of a reactive product(37-41). Studies of drug metabolism provide indirect evidence of enzyme induction in at least 50% of our patients with AP (1,42) who often show hypersecretion of hepatocyte-derived products in bile and/or serum. With the recent confirmation that the pancreas possesses the full machinery to process xenobiotics(43,44), it is no longer necessary to argue that reactive intermediates produced within the liver attack the pancreas by way of refluxed bile, or by the systemic circulation, although this may indeed happen in certain cases.

So, it seems that oxygen and/or organic free radicals detonate AP, but are they also responsible for the presumed "premature activation of zymogens"? This question has been addressed in the two best animal models of HPN. It has been shown that repeated dosing with allopurinol(45), or with a combination of superoxide dismutase and catalase(46) ameliorates AP in the DL-ethionine model. However, we have not found these substances useful in two ductal models (47,48). In one of these, 50 µl of a lysolecithin/sodium deoxycholate mixture is infused retrogradely during 2 min into the pancreatic duct of a rat. Haemorrhages are apparent within 3 min at which stage waves of PMNL have already invaded the interstitium: they quickly vanish, only to be replaced by numerous tissue macrophages at 24 h, by which time 60% of animals die(49). This peculiar behaviour of PMNL, and the dramatic recovery that followed the administration of N-acetylcysteine(NAC) to our patient who was dying from gall stone-related AP, complicated by sepsis and shock lung(50), led us to investigate the effect of this glutathione precursor and also of 5 aminosalicylic acid(5-ASA). The latter drug, in a dose of 10 µg/ml or less, markedly reduces the production of of H_2O_2, $OH^{\cdot}$ and singlet oxygen from activated PMNL; and also H_2O_2 from xanthine oxidase(51): at a 10-times higher dose it protects α_1PI from destruction by a concentration of HOCl that is found near phagocytosing PMNL(52). When a combination of NAC and 5-ASA was given before and soon after induction of AP by the lysolecithin brew, the 24 h mortality fell to 33% (48), but without a corresponding reduction in fat necrosis or HPN. The apparent corollary, that $O_2^{-\cdot}$, $OH^{\cdot}$ and HOCl do not directly activate pancreatic zymogens, but that the drugs acted indirectly by preventing further production of these potent species and/or by refurbishing cellular glutathione stores, receives support from our studies on human pancreatic juice(53). These findings, when considered alongside histological images of PMNL invasion of the interstitium and ductal debris in fatal cases of human(Figure 1) and experimental AP(12,20), reinforce our suspicion - as has been independently and cogently expressed by Rinderknecht recently(54) - that elastase and phospholipase A_2 derived from PMNL, and not (at least initially) from the pancreatic precursors of these enzymes, may cause the transformation to HPN. This concept is not as far-fetched as it may appear. Haemorrhagic and/or coagulative necrosis may accompany certain forms of experimental arthritis in which PMNL invasion is a characteristic feature, and it is suspected that these changes may be caused by the inadvertent extra-cellular discharge of PMNL secretions (55). This occurs when the primary chemotactic stimulant is overwhelming: detergents and crystals (eg calcium urate) are especially prone to provoke this response, so it would be surprising if bile could not do the same.

Rinderknecht(54) reminds readers that the chemical arsenal of activated PMNL includes: oxygen metabolites including HOCl (produced when myeloperoxidase interacts with H_2O_2 in the presence of CL^-); proteases (elastase, phospholipase A_2, collagenases, cathepsins, etc); extremely bioactive (eg LKB_4) products of the arachidonic acid cascade; cachectin; platelet and plasminogen-activating factors, etc. He also stresses that trypsin is not a prerequisite for the "multiple organ failure syndrome" (eg in burns, sepsis, etc), and that plasmin and thrombin (which become prematurely activated within the bloodstream in adverse conditions) can activate trypsin and phospholipase A_2, respectively. Thus, the eventual activation of pancreatic zymogens - by this and/or other mechanisms (vide supra) - would compound the injury and sound the death knell.

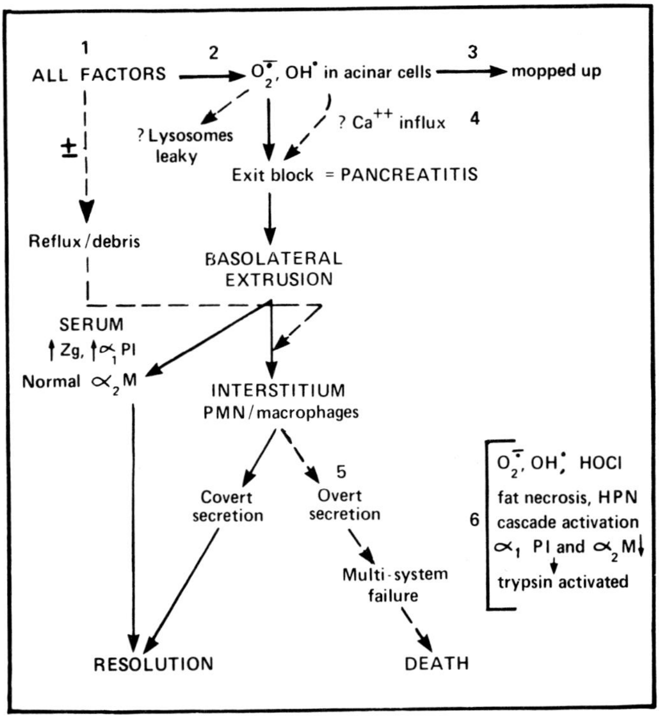

Fig 2

Figure 2 depicts these evolving concepts whilst also indicating (by numbers) the points at which future therapeutic research should be directed - bearing in mind that most clinical episodes of AP resolve spontaneously, and that it is difficult to predict at the outset which patient will develop HPN. Considering that patients seldom report to hospital until some 10 h after their first symptom, the only realistic strategy may be to intercept a potential vicious cycle by boostering cellular antioxidant stores and preventing further production of the most vicious of oxidants, $OH^{\cdot}$ and HOCl. Hopefully, this would buy time whilst the inciting agent passes (eg a gall stone migrates into the duodenum). Once the various cascades have been activated, the available options are heroic rather than practical or efficacious(1). Whereas cholecystectomy prevents further attacks of gall stone-related AP, patients with idiopathic disease often suffer further attacks: our recent work suggests that, as in the case of CP (vide infra), antioxidant supplements - with or without a cytochrome P448 inhibitor - may prevent this progression(5,6)

CHRONIC PANCREATITIS

Traditional Viewpoint

The disease manifests as a typical attack of pancreatitis but after full clinical recovery, in contrast to AP, pancreatic structure remains irreversibly altered (patchy loss of acini, ductal distortion, ductal plugs that may calcify, intra- and inter-lobular fibrosis). These changes increase with successive, typically self-limiting, attacks of pancreatitis that punctuate the clinical course of the disease(1). In developed countries CP is usually regarded as synonymous with chronic alcoholism and it has been suggested that high protein/high fat diets, and possibly cigarette smoke, magnify the risk. Traditional teaching emphasises the ductal lesions, regarding these as being caused by secretion of protein/calcium/mucin-rich juice which may be partially activated (because of reduced trypsin inhibitor and/or lysosomal hyperactivity), and contains reduced amounts of a "stone protein" that normally prevents calcium precipitation. There are several flaws in this concept. 1) The administration of alcohol to small animals for long periods does not produce CP. 2) The ductal-initiation theory fails to rationalise ultrastructural and functional studies which indicate that the acinar cells, whipped into a state of hyperactivity, bear the brunt of the early attack. 3) The disease can occur in teetotallers (as in 30% of our patients), whilst amongst those who drink alcohol regularly, some people develop symptoms after drinking only 20 g daily for $<$ 2 years and subsequent abstinence does not guarantee freedom from further attacks. 4) The high incidence of CP in certain tropical zones is not due to alcohol.

The Manchester Hypothesis (Figure 3)

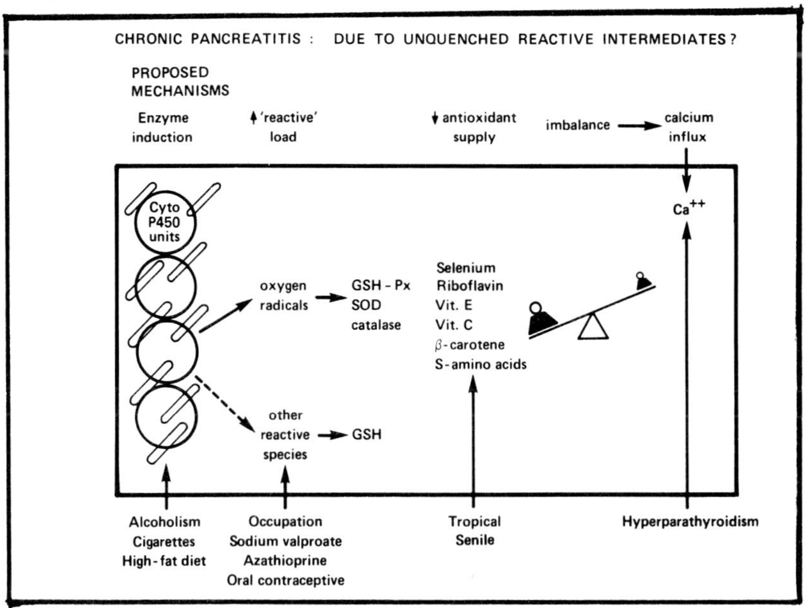

The aetiological factors might act at more than one of the proposed points in Fig 3: for example, the chronic consumption of alcohol leads to enzyme induction, whilst depleting glutathione stores and increasing the production of reactive intermediates from chemicals to which animals may have been concurrently exposed(39,40). With regard to the last item, the regular inhalation of halogenated hydrocarbons - chemicals that are known to yield reactive intermediates upon processing by the P450 system - may be especially relevant (37): whereas ingested lipophilic chemicals would pass preferentially to the liver which is best equipped to deal with them, the pancreas and liver would be equally exposed to inhaled agents that arrive through the abdominal arteries. Amongst the myriad potential effects of unquenched free radicals(23-25,28,38,55), two that may be most pertinent in CP could be their propensity to injure microtubules and, once they enter the tissue extracellular spaces that contain such low concentrations of antioxidant enzymes(33), to degrade cellular supporting structures/provoke fibrosis. Thus, the provision of ample antioxidants to enzyme-induced hamsters over a six month period resulted in pancreatic lipoatrophy but without pancreatitis or fibrosis(57).

We have adduced evidence for these theories. 1) Preliminary immunocytochemical studies of surgically resected specimens of pancreas and liver from our patients show induction of a range of drug metabolising enzymes within hepatocytes, pancreatic acinar and/or ductal cells(eg 56). These findings confirm our deductions from studies of drug metabolism(1,42, 58); serum/bile analysis(42,59,60) and hepatic morphology(61); and they also rationalise the expansion of the endoplasmic reticulum in pancreatic acinar cells, along with protein hypersecretin in the earliest stages of CP(1). 2) The idea that antioxidant stores have been compromised is suggested by (indirect) evidence of glutathione transferase B induction(1,42) and the large amounts of bilirubin in bile(61) - changes akin to those that occur when selenium deficient animals are exposed to enzyme inducers(62). Further support is gained from the findings of increased levels of ferroxidase I activity(1,42) and "molar ratios"(Figure 3), along with reduced levels of one or more antioxidant(62,63) in serum; increased amounts of free radical markers, including 9,11 linoleic acid(64) in bile; the changed composition of pancreatic juice which contains increased levels of lysosomal enzymes, lactoferrin and mucin; and the not infrequent presence of microvesicular fat and/or masses of lipofuscin within hepatocytes, acinar and ductal cells(1,56,60). Thus, the liver and pancreas in patients with CP show the same pattern of abnormalities but pancreatic pain dominates the clinical picture. 3) Inadequate diets seem to be the main reason why antioxidant supply runs dry in patients with CP (65,66), but excessive consumption of antioxidants due to repetitive exposure to xenobiotics seems to be another important factor(62,63): malabsorption of antioxidants, when pancreatic exocrine failure supervenes, would compound these problems. 4) Disturbances in the normal relationship between antioxidant demand via the cytochrome P450 system and antioxidant supply can be guaged by appropriate studies(Figure 4). In general, the same pattern of abnormalities is found in patients with CP and AP (especially idiopathic disease), but the degree of (relative) antioxidant lack tends to be much higher in the former (65,67). 5) Perhaps the strongest argument in support of the hypothesis is that an antioxidant cocktail prevented painful attacks in our pilot study(56,68): a placebo-controlled trial is in progress.

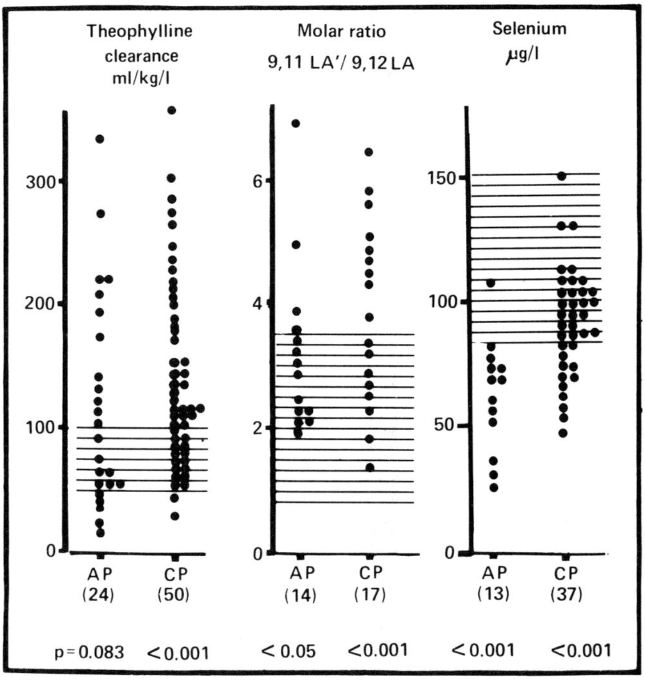

Figure 4. Markers of enzyme induction, free radical activity, and antioxidant status in patients with AP or CP. The hatched lines are reference ranges, and significance values indicate differences from control data. Note: the selenium levels in AP were admission values (with one exception, 110 ug/l), whilst only a third of the levels in CP were in an attack (lowest points, ref 62).

OVERVIEW

An attack of pancreatitis, whether due to AP or (an

exacerbation of) CP, represents a "road-block" within acinar cells, causing products to be diverted by the endocrine route. This phenomenon seems to be initiated by unquenched free radicals derived from xanthine oxidase in some patients and the cytochrome P450 complex in other cases: there is no reason why other potential biological sources of free radicals should not be able to precipitate the problem. The entry of acinar secretions (? along with free radical oxidation products) into the interstitium leads to oedema and chemotaxis - changes that are most striking when debris (bile, proteinaceous material, cell casts, etc) is present within the pancreatic duct. It is proposed that the extracellular secretions from activated PMNL cause the metamorphosis to HPN: this sequence is rare, after the first attack, in CP. The degree of imbalance between the oxidant load and antioxidant supply in the pancreas; the frequency/duration of imbalance episodes; the site and source of free radicals; and their nature - whether oxygen metabolites, organic free radicals, or both - may collectively dictate whether AP or CP ensues (the former is characterised by a "restitio ad integram"). These theories, if correct, offer scope for specific methods to control these conditions. Potentially more important is the possibility that CP and idiopathic AP may be preventable by appropriate dietary advice (57,55) and antioxidant supplementation in groups who are placed at risk through occupational chemicals, prescriptions, addictions, and other aspects of our modern lifestyle(59).

References

1. JM Braganza: Recent Advances in Gastroenterology 6, ed R Pounder (Churchill Livingstone, 1986) pp 251-280

2. LD Isenman, SS Rothman: Proc Natl Acad Sci USA:74(1977) 4068-4072

3. D Celener, R Kurtbart, E Espinal, C Perec, O Tiscornia: International meeting of pancreatology, Sao Paulo (1986)

4. RJL Anderson, JM Braganza, RM Case, AK Khullar, FS Steven: J Physiol 342(1983)11-12P

5. JW Brodrick, MC Geokas, C Largman, M Fassett, JH Johnson: Am J Physiol 237(1979)E474-E480

6. H Rinderknecht: Dig Dis Sci 31(1986)314-321

7. MC Geokas, C Largman, JW Brodrick, JH Johnson: Am J Physiol 236(1979)E77-E83

8. AK Foulis: Recent Advances in Histopathology,eds PR Anthony RNM MacSween (Churchill Livingstone,1984) pp188-196

9. PR Durie, KJ Gaskin, JE Ogilvie, CR Smith, GG Forstner, C Largman:Jou Paediatr Gastroenterol 4(1985)199-207

10. MC Geokas, C Largman, PR Durie, JW Brodrick, S Ray, M O'Rourke, J Vollmer. Am J Pathol 105(1981)31-39

11. MC Geokas, C Largman, JW Brodrick, M Fassett: Am J Physiol

238(1980) G238-G246

12. KR Wedgwood, G Adler, H Kern, HA Reber: Dig Dis Sci 31 (1986) 1081-1083

13. P Coleman, SG Barnwell, PJ Lowe:Gut 25(1984)A1176

14. PJ Broe, GD Zuidema, JL Cameron: Surgery 91(1982) 377-382

15. T Kimura, GD Zuidema, JL Cameron: Surgery 88(1980) 661-666

16. PJ Broe, JL Cameron: Ann Surg 195(1982) 566-573

17. G Adler, G Rohr, HF Kern: Dig Dis Sci 27(1982) 993-1002

18. WM Keynes, RG Keith, eds:The pancreas(William Heinemann, 1981) p 280

19. H Koike, ML Steer, J Meldolesi: Am J Physiol 242(1982) G297-G307

20. B Lombardi, LW Estes, DS Longnecker: Am J Pathol 79(1975) 465-480

21. KN Rao, J Tuma, B Lombardi: Am J Pathol 70(1976) 720-726

22. PS Barie, MV Tahamont, AB Malik: Am Rev Respir Dis 126 (1982) 904-908

23. RF del Maestro: Free radicals in molecular biology, ageing and disease, eds D Armstrong, RS Sohal, RG Butler, TF Slater (Raven Press, 1984) pp87-102

24. DG Weissman, ed: Advances in Inflammation Research (Raven Press, 1983)

25. J Travis: Am Rev Respir Dis 135(1987) 773-774

26. SR Cairns, TJ Peters: Clin Sci 67(1984) 337-345

27. G Poli, E Albano, F Biasi, E Chiarpotto, MU Dianzani: FEBS Lett 160(187-190)1983

28. RD Irons, DA Neptun: Arch Toxicol 45(1980)297-305

29. H Sanfey, MG Sarr, GB Bulkley, JL Cameron: Acta Physiol Scand suppl 548(1986)109-118

30. H Sanfey, PJ Broe, JL Cameron: Surgery 150(1985)297-300

31. MG Sarr, GB Bulkley, JL Cameron: Surgery 101(1987)292-296

32. DW Christ, GB Bulkley, JL Cameron, Surg Forum 37(1986) 151-153

33. B Halliwell, JMC Gutteridge: Arch Biochem Biophys 246 (1986)501-504

34. KC Guice, DE Miller, KT Oldham, CM Townsend, JC Thompson: Am J Surg 151(1986)163-169

35. T J-K Toung, M Sendak, RJ Traystman, MC Rogers, GB Bulkley, JC Cameron: Circulation 74(1986)II-91

36. G Dobrilla, F Chilovi, S Amplatz: Acute pancreatitis,eds PA Banks, G Bianchi Porro(Masson Italia Editori,1984)pp 23-41

37. JM Braganza, JE Jolley, WR Lee: Internat J Pancreatol 1 (1986)9-19

38. B Halliwell, JMC Gutteridge: Free radicals in biology and medicine(Oxford University Press,1985)

39. A Sato, T Nakajima, Y Koyama: Br J Ind Med 37(1980) 382-386

40. Strubelt: Biochem Pharmacol 29(1980)1445-1449

41. JH Lewis, HJ Zimmerman, CT Garrett, E Rosenberg: Hepatology

2(1982)870-873

42. JM Braganza: Proceedings of International symposium on pancreatic cancer, Padova 1986, eds R Naccaratto, R Burlino (1987) In Press

43. P Weibken, BM Schaeffer, DS Longnecker, TJ Curphy: Drug Metab Dispos 1204(1984)427-431

44. TT Kawabata, DG Wick, FP Guengerich, J Baron: Cancer Res 44(1984):215-223

45. H Degertekin, K Van Meter, A Ertan, K Akdamar, R Yates: Gastroenterology 84(1984)A1059

46. S Bank, A Sandberg, J Aaron, R Greenwald, S Lendvai Dig Dis Sci 30(1985)962

47. RJL Anderson, IJM Jeffrey, PM Kay, JM Braganza: Internat J Pancreatol 1(1986)237-248

48. S Uden, A Blower, R MacMahon, TV Taylor, JM Braganza: abstract submitted

49. AL Blower, TV Taylor: Surgical Research Society Meeting, Leeds(1987)

50. JM Braganza, AM Holmes, AR Morten, L Stalley, R Ku, R Kishen: Lancet i(1986):914-915

51. Y Kiyachi, A Yoshioka, S Imamura, Y Nima: Gut 28(1987) 190-195

52. OI Aruoma, M Wasil, B Halliwell, BM Hoey, J Butler: Biochem Pharmacol(1987) In Press

53. PM Guyan, JM Braganza, J Butler: Proceedings of the Free Radical Society Meeting, St Andrews(1987) In press

54. H Rinderknecht: Internat J Pancreatol(1987) In Press

55. G Weissmann, C Serhan, HM Korchak, JE Smolen: Ann NY Acad Sci 389(1982)11-24

56. JM Braganza, IJM Jeffrey, J Foster, RF McCloy: Pancreas (1987)In Press

57. GL Parker, S Branigan, JB Houston, JM Braganza: Gut 27(1986) A603

58. DWK Acheson, P Rose, JB Houston, JM Braganza, Clin Chim Acta 153(1985)73-84

59. M Taj, P Rose, LP Hunt, GN Smith, JM Braganza: Internat J Pancreatol 1(1986)309-326

60. NH Haboubi, HH Ali, JM Braganza: Mt Sinai J Med 53(1986) 380-388

61. JM Braganza, P Kay, GH Kay, KP Hunt: Mt Sinai J Med 53 (1986)389-395

62. JM Braganza, C Hewitt, JP Day: Trace Metals in Medicine (1987) In Press

63. J Martensson, T Bolin: Am J Gastroenterol 81(1986)1179-1184

64. GN Smith, M Taj, JM Braganza: J Free Rad Biol Med(1987) In Press

65. P Rose, E Fraine, LP Hunt, DWK Acheson, JM Braganza: Hum Nutr Clin Nutr 40C(1986)151-164

66. S Uden, DWK Acheson, J Reeves, H Martin, LP Hunt, S Brown, JM Braganza: Meeting of the Medical Research Society, Guildford(1987)Abstract In Press

67. JM Braganza, E Fraine, H Martin, P Rose: International Meeting of Pancreatology, Sao Paulo(1986)

68. JM Braganza, A Thomas, A Robinson: Internat J Pancreatol (1987)In Press

69. TL Dormandy: Lancet ii(1983)1010-1014

Acknowledgements

I am grateful to the Wellcome Trust, the British Digestive Foundation, The North West Regional Health Authority, Smith, Kline & French Ltd., and the Central District Research Grants Committee for supporting our various studies. I appreciate the collaboration of experts, the contributions of research associates, the assistance of secretaries and chartists, and, above all, the endurance and faith of our patients.

Addendum

Since this paper was submitted, two reports have appeared in support of the ideas presented regarding AP. Sodium deoxycholate has been shown to increase the formation of reactive oxygen species by colonic mucosa (PA Craven, J Pfanstiel, R Saito, FR DeRuberts: Gastroenterology 92(1987) 1998-2008). These species seem to initiate pancreatic oedema but they do not induce haemorrhagic necrosis in the DL-ethionine model of AP (PL Rutledge, AK Saluja, RE Powers, ML Steer: Gastroenterology 93(1987)41-47).

FREE RADICALS: CHEMISTRY, PATHOLOGY AND MEDICINE
edited by Catherine Rice-Evans & Thomas Dormandy, © 1988 Richelieu Press, London

Some New Aspects on Free Radical Reactions in Red Cell Pathology*

Michael R. Clemens, H. Einsele, C. Ladner
and
H. D. Waller
Eberhard-Karls-Universität Tübingen
Medizinische Klinik und Poliklinik
Otfried-Müller-Strasse 10, D-7400 Tübingen 1
Federal Republic of Germany

Abstract. The role of the membrane lipid composition in lipid peroxidation and oxydative haemolysis, and its relationship to red cell aging is examined. The following conclusions can be drawn:

1. The amount of cholesterol in red cell membranes is negatively correlated to the susceptibility of red cells to lipid peroxidation. A close relationship exists between the content of polyunsaturated fatty acids in the membrane - in particular of arachidonic and docosahexaenoic acid - and the susceptibility of red cells to lipid peroxidation. These findings are valid for normal and abnormal red cells.
2. The degree of lipid peroxidation and haemolysis upon peroxide stress of red cells is closely related.
3. Young cells are much more vulnerable to peroxides than old cells despite higher activities of "antioxidant" enzymes in young cells. This is evident in experiments performed with in vitro and in vivo-aged red cells and can be attributed to an elevated amount of polyunsaturated fatty acids in young cells.
4. A higher susceptibility of red cells to lipid peroxidation has been observed in haemolytic states. On the basis of our results this is probably a secondary abnormality which seems not to be involved in the pathogenesis of a variety of haemolytic diseases except those accompanied by a decompartmentalisation of cell constituents (e.g. sickle cell anaemia).
5. Uric acid known to be a powerful radical scavenger, inhibits red cell lipid peroxidation. Further radical scavengers did not protect red cells and haemolysates against lipid peroxidation. The oxygen species initiating red cell lipid peroxidation is not yet identified.

*Dedicated to Prof. Dr. G.W. Löhr (Freiburg) on the occasion of his 65th birthday

Introduction

Oxygen radicals and other oxygen-derived species have been implicated as important causative agents of aging and of several human diseases. There is little doubt that free radical reactions participate in the pathogenesis of a variety of haematological diseases. However, until now it is unclear whether free radicals cause any human disease.

In the present paper we will focuse on the role of radical reactions in red cells and its potential involvement in some haemolytic states. For red cells it might be expected to be highly vulnerable to oxygen radicals. In fact, several studies have shown that normal red cells are exceptionally resistant to oxidative damage (see review 1). This can be partly ascribed to cellular antioxidants and enzymes, but it mainly reflects an effective structural compartmentalisation of cell constituents.

The administration of certain drugs, however, stimulates the generation of activated oxygen species, which may overwhelm the protective mechanisms and may induce membrane lipid peroxidation and consequently haemolysis. This is particularly likely if the defenses of the red cells are already defective due to a hereditary disorder. Susceptibility to lipid peroxidation of red cells is greatly increased in most haemolytic states, which may partly be a result of a decompartmentalisation of cell constituents e.g. in unstable haemoglobin diseases. The importance of haem-iron, antioxidants and enzymes in oxidative damage of red cells has been extensively investigated and reviewed (2-6). This paper will indicate the role of the membrane lipid composition in oxidative haemolysis, its relationship to red cell aging, and which oxygen species might be involved in oxidative damage to red cell membranes and in the consequent haemolysis.

Materials and Methods

Materials. All chemicals used were analytical grade. Celluloses, uric acid, Percoll[R] and albumin were purchased from Sigma (Munich, F.R.G.), sodium citrate from Braun (Melsungen, F.R.G.), desferrioxamine from Ciba-Geigy (Wehr, F.R.G.), phenylhydrazine from EGA-Chemie (Steinheim, F.R.G.), Sephadex G-25 from Pharmacia (Uppsala, Sweden), boron trifluoride 14% in methanol from Serva (Heidelberg, F.R.G.) and all other chemicals from Merck (Darmstadt, F.R.G.).

For red blood cell analysis blood was mixed with sodium citrate (3.13%) as anticoagulant (1:5, v/v). Red cells were separated from plasma with microcristalline and alpha-cellulose (1:1, w/w) (7). Isoosmotic phosphate-buffered saline (1 vol KH_2PO_4/ K_2PO_4 and 9 vol 0.15 M NaCl) (PBS) was used for washing red cells three times and for suspending the red cells in the incubation medium or prior to lipid extractions. The age-dependent separation of red cells was performed by using Percoll, meglamine diatrizoate 60%, NaCl and water (11). Red cell membranes were isolated as described by Burton et al. (8).

Blood Donors. Controls comprised healthy individuals (age range 22-40 years). Clinical, haematological and biochemical examinations revealed no abnormalities in any of these controls.

Alcoholics studied were admitted either to the Universitäts-Nervenklinik Tübingen for ethanol withdrawal and psychotherapy (patients without liver cirrhosis) or to our hospital for treatment of decompensated liver cirrhosis. Futher clinical and biochemical data of these patients were reported elsewhere (9, 10).

Lipid Peroxidation. The susceptibility to lipid peroxidation of red cells (haematocrit 2.5%) was estimated with 10 mM hydrogen peroxide, and 0.25 or 1 mM azide, respectively, in PBS pH 7.4 for 2 hours at 37° (12). The degree of lipid peroxidation was measured by analysing the release of volatile hydrocarbons into a head space vial system by gas chromatography using a Porasil C column (13).

Haemolysis. Two hours after initiation of the peroxidative reaction, haemolysis was determined photometrically at 540 nm by measuring the haemoglobin concentration in the supernatant of red cell suspensions after conversion to cyanomethaemoglobin.

Lipid Analysis. Cholesterol in red cell membranes was estimated as described by Ott et al. (14). For the analysis of fatty acids, lipid extraction was performed by the method of Jaeger et al. (15). A solution of chloroform/methanol 2:1 (v/v) was added to membranes in a ratio of 20:1. The sample was filtered and non-lipid material removed by the addition of 2.0 g Sephadex G-25. Transmethylation of fatty acids was performed with 0.5 ml BF_3 in methanol. Fatty acid methylesters were extracted in hexane, analysed by gas chromatography using various columns (FFAP or SP 2100 see ref. 10; SP 2330 see ref. 16) and were identified and calculated as described previously (10, 16).

Antioxidants. In experiments performed with antioxidants, red cells and haemolysates (containing red cell membranes) were preincubated for 15 min with the antioxidant prior to the initiation of peroxidation.

In vitro Aging. For in vitro determination of lipid peroxidation during long-term incubation, red cells were suspended in PBS (haematocrit 2.5%). The gas phase of the head space vials used contained synthetic air. To prevent bacterial growth, sterile conditions were used and piperacillin and cefoxitin were added to the suspension. Bacterial cultures set up at the end of the incubation period were negative in all cases. Energy supply was provided by the addition of 20 mM glucose to the incubation medium.

Methaemoglobin. Red cell content of methaemoglobin was determined according to Henry et al. (17).

Enzyme activities. Red cell enzyme activities were measured as described previously (18).

Results

Lipid Composition and Lipid Peroxidation of Red Cells

Normal red cells. A significant positive correlation was found between the proportion of arachidonic acid and docosahexaenoic acid in red cell membranes, and the degree of peroxide-induced lipid peroxidation monitored by the release of pentane and ethane from red cells, respectively (Fig. 1 and 2). An evident negative correlation exist between the cholesterol content of red cell membranes and the peroxide-induced formation of pentane (Fig. 3). There was no correlation between the degree of lipid peroxidation and the content of linoleic acid in the membrane.

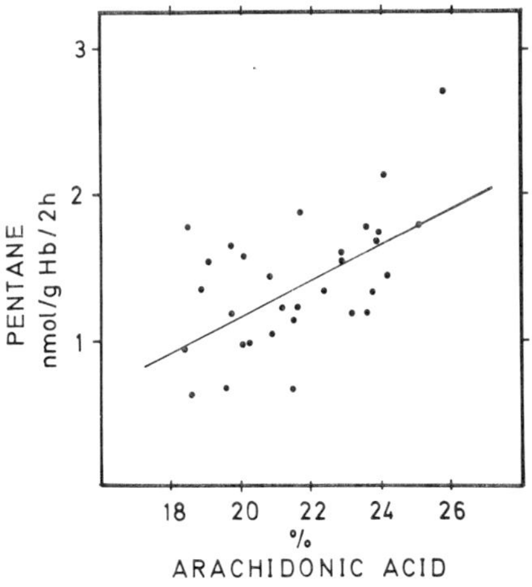

Figure 1: Relationship between lipid peroxidation of red cells (monitored by the release of pentane) and the fatty acid composition of their membranes (arachidonic acid). $r = +0.57$, $p<0.01$

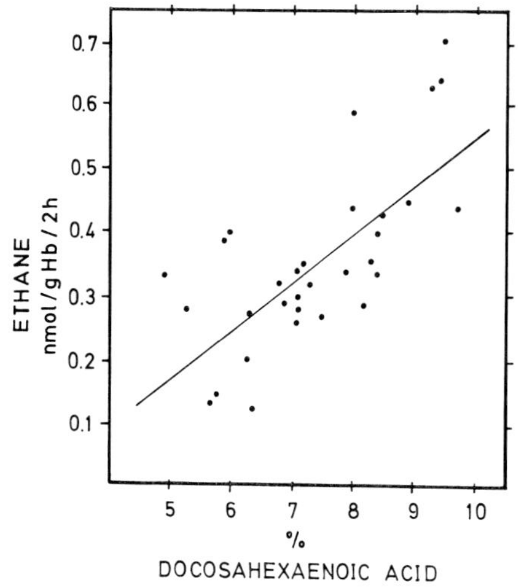

Figure 2: Relationship between lipid peroxidation of red cells (monitored by the release of ethane) and the fatty acid composition of their membranes (docosahexaenoic acid). $r = +0.76$, $p < 0.01$. Pentane is released during the peroxidation of n-6 fatty acids (e.g. arachidonic acid; C 20:4 n-6), see Fig. 1; ethane is released during the peroxidation of n-3 fatty acids (e.g. docosahexaenoic acid; C 22:6 n-3).

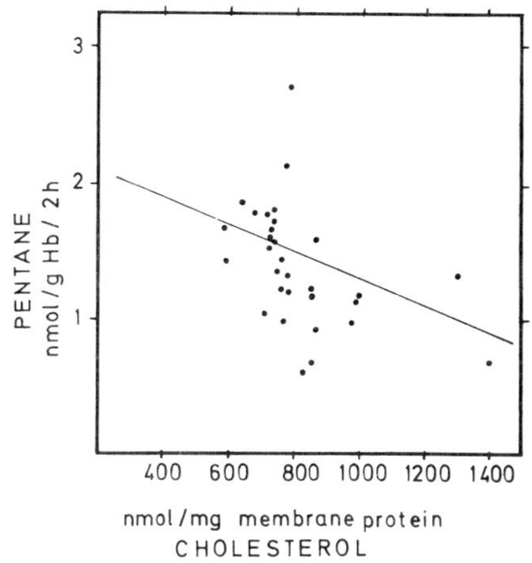

Figure 3: Relationship between the pentane release during peroxide-induced red cell lipid peroxidation and the cholesterol content of their membranes. $r = -0.44$, $p<0.05$

Red cells from alcoholics. Fatty acid analysis of red cells from alcoholics without liver cirrhosis showed a significant deficiency of linoleic acid (controls (n=8): 17.0 ± 1.4 %; alcoholics (n=16): 12.1 ± 1.3 %; $p<0.001$). The peroxide-induced lipid peroxidation was not different from controls (controls: 0.88 ± 0.08 nmol/g Hb/ 2h; alcoholics: 0.91 ± 0.09 nmol/g Hb/ 2h). However, red cell susceptibility to lipid peroxidation was found to be clearly decreased in alcoholics with liver cirrhosis (controls (n=13): 0.80 ± 0.12 nmol/ gHb/ 2h; liver cirrhosis (n=17): 0.35 ± 0.16

nmol/ gHb/ 2h), whose red cells contained a significantly decreased proportion of arachidonic acid (controls: 17.9 ± 1.6 %; liver cirrhosis: 14.6 ± 2.8; $p<0.001$) (for further results see Ref. 44). These results have been confirmed by experiments performed with red cell obtained from Sprague-Dawley rats: solely the content of arachidonic acid in their membranes was close correlated to the susceptibility of red cells to peroxide-induced lipid peroxidation (44).

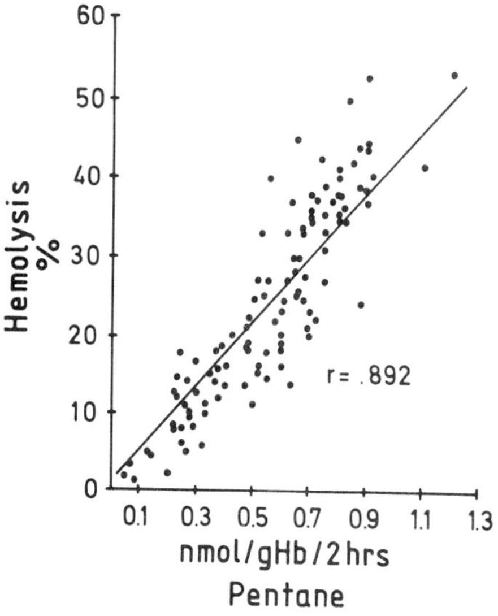

Figure 4: Relationship between the degree of peroxide-induced lipid peroxidation of red cells obtained from healthy individuals and patients suffering from various disease, and the degree of hemolysis after the 2 hour incubation procedure (10 mM H_2O_2 in the presence of 0.25 mM azide, 37 °C). $p<0.001$

Haemolysis and Lipid Peroxidation

A highly significant correlation was found between haemolysis and the degree of lipid peroxidation evaluated by the release of pentane and ethane upon incubation of red cells from normals and patients with a variety of diseases (pentane shown in Fig. 4; r = 0.89). In patients with liver cirrhosis red cells were often extremely resistant to peroxide-induced lipid peroxidation due to a decreased content of arachidonic acid and an increased content of cholesterol in their membranes. In those cells there was found again a close correlation between lipid peroxidation monitored by the pentane formation, and hemolysis upon peroxide incubation of red cells (r = +0.83; p<0.001; Fig. 5).

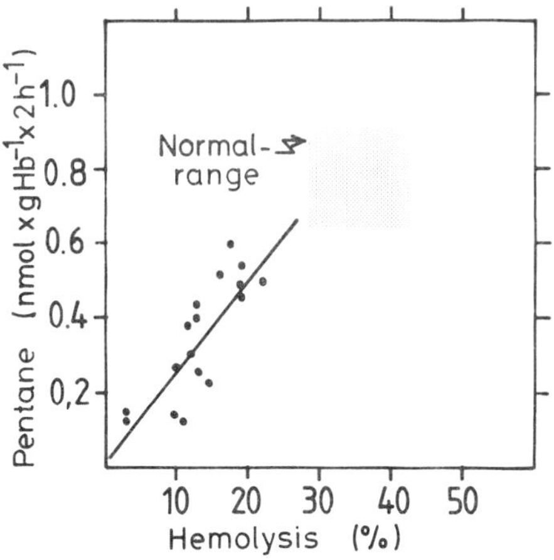

Figure 5: Relationship between the degree of peroxide-induced lipid peroxidation of red cells from alcoholics with liver cirrhosis and the degree of hemolysis after the 2 hour incubation procedure (conditions see Fig. 4). r = +0.83, p<0.001

Cell Age, Lipid Composition and Lipid Peroxidation

In order to test, whether young cells are better protected against oxidant stress than old cells, we measured the susceptibility of red cells to hydrogen peroxide-induced lipid peroxidation upon age centrifugation, which leads to various concentrations of reticulocytes in the cell suspensions examined. The reticulocyte-enriched fractions were found to be in the upper 4 fractions, the old cells were found in the bottom. Fig. 6 and 7 clearly show that the higher the reticulocyte count in each fraction, the more susceptible are the red cells to lipid peroxidation. Analyzing the main substrate for peroxidation, e.g. the polyunsaturated fatty acids in the membrane, we found that cells from reticulocyte-rich suspensions contain higher amounts of arachidonic and docosahexaenoic acid than older cells from the bottom (Table 1). This is consistent with previous studies (34).

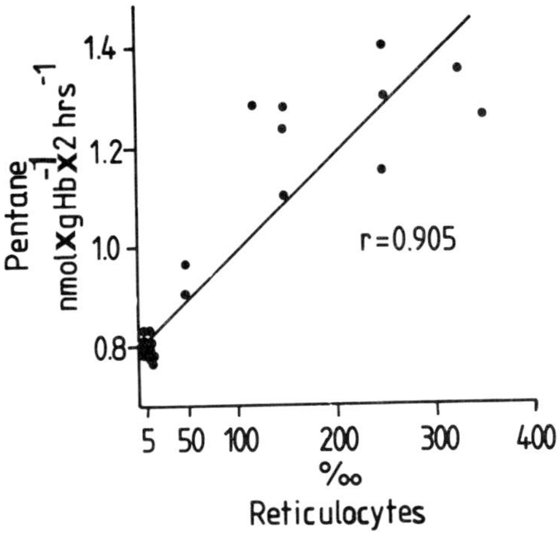

Figure 6: Relationship between red cell age and susceptibility to lipid peroxidation monitored by the release of pentane from n-6 fatty acids. Reticulocyte-enriched red cell preparations were obtained by centrifugation of normal cells in a density gradient. Lipid peroxidation was induced by the addition of H_2O_2 10 mM in the presence of azide 0.25 mM, incubation at 37°C for 2 hours. $p<0.001$

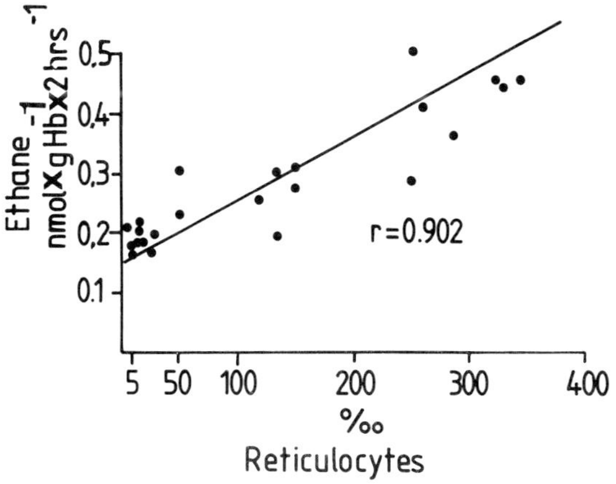

Figure 7: Relationship between red cell age and susceptibility to lipid peroxidation monitored by the release of ethane from n-3 fatty acids. For red cell preparation and incubation procedure see Fig. 6

Table 1
Fatty acid composition of membranes from young and old red blood cells separated by density centrifugation

No.	Cell Age	C 20:4 n-6(%)	C 22:6 n-3 (%)
1	young	19.2	8.9
	old	15.9	6.0
2	young	17.4	8.6
	old	16.4	7.8
3	young	24.5	7.9
	old	20.6	7.7
4	young	18.2	10.8
	old	17.1	9.8
5	young	25.6	7.4
	old	22.0	5.2
6	young	20.7	6.6
	old	18.1	5.7

In Vitro Aging of Red Cells and Lipid Peroxidation

Red cells incubated in the presence of synthetic air at 37°C for a longer period of time were found to develop a time-dependent decrease of activities of glucose-6-phosphate dehydrogenase and glutathione reductase as well as of the reduced glutathione content (Table 2).

Table 2
Decrease of enzyme activities during in vitro aging

Incubation time	Glutathione mmol/L RBC	G-6-PD IU/10^{11} RBC	GR
before incubation	20.0±2.2	12.0±1.2	18.0±1.8
4 days	5.0±2.1	11.0±2.1	14.5±1.5
7 days	0.2±0.1	5.0±0.8	0.9±0.2

Normal red blood cells were incubated at 37°C in the presence of glucose 20 mM. Mean ± S.D., n = 3

Concurrently with decreased enzyme activities, volatile hydrocarbons as an index of lipid peroxidation, haemolysis and methaemoglobin formation were observed (Fig. 8). Analysis of the fatty acid composition of red cell membranes prior and after long-term incubation procedures revealed a significant reduction in the amount of polyunsaturated fatty acids (Table 4). Parallel to this finding red cells after long-term incubation show a markedly reduced release of hydrocarbons when incubated with hydrogen peroxide compared with freshly drawn cells, i.e. in vitro aged red cells are extremly less susceptible to lipid peroxidation than freshly drawn cells despite the lower activities of protective enzymes (Table 3).

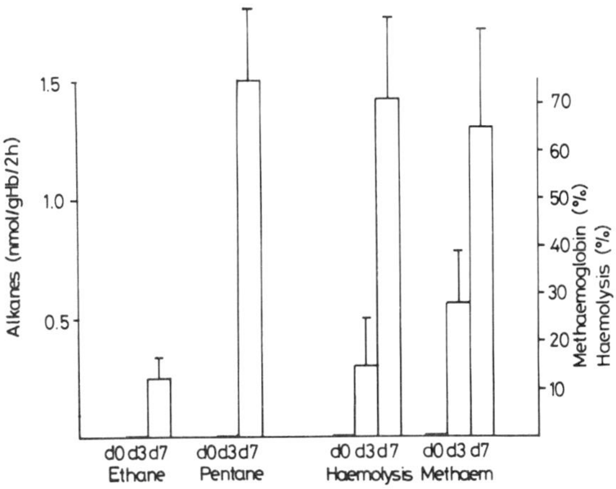

Figure 8: Lipid peroxidation, haemolysis and methaemoglobin formation during in vitro aging of normal red cells. Haematocrit 2.5 %, gas phase of head space vials filled with synthetic air; incubation at 37°C, analyses before incubation (d 0), after 3 days (d 3) and after 7 days (d 7) incubation. n = 15, S.D. given.

Table 3

Decreased susceptibility to peroxide-induced lipid peroxidation of in vitro aged normal red blood cells

	n	Ethane	Pentane
Controls	38	0.19±0.06	0.79±0.14
in vitro aged RBC	8	0.05±0.04	0.10±0.11

In vitro aging of red cells for 7 days at 37°C in the presence of glucose 20 mM. Susceptibility of red cells to lipid peroxidation was estimated in normal (controls) and in vitro aged cells upon hydrogen peroxide incubation (10 mM) in the presence of azide (0.25 mM) at 37°C for 2 hours. Alkanes expressed as nmol/gHb/2h.

Table 4
Polyunsaturated fatty acid composition of red cell membranes obtained before and after in vitro aging

Time	C 18:2 n-6	C 20:4 n-6	C 22:6 n-3
before	9.5+0.8	24.0+0.9	7.0+1.3
after	7.0+1.1	18.2+1.2	4.0+1.0

Conditions for in vitro aging: incubation of red cells for 7 days (haematocrit 2.5%) at 37°C in the presence of glucose 20 mM, gas phase of head space vials filled with synthetic air. Fatty acids in % of all fatty acids analysed. Mean ± S.D., n=6

Free Radical Scavengers and Red Cell Lipid Peroxidation

Experiments with red cells. After preincubation of red cells with various radical scavengers (mannitol, ethanol, thiourea, albumin, desferrioxamine) and an incubation of the cells with hydrogen peroxide in the presence of azide and the individual scavenger, the degree of lipid peroxidation did not differ from those experiments without scavengers. In contrast, uric acid was able to reduce significantly the hydrocarbon release (Table 5). Additionally the fatty acid composition of red cell membranes was analysed before and after hydrogen peroxide exposure of red cells, either in the presence or absence of mannitol, uric acid and desferrioxamine (Table 6). Results obtained by measuring the degree of lipid peroxidation by the hydrocarbon release were confirmed: solely uric acid was able to reduce fatty acid breakdown upon hydrogen peroxide incubation of red cells. Comparable results were obtained by measuring the haemolysis upon peroxide-induced lipid peroxidation.

Experiments with haemolysates. To exclude the possibility of the added scavenger not being able to permeate the red cell membrane, haemolysed red cells in the same haemoglobin concentration as the erythrocytes in the previous experiments were prepared and supplied with the same concentrations of radical scavengers. As shown in Table 7, mannitol, ethanol, thiourea and albumin were not able to decrease the degree of lipid peroxidation. However, hydrocarbon release from haemolysates was markedly decreased using uric acid and desferrioxamine. The extreme antioxidant effect of desferrioxamine was confirmed by analysing the hydrogen peroxide-induced fatty acid breakdown in haemolysed red cells, which was almost completely inhibited in the presence of the desferrioxamine (Table 8).

Table 5
Free radical scavengers and red cell lipid peroxidation

Addition	n	Pentane	Ethane
		nmol/g Hb/ 2h	
none	44	0.84±0.19	0.21±0.04
mannitol 100 mM	4	0.68±0.22	0.24±0.06
ethanol 100 mM	4	0.74±0.24	0.20±0.01
thiourea 2 mg/ml	3	0.70±0.25	0.28±0.07
albumin 1 mg/ml	2	0.80±0.20	0.23±0.06
uric acid 2 mg/ml	3	0.44±0.18	0.10±0.06
desferrioxamine 0.5 mg/ml	4	0.78±0.21	0.20±0.05

Incubation of red cells (haematocrit 2.5%) with 10 mM hydrogen peroxide in the presence of 0.25 mM azide in PBS pH 7.4 at 37°C for 2 hours. Addition of scavengers 15 min before inducing lipid peroxidation by hydrogen peroxide. Mean ± S.D.

Table 6
Free radical scavengers and breakdown of polyunsaturated fatty acids during red cell lipid peroxidation

	n	C 18:2 n-6	C 20:4 n-6	C 22:6 n-3
before incubation	8	14.9±2.1	20.3±0.8	8.5±0.9
after incubation	10	13.1±1.3	10.1±1.4	2.1±1.0
+ uric acid 2 mg/ml	4	11.9±1.0	15.9±1.1	4.7±1.1
+ desferri-oxamine 0.5 mg/ml	4	15.1±2.3	11.1±1.8	2.5±1.2

Experimental conditions see Table 5. Fatty acid analyses performed after 2 hours incubation of red cells. Fatty acids in % of all fatty acids analysed. Mean ± S.D.

Table 7
Free radical scavengers and lipid peroxidation of haemolysates

Addition	n	Pentane	Ethane
none	6	0.88±0.22	0.22±0.05
mannitol 100 mM	3	0.79±0.23	0.19±0.04
ethanol 100 mM	3	0.89±0.20	0.23±0.05
thiourea 2 mg/ml	3	0.76±0.21	0.27±0.04
albumin 1 mg/ml	2	0.84±0.21	0.24±0.06
uric acid 2 mg/ml	3	0.20±0.08	0.12±0.02
desferrioxamine 0.5 mg/ml	4	0.10±0.08	0.08±0.02

Incubation of haemolysates containing membranes (haemoglobin concentration 0.8 g/dl) with 10 mM hydrogen peroxide in the presence of 0.25 mM azide in PBS pH 7.4 at 37°C for 2 hours. Addition of scavengers 15 min before inducing lipid peroxidation with hydrogen peroxide. Mean ± S.D. Alkanes as nmol/gHb/2h.

Table 8
Free radical scavengers and breakdown of polyunsaturated fatty acids during lipid peroxidation of haemolysates

	n	C 18:2 n-6	C 20:4 n-6	C 22:6 n-3
before incubation	4	13.1±1.2	21.8±2.5	8.6±1.8
after incubation	4	13.8±1.4	11.1±2.3	2.8±1.0
+ desferri-oxamine 0.5 mg/ml	4	12.6±1.0	19.3±0.5	7.6±0.7

Experimental conditions see Table 7. Fatty acid analyses performed after 2 hours incubation of haemolysates. Fatty acids expressed as percentage of all fatty acids analysed. Mean ± S.D.

Discussion

It has been suggested that the susceptibility of membranes to peroxidation is affected by the proportion of polyunsaturated fatty acids in the membrane (2,19-22). For example, the increase of polyunsaturated fatty acid content of microsomes enhances their sensitivity to peroxidation (23). The relationship between haemoglobin and membrane oxidation using liposomes containing haemoglobin as a model (haemosomes) has recently been studied (24). In this model oxidation of both haemolysate and purified haemoglobin occur much more rapidly in liposomes prepared from unsaturated than from saturated lipids. In the same model increased cholesterol concentrations in the haemosomes results in the almost complete elimination of haemoglobin-induced lipid peroxidation (25).

In other words, the content of cholesterol as well as the content of polyunsaturated fatty acids in the plasma membrane may influence susceptibility to peroxidation: 1. cholesterol may probably act both by intercepting some radicals and by altering the internal structure of the membrane through the interaction of its large hydrophobic ring with fatty-acid side-chains; 2. the number of unsaturation in fatty acid chains is of relevance for its vulnerability to free radical attack.

Our here presented results confirm that red cell susceptibility to lipid peroxidation is dependent both on the amount of cholesterol and of polyunsaturated fatty acids, either in normal red cells or in cells from patients with an abnormal lipid metabolism. It has to be qualified that a positive correlation between the content of substrate and the degree of lipid peroxidation existsobviously for arachidonic and docosahexaenoic acid, but not for linoleic acid. This could be attributed to the fact that mainly arachidonic and docosahexaenoic acid but not linoleic acid break down during red cell membrane peroxidation (Table 4,6, and 8), whereas linoleic acid is of minor importance.

In the light of these findings it is not surprising that young cells containing more polyunsaturated fatty acids, are much more vulnerable to peroxide-induced lipid peroxidation than old cells, which has been supported by our results both for the situation in vitro and in vivo. There is evidence that under the conditions chosen here, the significantly higher activity of "antioxidant" enzymes and higher amount of glutathione in young cells (Table 2, Ref. 18, 26 and 43), seem to be of minor importance and can be overwhelmed by an elevated amount of substrate for lipid peroxidation. This has to be taken into account while evaluating the results obtained with red cells in haemolytic states (27,28), which revealed an increased susceptibility to lipid peroxidation of those cells, estimated by the formation of thiobarbituric acid-reactive material (29). Thus, a higher vulnerability of red cells to lipid peroxidation in haemolytic states is probably a secondary abnormality and, therefore, not involved in the pathogenesis of a variety of haemolytic diseases. The latter may not be valid for haemolytic diseases with abnormalities in the compartmentalisation of the cell constituents (e.g. sickle cell anaemia, thalassemia, unstable haemoglobin disease). The increased sensitivity to lipid peroxidation and to oxidative haemolysis of red cells in haemolytic diseases may trigger and intensifying haemolytic diseases.

The possibility that autoxidation of oxyhaemoglobin to methaemoglobin involves the generation of a superoxide radical had been previously shown (30-32). The enzymatic and spontaneous dismutation of superoxide radicals yields hydrogen peroxide. The presence of both $\cdot O_2^-$ and H_2O_2 creates the possibility for the generation of hydroxyl radicals ($OH^\cdot$). The latter reaction can be catalysed by iron or other transition metals and allows the reaction to proceed at a much faster rate. At this point the availibility of reducing equivalents is critical, since hydroxyl radicals have the potential to initiate lipid peroxidation (see review Ref. 1). A recent study showed that iron bound to haemoglobin is unlikely to produce hydroxyl radicals in free solution, but that peroxides can cause a release of iron from the protein. This released iron, which can be bound by transferrin or desferrioxamine, is likely to be the true promotor of hydroxyl radical generation in red cells and contribute to the peroxidation in haemoglobin-containing systems (33). Thus, haemoglobin may act as an iron supplying agent but not as a promotor of red cell lipid peroxidation per se. This has been confirmed for red cells by our experiments performed with haemolysates and desferrioxamine (Table 5 and 6), showing that desferrioxamine can almost completely block the hydrogen peroxide-induced lipid peroxidation and fatty acid-breakdown, which requires haemoglobin in its reduced ligand state (35).

Uric acid has been hypothesized to provide antioxidant protection for red cells, and evidence from several studies and from the here reported results supports the idea that uric acid functions as a free radical scavenger (36-42). Despite the fact that uric acid is a powerful scavenger of singlet oxygen and of hydroxyl radicals, the initiating species in hydrogen-peroxide-induced red blood cells lipid peroxidation is not clearly identified: a variety of further hydroxyl radical scavenger was not able to inhibit red cell and haemolysate lipid peroxidation. Further studies are needed to clarify the initiating process in hydrogen peroxide-induced lipid peroxidation and red cell aging.

Acknowledgements

For the excellent technical assistance we thank U. Wegner. This work was supported in part by a grant of the Hoechst AG, Klinische Forschung, Werk Albert, Wiesbaden.

References

1. MR Clemens and HD Waller: Chem Phys Lipids (1987) in press

2. TL Dormandy: Br J Haematol 20 (1971) 457

3. EC Gordon-Smith and JM White: Br J Haematol 26 (1974) 513

4. RW Carrell, CC Winterbourn and EA Rachmilewitz: Br J Haematol 30 (1975) 259

5. D Chin, B Lubin and SB Shohet: Free Radicals in Biology Vol. V, Academic Press, New York (1982) 115

6. RP Hebbel: J Lab Clin Med (1986) 401

7. E Beutler, C West and KG Blume: J Lab Clin Med 88 (1976) 328

8. GW Burton, KU Ingold and KE Thompson: Lipids 16 (1981) 946

9. MR Clemens, H Einsele, H Remmer and HD Waller: Clin Chim Acta 145 (1985) 283

10. MR Clemens, HW Schied, W Daiss and HD Waller: Klin Wochenschr 64 (1986) 181

11. L Vettore, MC DeMatteis and P Zampini: Am J Hematol 8 (1980) 291

12. MR Clemens, H Einsele, H Frank, H Remmer and HD Waller: Biochem Pharmacol 32 (1983) 3877

13. MR Clemens and H Remmer: Blut 45 (1982) 329

14. P Ott, Y Binggeli and U Brodbeck: Biochim Biophys Acta 685 (1982) 211

15. H Jaeger, HU Kloer, H Ditschuneit and H Frank: Application of glass capillary gas chromatography, Marcel Dekker, New York (1981) 395

16. MR Clemens, M Müh-Zange and J Werringloer: IRCS Med Sci 13 (1985) 989

17. R Henry, DC Cannon and JW Winkleman: Methemoglobin in Clinical Chemistry, Principles and Practice. Harper and Row, New York (1974) 1149

18. HD Waller, FJ Birke, FJ Tigges and HC Benöhr: Klin Wochenschr 52 (1974) 179

19. TL Dormandy: Ann R College Surg 62 (1980) 188

20. TL Dormandy: Lancet i (1978) 647

21. M Tien and SD Aust: Biochim Biophys Acta 712 (1982) 1

22. JMC Gutteridge: Biochem Biophys Res Commun 74 (1977) 529

23. HE May and PB McCay: J Biol Chem 243 (1968) 2296

24. J Szebeni, CC Winterbourn and RW Carrell: Biochem J 220 (1984) 685

25. J Szebeni and K Toth: Biochim Biophys Acta 857 (1986) 139

26. G Bauer, A Jung and A Wendel: Klin Wochenschr 60 (1982) 867

27. J Stocks, EL Offerman, CB Modell and TL Dormandy: Br J Haematol 23 (1972) 713

28. J Stocks, M Kemp and TL Dormandy: Lancet i (1971) 266

29. J Stocks and TL Dormandy: Br J Haematol 20 (1971) 95

30. R Wever, B Oudega and BF van Geldern: Biochim Biophys Acta 302 (1973) 475

31. HP Misra and I Fridovich: J Biol Chem 247 (1972) 6960

32. CC Winterbourn, BM Grath and RW Carrell: Biochem J 155 (1976) 493

33. JMC Gutteridge: FEBS Lett 201 (1986) 291

34. GB Phillips, JT Dodge and C Home: Lipids 4 (1969) 544

35. MR Clemens, H Einsele, H Remmer and HD Waller: Biochem Pharmacol 34 (1985) 1339

36. BN Ames, R Cathcart, E Schwiers and P Hochstein: Proc Natl Acad Sci USA 78 (1981) 6858

37. RC Smith and L Lawing: Arch Biochem Biophys 223 (1983) 166

38. RC Smith and V Nunn: Arch Biochem Biophys 232 (1984) 348

39. N Ogino, S Yamamoto, O Hayaishi and T Tokuyama: Biochem Biophys Res Commun: 87 (1979) 184

40. C Deby, G Deby-Dupont, FX Noel and L Lavergne: Biochem Pharmacol 30 (1981) 2243

41. C Deby and G Deby-Dupont: Agents Actions 11 (1981) 651

42. AM Cohen, RE Aberdroth and P Hochstein: FEBS Lett 174 (1984) 147

43. GW Löhr and HD Waller: Klin Wochenschr 37 (1959) 833

44. MR Clemens, H Einsele and H Remmer: Arch Toxicol 60 (1987) 167

FREE RADICALS: CHEMISTRY, PATHOLOGY AND MEDICINE
edited by Catherine Rice-Evans & Thomas Dormandy, © 1988 Richelieu Press, London

Are Reactive Oxygen Species Involved in Cerebral Malaria?

Nicholas H. Hunt, Cassandra M. Thumwood, Ian A. Clark* and William B. Cowden, Departments of Experimental Pathology, John Curtin School of Medical Research, and *Zoology, Faculty of Science, Australian National University, Canberra, ACT 2601, Australia.

1. Introduction

People infected with Plasmodium falciparum may develop cerebral malaria, a frequently fatal complication of the disease for which there is no effective direct treatment. The neurological symptoms include behavioural changes, convulsions and coma. Post-mortem examinations show haemorrhages, oedema, endothelial cell damage and mononuclear cell infiltration (1).

The difficulties inherent in studying the early stages of human cerebral malaria have led to the development of animal models. One of these is P. berghei (ANKA strain) which, in certain strains of mice, leads to cerebral symptoms, breakdown of the blood-brain barrier and a high mortality rate (2-4). Although this system does not reproduce perfectly the pathology of human cerebral malaria, it is useful for gaining insights which may be applicable to the human disease.

The immune system appears to be involved in the pathological process in murine cerebral malaria (5,6), and there are some indications that this may also be the case in humans (7,8). Mononuclear cells marginate in cerebral vessels and infiltrate the brain parenchyma in the mouse model (4). We previously have suggested involvement of reactive oxygen species (ROS), released by phagocytes, in the host response against malaria parasites and the concomitant pathological changes (9,10). Damage to cerebrovascular endothelial cells is a feature of murine malaria (2,4), and ROS can harm endothelial cells in vivo and in vitro (11,12). Thus, in the rodent model, activation of phagocytes might lead to

---

Acknowledgements. This investigation received support from the malaria component of the UNDP/World Bank/World Health Organization Special Programme for Research and Training in Tropical Disease and the National Health and Medical Research Council of Australia. We thank Dr G. Grau for helpful discussions and Mrs K. Rabl for expertly typing the manuscript.

oxidative damage to endothelium and thereby breach the blood-brain barrier.

In human cerebral malaria, it has been argued that the pathology is a consequence of plugging of microvessels by parasitized erythrocytes and the regional anoxia that would follow this phenomenon (13). Spontaneous clearing of these blockages would allow reperfusion of previously ischaemic areas. The damage seen in ischaemia-reperfusion injury is thought to be caused by ROS generated during the period of reperfusion with solutions containing oxygen, or oxygenated blood (14). Activated phagocytes may contribute to a second phase of this process (15). Since blockage of microvessels is infrequent in the rodent model (4), this scenario is unlikely to occur there, although it may be relevant to human cerebral malaria.

A substantial body of work in animal models of trauma has shown that ROS derived from arachidonic acid metabolism in brain parenchyma are a direct cause of cerebrovascular abnormalities (16,17). Thus, there are several reasons for hypothesizing that ROS might be involved in the pathology of cerebral malaria. We have studied this by administering the antioxidants butylated hydroxyanisole (BHA) and ethoxyquin (EQ) to CBA/H and A/J mice infected with P. berghei ANKA.

2. Methods

CBA/H and A/J mice bred at the John Curtin School of Medical Research breeding establishment were infected i.p. with 10^6 parasitized erythrocytes obtained from animals with P. berghei ANKA infection. The parasite was from Dr G. Grau, WHO Immunology Research and Training Centre, University of Geneva. Development of the parasite was assessed by examining thin blood smears.

The antioxidant-supplemented diets were prepared from standard laboratory pellets ground to a coarse powder. To 925g of this was added 75g BHA dissolved in 750ml acetone. For EQ, 50g in 750ml acetone was added to 950g food powder. The acetone was driven off in a Büchi rotary evaporator and the mixtures were then added to a further 9kg of ground food, thoroughly dispersed and the whole pelleted. Control food was treated the same way, without BHA or EQ.

For administration by osmotic pump, the antioxidants were dissolved in polyethylene glycol (PEG) 200 at 0.7g/ml; 0.25ml of this solution was placed in the pumps (model 2002 Alza, Palo Alto, CA, USA) which were implanted intra-abdominally under avertin (1.5ml/100g body weight) anaesthesia. These pumps release 0.5μl/hr for >14 days.

Permeability of the blood-brain barrier was determined by two methods (4): injection of ^{125}I-bovine albumin, or ^{51}Cr-labelled mouse red blood cells, with subsequent perfusion and sampling of brain tissue or cerebrospinal fluid (CSF). Data were analyzed with a computer programme (18) which allows assessment by parametric or non-parametric tests, as appropriate.

3. Results

CBA mice were fed the diets containing antioxidants for 7 days prior to infection with *P. berghei* ANKA. This was to allow recovery from weight loss, which is most pronounced for the 2-3 days after the animals are placed on the diets and may be due to a reduction in food intake. Mice fed the BHA diet were ~5% lighter than controls during the early part of the infection, whereas EQ-fed mice were ~10% lighter (not shown). The increase in parasitaemia which occurs up to day 10 in control-fed mice, at which time most have died (Fig. 1), was delayed by about 2 days in mice fed diets containing BHA or EQ (not shown).

Ninety-seven percent of CBA/H mice fed the control diet died with cerebral symptoms and low parasitaemias (<15%) on days 6-11 after infection (Fig. 1). Feeding diets containing BHA or ethoxyquin reduced the numbers of mice dying during this period to 27% and 7%, respectively. Comparable results were obtained with A/J mice (not shown). All of the BHA-fed mice which survived the early phase (days 6-11) died between days 25-30 (Fig. 1) at high parasitaemias and without cerebral symptoms; haemolysis due to lysis of red cells by the parasite was the likely cause of death. Similarly, most of the EQ-fed mice died between days 20-40 (Fig. 1), but in about 30% of cases the infection resolved and the mice survived.

The anti-oxidants also were administered in osmotic pumps; CBA/H mice were infected 2 days after implantation of pumps containing PEG (vehicle), BHA or EQ. BHA once again caused 5% reduction in body weight, but EQ did not (not shown). Both antioxidants delayed the increase in parasitaemia by 1-2 days (not shown). All mice without pumps, and 92% of those with pumps containing PEG alone, died on days 7-9 (Fig. 2) with cerebral symptoms and low parasitaemias (<15%). However, 69% of mice treated with BHA and 33% of mice treated with EQ survived this period and were still alive at day 19 when the experiment was terminated (Fig. 2).

Albumin is usually well excluded from the brain parenchyma, and saline perfusion reduced to very low values the amount of injected ^{125}I-BSA retained in various areas of the brain in A/J

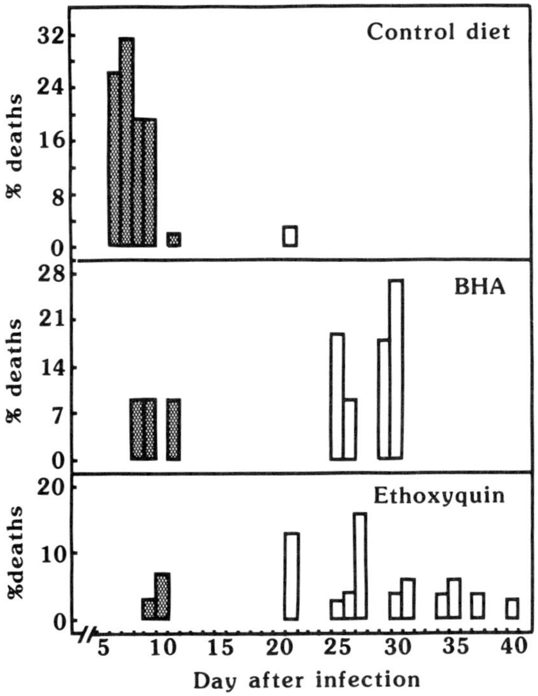

Fig 1. Effects of diets containing 0.75% (w/w) BHA or 0.5% (w/w) EQ on mortality in CBA/H mice infected with P. berghei ANKA. Shaded bars represent animals dying with cerebral symptoms. For control diet, N = 58; for BHA, N = 11; for EQ, N = 30.

mice (Table 1). However, in mice with cerebral involvement during P. berghei ANKA infection (around day 8 after infection) there was up to a 17-fold increase in tissue retention of ^{125}I. This was greatly reduced in mice fed the diet containing BHA (Table 1). Essentially the same pattern was seen when ^{51}Cr-labelled erythrocytes were injected (Table 2): ^{51}Cr retention was significantly increased in P. berghei-infected mice, but this change was largely attenuated in the mice fed the diet containing BHA. The appearance of ^{125}I and ^{51}Cr in the cerebrospinal fluid (CSF) of mice with cerebral involvement was also greatly reduced when mice were maintained on the BHA diet (not shown).

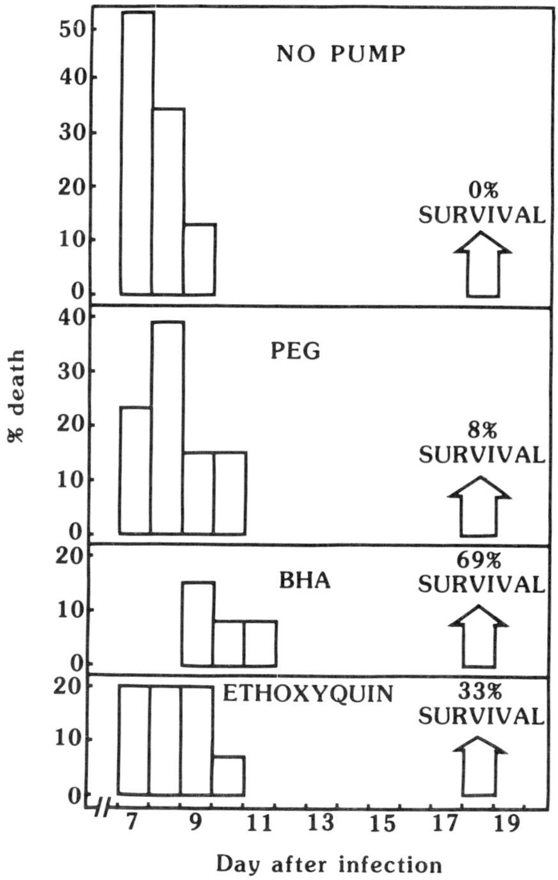

Fig. 2. Effects of agents delivered in osmotic pumps on mortality in CBA/H mice infected with *P. berghei* ANKA. Pumps were implanted 2 days before infection with 10^6 parasitized erythrocytes. For no pumps, N = 15; for PEG, N = 13; for BHA, N = 13; for EQ, N = 15.

Animals	^{125}I (cpm) in tissue Olf	Cr	Cb	St
Uninfected	185±3	702±85	257±57	227±25
Infected plus control diet	1,393±208	13,826±1,938	3,335±695	3,777±580
Infected plus BHA diet	598±51	3,621±441	992±113	997±138

Table 1. Retention of radioactivity in brain tissue after injection of ^{125}I-BSA. 1 hr after injection of radioactivity, blood was purged from the vessels with saline, areas of the brain removed and radioactivity counted. Olf = olfactory lobe; Cr = cerebrum; Cb = cerebellum; St = brain stem. Values are mean ± SEM from 10-16 mice per group.

Animals	^{51}Cr (cpm) in tissue Olf	Cr	Cb	St
Uninfected	32±3	88±20	36±10	35±9
Infected plus control diet	121±20	1,172±278	395±110	381±105
Infected plus BHA diet	69±8	243±29	78±13	88±14

Table 2. Retention of radioactivity in brain tissue after injection of ^{51}Cr-labelled erythrocytes. Procedures and terms as in Table 1.

4. Discussion

Administration of BHA or EQ in the diet or through osmotic pumps greatly reduces cerebral malaria in mice infected with *P. berghei* ANKA. These antioxidants also inhibit the movement into brain tissue of albumin and erythrocytes, events which accompany the neurological symptoms and mortality of murine cerebral malaria.

It is conceivable that the antioxidants influence the outcome of the disease by a direct effect on the malaria parasite, because they slow the rate of increase of the parasitaemia. However, the actual degree of parasitaemia is not the important factor, because mice which do not die of cerebral malaria eventually are subjected to much higher erythrocyte occupancy without developing central nervous system complications. It is also possible that the weight loss induced by the treatments significantly influences the outcome of the disease. However, BHA did not induce weight loss in every experiment and EQ in pumps had little effect on weight but still reduced mortality.

BHA has proved useful in vivo as an antioxidant (19,20), on the basis of its known ability to act as a resonance-stabilized free radical scavenger (21). The other antioxidant, ethoxyquin, when given in the diet increases the longevity of mice (22). The actions of these agents in the mouse cerebral malaria model are consistent withour earlier suggestion (9,10) that oxygen-derived free radicals may play a role in the immunopathology of malaria. This is further strengthened by our preliminary observation that daily injections of PEG-coupled superoxide dismutase and catalase also prevent murine cerebral malaria (not shown).

It has been suggested that human cerebral malaria has an inflammatory component (23), including mononuclear cell infiltration (1) which is also seen in the murine model (2-4). The $L3T4^+$ subset of T lymphocytes is essential for the development of central nervous system pathology in mice infected with *P. berghei* ANKA (6) and these cells are detectable in the CSF near the end of this period of pathological change (Thumwood, unpublished). A role for soluble factors of monocyte/macrophage origin is indicated by the ability of antibodies against tumour necrosis factor (TNF) to completely prevent neurological symptoms and related mortality in the same mouse model of cerebral malaria (24).

TNF is a peptide released by monocytes and macrophages which has a wide spectrum of activities (25,26) and appears to mediate the pathological changes associated with endotoxaemia (27). The nature of its activities strongly suggests that it is an important determinant of malaria pathology (see discussion in 28-30). Of particular relevance to cerebral malaria may be its ability to disorder endothelial cells directly (31), to express HLA antigens (32) and to cause them to express a protein which allows adherence of neutrophils (33). TNF potentiates the release of ROS by monocytes (34) and, as mentioned previously, endothelial cells can be damaged by ROS both *in vivo* and *in vitro* (11,12).

5. Summary and conclusions

A possible sequence of events in cerebral malaria is as follows. Malarial antigen activates T lymphocytes. These cells, probably the L3T4$^+$ subset, produce IFN- which can promote the release of TNF (35) and ROS (36) by monocytes and macrophages. TNF damages endothelial cells directly (31) or by causing adherence to them of activated, ROS-generating phagocytes. Endothelial cell damage results in breakdown of the blood-brain barrier and movement of albumin and erythrocytes into the central nervous system. Antioxidant treatment could interfere with this process in at least two ways: firstly, at the effector stage by scavenging oxygen-derived free radicals released by activated phagocytes, thus protecting endothelial cells from damage; or, secondly, by inhibiting the proliferation of T cells, as BHA is known to do *in vitro* (37). However, we have been unable to find any effect of feeding the BHA diet on the phenotype or proliferation index of lymph node cells in *P. berghei* ANKA-infected CBA/H mice (Thumwood, unpublished). A number of potential or actual interactions between antioxidants and the immune system have been reviewed recently (38), including the ability of these molecules to interfere with leukocyte arachidonic acid metabolism. The use of other agents, such as PEG-coupled superoxide dismutase and catalase, or inhibitors of arachidonate cyclo-oxygenase and lipoxygenase, should allow us to identify the mechanism of action of antioxidants in the murine cerebral malaria system.

References

1. Toro, G. & Roman, G. Arch. Neurol. 35 (1978), 271.
2. Mackey, L.J., Hochmann, A., June, C.H., Contreras, C.E. & Lambert, P.H. Clin. Exp. Immunol. 42 (1980), 412.
3. Rest, J.R. Trans. Roy. Soc. Trop. Med. Hyg. 76 (1982), 410.
4. Thumwood, C.M., Hunt N.H., Clark, I.A. & Cowden, W.B. Manuscript submitted.
5. Finley, R.W., Mackey, L.J. & Lambert, P.-H. J. Immunol. 129 (1982), 2213.
6. Grau, G.E., Piguet, P.-F. Engers, H.D., Louis, J.A., Vassalli, P. & Lambert, P.-H. J. Immunol. 137 (1986), 2348.

7. Adam, C., Ceniteau, M., Gougerot-Picidalo, M., Veroust, P., Lebras, J., Gibert, C. & Morel-Maroger, L. Infect. Immun. 31 (1981), 530.
8. Cohen, S. & Lambert, P.-H. Immunology of Parasitic Infections, ed. S. Cohen & K.S. Warren. Oxford, Blackwell Scientific Publications (1982), 422.
9. Clark, I.A., Hunt, N.H. & Cowden, W.B. Adv. Parasitol. 25 (1986), 1.
10. Clark, I.A. & Hunt, N.H. Infect. Immun. 39 (1983), 1.
11. Sacks, T., Modlow, C.F., Craddock, P.R., Bowers, T.K. & Jacob, H.S. J. Biol. Chem. 258 (1978), 12534.
12. DelMaestro, R.F. Canad. J. Physiol. Pharmacol. 60 (1982), 1406.
13. MacPherson, G.G., Warrell, M.J., White, N.J., Looareesuwan, S. & Warrell, D.A. Am. J. Pathol. 119 (1985), 385.
14. McCord, J.M. & Roy, R.S. Canad. J. Physiol. Pharmacol. 60 (1982), 1346.
15. Lucchesi B.R. & Mullane, K.M. Ann. Rev. Pharmacol. Toxicol. 26 (1986), 201.
16. Demopoulos, H.B., Flamm, E.S., Pietronigro, D.D. & Seligman, M.L. Acta Physiol. Scand. Suppl. 492 (1980), 91.
17. Kontos H.A. Circ. Res. 57 (1985), 508.
18. Blaker, W.D Comput. Biol. Med. 17 (1987), 37.
19. Dillard, C.J., Downey, J.E. & Tappel, A.L. Lipids 19 (1984), 127.
20. Cowden, W.B., Lewis-Hughes, P.H. & Clark, I.A. Biochem. Pharmacol. 34 (1985), 3601.
21. Emanuel, N.M. & Lyaskovskaya, Y.N. The Inhibition of Fat Oxidation Processes. Translated K.A. Allen. Oxford, Pergamon Press (1968), 247.
22. Comfort, A., Youhotsky-Gore, I. & Pathmanathan, K. Nature 229 (1971), 254.
23. Maegraith, B. & Fletcher, A. Adv. Parasitol. 10 (1972), 49.
24. Grau, G.E., Fajardo, L.F., Piguet, P.F., Allet, B., Lambert, P.-H. & Vassalli, P. Science (1987) in press.
25. Beutler, B. & Cerami, A. N. Engl. J. Med. 316 (1987), 379.
26. Le, J. & Vilcek, J. Lab. Invest. 56 (1987), 234.
27. Beutler, B., Milsark, I.W & Cerami, A. Science 229 (1985), 869.
28. Clark, I.A., Cowden, W.B., Butcher, G.A. & Hunt, N.H. Am. J. Pathol. (1987) in press.
29. Clark, I.A. Ann. Trop. Med. Parasitol. (1987) in press.
30 Clark, I.A., Chaudhri, G., Thumwood, C.M., Hunt, N.H. & Cowden, W.B. In Free Radicals, Oxidant Stress and Drug Action, ed. C. Rice-Evans, London, Richelieu Press, (1987).

31. Sato, N., Goto, T., Haranaka, K., Satomi, N., Nariuchi, H., Mano-Hirano, Y. & Sawasaki, Y. J. Natl. Canc. Inst. 76 (1986), 1113.
32. Collins, T., Lapierre, L.A., Fiers, W., Strominger, J.L. & Poher, J.S. Proc. Natl. Acad. Sci. USA 83 (1986), 446.
33. Pohlman, T.H., Stanness, K.A., Beatty, P.G., Ochs, H.D. & Harlan, J.M. J. Immunol. 136 (1986), 4548.
34. Shparber, M. & Nathan, C.F. (1986). Autocrine activation of macrophages by recombinant tumour necrosis factor (rTNF) but not recombinant interleukin-1 (rIL-1). Blood 68, 86a.
35. Collart, M.A., Belin, D., Vasali, J.-D., De Kassado, S. & Varsalli, P. J. Exp. Med. 164 (1986), 2113.
36. Nathan, C.F., Murray, H.W., Wiebe, M.E. & Rubin, B.Y. J. Exp. Med. 158 (1983), 670.
37. Chaudhri, G., Clark, I.A., Hunt, N.H., Cowden, W.B. & Ceredig, R. J. Immunol. 137 (1986) 2646.
38. Ball, S.S., Weindruch, R. & Walford, R.L. In Free Radicals, Aging and Degenerative Diseases. Ed. J. Johnson Jr. Liss, New York (1986), 427.

REACTIVE OXYGEN SPECIES IN BIOCHEMICAL SYSTEMS

FREE RADICALS: CHEMISTRY, PATHOLOGY AND MEDICINE
edited by Catherine Rice-Evans & Thomas Dormandy, © 1988 Richelieu Press, London

Protein Kinase C Participates in the Regulation of Lipid Peroxidation in Biological Membranes.

V.E.Kagan*, R.A. Bakalova*, E.A. Serbinova*, G.M. Koynova*, G.N. Baldenkov**, V.A. Tkachuk**, Ts.S. Stoytchev*.
*Institute of Physiology, Bulgarian Academy of Sciences, Sofia 1113, Bulgaria, **National Research Center for Cardiology, Academy of Medical Sciences USSR, Moscow.

Introduction.

It is now accepted that low steady-state concentrations of lipid peroxidation (LPO) products play an essential part in cell metabolism and are necessary for the physiological modifications of biomembranes (1,2). Excessive generation of the products, on the other hand, underlies membrane damage in many pathological conditions (1-8). In other words, destruction of membrane structures in many pathological states is probably due not to the appearance of new pathobiochemical pathways but to the uncontrolled increase or imbalance of normal reactions.

The main components of the antioxidant system which controls the intensity of LPO reactions at a normal steady-state level are well known: they include water-soluble and lipophilic free radical scavengers (1,2,9,10), transition metal chelators (1,4,10,11) and specific antioxidant enzymes (superoxide dismutase, catalase, glutathione peroxidase) (1,10,12,13). Recently, the participation of calcium ions in this regulation has been demonstrated (1,14-17), pointing to the possible role of secondary messengers.

It is well established that modifications of the physico-chemical properties of biomembranes due to the action of activated oxygen species and LPO products can lead to severe disturbancies of calcium homeostasis and to changes in specific receptor-enzymic complexes (adenylate cyclase, guanylate cyclase, phosphatidylinositol specific phospholipase C) (18-20). However, it is still unclear what pathways produce secondary messengers in LPO reactions in the cell, and how important these are in co-ordination the activity of different components of the antioxidant system.

The present experiments were designed with these questions in mind. The findings illustrate the influence of the effectors of the two main regulatory system - cAMP-dependent and the other phosphatidylinositol-dependent - on LPO reactions in membrane structures of liver and heart. For this purpose we used forskolin (FK), an activator of the adenylate cyclase system, as well as phorbol-12-myristate-13-acetate (PMA), a structural analogue of diacylglycerol and an activator of protein kinase C (PK-C), which participates in phosphatidylinositol- dependent transmembrane signal transduction (21-24).

Materials and methods.

Male Wistar rats weighting 150 - 180 g were used. Liver microsomal fractions were isolated by the differential centrifugation (10 000 g supernatant for 60 min at 100 000g, "VAC-602" ultracentrifuge, GDR) of homogenates obtained after liver perfusion with ice-cold 1.15% KCl. In some cases microsomes were obtained after perfusion of liver with ice-cold 20 mM Tris-HCl buffer (pH=7.4), and after preincubation of liver homogenate with the substances studied (10 min at 30°C). Rat heart microsomal fractions were also isolated by differential centrifugation according to the procedure of Levitsky at . al. (25). Total extracts of liver lipids were obtained by the Folch method (26). Liposomes were prepared by sonication (Soniprep MSE-150W desintegrator at 12W) of liver lipid suspensions in a medium containing Tris-HCl buffer (pH=7.4), 20 mmoles/1 until complete clearance (27). Protein concentration was assayed with the biuret reagent using bovine serum albumin as a standard. LPO was induced in the following systems:-

1. In heart homogenates after a 10-minute preincubation at 30°C in a medium which contained $FeCl_2$, 80 μmoles/1, ascorbate, 1 mmol/1, Tris-HCl buffer 20 mmol/1 (pH=7.4), and protein, 1mg/ml.
2. In liver homogenates after a 10-minute preincubation at 30°C in a medium which contained $FeCl_2$, 20 μmol/1, ascorbate, 0.5mmol/1, Tris-HCl buffer (pH=7.4), 20 mmol/1, and protein, 1 mg/ml.
3. In microsomal membranes from heart or liver after a 10-minute preincubation at 30°C in a medium which contained $FeCl_2$, 20 μmol/1, ascorbate (or NADPH), 0.5mmol/1 Tris-HCl buffer (pH=7.4), 20mmol/1, and protein, 1mg/ml.
4. In liver lipid liposomes after a 10-minute preincubation at 30°C in a medium which contained $FeCl_2$, 100 μmol/1, ascorbate, 0.5 mmol/1, Tris-HCl buffer (pH=7.4), 20 mmol/1, and 1 mg/ml of lipids.

All incubations were conducted with constant stirring at 37°C. The accumulation of LPO products was followed spectrophotometrically using the 2-thiobarbituric acid

(TBA) method (Perkin-Elmer 552 spectrophotometer) at 535 nm (28). The amounts of TBA-reactive products were calculated using the molar extinction $1.56 \times 10^5\ M^{-1}.cm^{-1}$ (28). The pharmacodynamics of PMA action on LPO were studied in mice. PMA was injected intraperitoneally in a dose of 10 µmol /kg weight. At defined time intervals after the injection, the animals were killed, the required organs were removed and homogenates were prepared as described. LPO was induced in tissue homogenates as mentioned above.

The reagents used were Tris-HCl, bovine serum albumin (Reanal), ascorbate, d,l-alpha-tocopherol (α-T) (Serva), NADPH, 2-TBA (Sigma), PMA, 4α-phorbol-12,13-didecanoate (4α-PDD), FK, Polymyxin B (Px B) (Merck).

Results and discussion.

Fig. 1 shows the dependence of (Fe^{2+} + ascorbate)-induced LPO inhibition on the concentrations of PMA added to the microsomal suspensions, rat liver homogenates or liposomes.

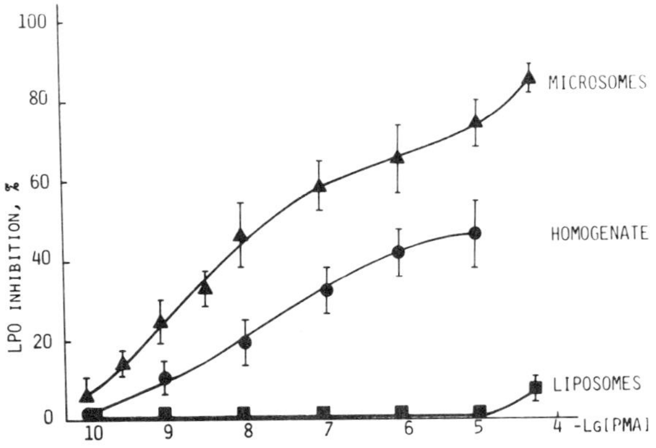

EFFECTS OF PHORBOL 12-MYRISTATE 13-ACETATE (PMA) ON (Fe^{2+} + ASCORBATE)-INDUCED LPO IN RAT LIVER MICROSOMES, HOMOGENATE AND LIPOSOMES (FROM LIVER LIPIDS).

Fig. 1

At concentrations of 10^{-9}-10^{-8} mol/1, the inhibitory effect of PMA in microsomal membranes was 25-45% and in liver homogenate 10-20%. In liposomes PMA did not inhibit LPO up to a concentration of 10^{-5} mol/1. The inhibitory effect of PMA was far more pronounced than the effect of known free radical scavengers. For instance, α-T in the concentration range, 10^{-9}-10^{-8} mol/1 had virtually no effect on non-enzymic LPO in liver homogenates or microsomes (9); and it caused only 45-55% inhibition at concentrations as high as 10^{-6}-10^{-5} mol/1 (9). At these concentrations α-T posesses the same efficiency in inhibiting LPO in liposomes (9,29,30). This suggests that the mechanisms which underly inhibition of LPO by PMA and by α-T could be quite different. It is known that at concentrations of 10^{-9}-10^{-7} mol/1 PMA is able to activate PK-C. One can therefore suggest that LPO inhibition in the presence of PMA is due to PK-C activation. The data in Table 1 illustrate the action of PMA on liver microsomes isolated from homogenates preincubated for 10 min at 30°C with PMA at a concentration of 10^{-7} mol/1 (the saturating concentration for PK-C activation) (24,31,32).

Table 1

EFFECTS OF PHORBOL 12-MYRISTATE 13-ACETATE (PMA), FORSKOLIN (FK) AND ALPHA-TOCOPHEROL (α-T) ON (Fe^{2+} + ASCORBATE)-INDUCED LPO IN MICROSOMES, ISOLATED FROM PREINCUBATED RAT LIVER HOMOGENATES.*

INCUBATION TIME, MIN	LPO INHIBITION, %				
	PMA**	α-T*^	α-T##	FK##*	PMA+FK
5	72,0±3,0	8,0±0,5	53,0±0,5	0,0	21,0±1,0
10	69,0±2,0	0,0	48,0±5,0	0,0	19,0±0,7
20	31,0±0,5	0,0	0,0	0,0	21,0±2,2

*PREINCUBATION FOR 10 MIN AT 30°C.
**0,1μM PMA
*^0,1μM α-T
##100μM α-T
##*1μM FK

PMA decreases the LPO product concentration by 72% after 5 min whereas α-T at the same concentration produces only 8% inhibition. The dynamics of PMA-dependent LPO inhibition also differ from α-T-dependent inhibition. Even at much higher concentrations (10^{-4} mol/l) the inhibitory effect of α-T decreases sharply and is practically eliminated 20 min after LPO induction. This is presumably due to its utilization in the course of the reaction (9). By contrast, PMA-dependent LPO inhibition is fully retained during the first 10 minutes of incubation and decreases only to 43% 20 minutes after LPO induction (Table 1). If PK-C participates in LPO inhibition in the presence of PMA, then a slow decrease of the inhibitory effect could be explained by a dephosphorylation process. These considerations are, of course, only indirect indications of the possible involvement of PK-C in PMA-dependent LPO inhibition. To obtain more reliable evidence we used a PMA analogue, 4α-PDD, a very weak activator of PK-C (33,34), as well as the deesterified derivative of PMA, phorbol, which does not activate PK-C at all (33,34).

Table 2

EFFECTS OF PHORBOL 12-MYRISTATE 13-ACETATE (PMA), 4α-PHORBOL 12,13-DIDECANOATE (4α-PDD), PHORBOL, POLYMYXIN B (Px B) ON (Fe^{2+} + ASCORBATE)-INDUCED LPO IN MICROSOMES, ISOLATED FROM PREINCOBATED RAT LIVER HOMOGENATE.*

INCUBATION TIME, MIN	LPO INHIBITION, %				
	PMA**	4α-PDD#*	PHORBOL##	Px B##*	PMA+Px B##*
5	72,0±3,0	6,3±0,2	0,0	9,0±1,0	8,0±3,0
10	69,0±2,0	7,0±0,4	0,0	0,0	3,0±0,5

*PREINCUBATION FOR 10 MIN AT 30°C.
**0,1μM PMA, #*0,1μM 4α-PDD,
##0,1μM PHORBOL, ##*100μM Px B

The data in Table 2 show that at a concentration of 10^{-7} mol/l phorbol does not influence LPO in microsomal membranes, whereas the effect of 4α-PDD does not exceed 10%. When Px B, a non-specific inhibitor of PK-C (35), was added to liver homogenates the PMA-dependent inhibition of LPO in microsomes isolated from preincubated homogenates was significantly lower (Table 2). It should be noted that Px B (10^{-4} mol/l) did not inhibit LPO either in microsomes in the absence of PMA or in liposomes.

It can therefore be concluded that the PMA-dependent inhibition of LPO in rat liver microsomal membranes is due to activation of PK-C.

This means that PK-C can participate in the regulation of LPO in biomembranes. The inhibitory effect of PMA in microsomal suspensions is far less pronounced than in microsomal fractions isolated from homogenates preincubated with PMA (Table 1, Fig. 1). It is known that PK-C is a cytosolic (water-soluble) enzyme, which can be translocated to membranes on hormonal stimulation or on treatment with PMA (15,24,36-40). Evidently PK-C binding by membrane fractions in which LPO is induced can be responsible for the difference in the efficiency of PMA action under these two experimental conditions.

Activation of PK-C is a necessary prerequisite of hormonal signal transduction and of membrane-mediated regulation of cell functions and proliferation (41,42). In the phosphatidylinositol-dependent system of transmembrane signal transduction PK-C plays the role analogous to the role of protein kinase A (PK-A) in the system in which cAMP is used as a secondary messenger (15,24,43). For many physiological and metabolic processes such as cell division, contractility, the activation of neutrophils and the functioning of Ca-channels, the effects of PK-C and of PK-A are opposite or antagonistic (41-48). Does this hold true for LPO regulation? To answer this question we examined the effects of FK, an activator of adenylate cyclase, on LPO in microsomal membranes in the absence and in the presence of PMA.

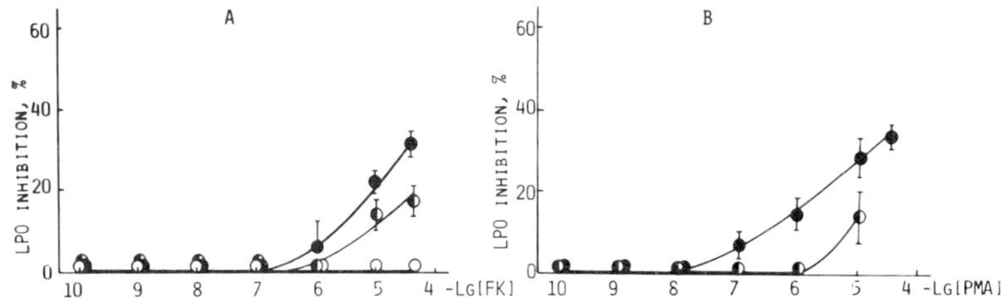

EFFECTS OF FORSKOLIN (FK) (A) AND PHORBOL 12-MYRISTATE 13-ACETATE(PMA) + FK (1μM) (B) ON (Fe^{2+} + ASCORBATE)-INDUCED LPO IN RAT LIVER MICROSOMES (●), HOMOGENATES (◐) AND LIPOSOMES (FROM LIVER LIPIDS) (○), IN VITRO.

Fig. 2

The data in Fig. 2A show that FK alone has virtually no influence on the accumulation of LPO products in microsomal suspensions, liposomes or liver homogenates. Simultaneous addition of FK (10^{-6} mol/l) and PMA (10^{-7} mol/l) to liver homogenates with subsequent incubation

(10 min at 30°C) causes sharp decrease in the inhibitory effect of PMA on (Fe^{2+} + ascorbate)-induced LPO in microsomes isolated from preincubated homogenates (Table 1). A similar effect was observed when FK and PMA were added directly to microsomal suspensions or to liver homogenates in which (Fe^{2+} + ascorbate)-dependent LPO was induced (Fig. 2B). Under the same experimental conditions PK-A-dependent processes prevent the inhibitory effects of PMA on LPO, suggesting the possibility of antagonistic interactions between cAMP- and phosphatidylinositol-dependent regulatory systems.

Similar results were obtained with rat heart homogenates and microsomes (Fig.3, Table 3). The inhibitory effect

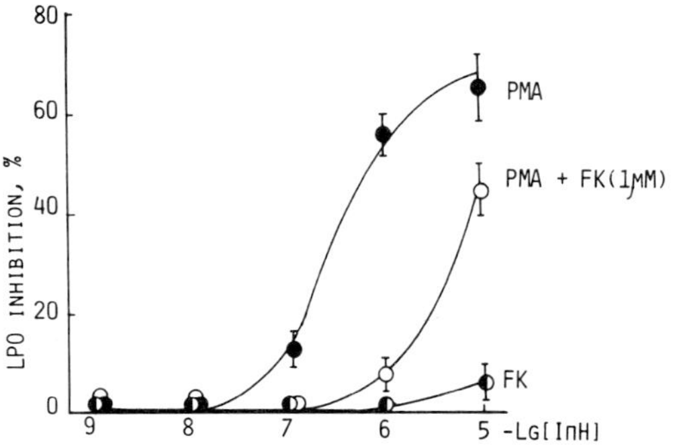

EFFECTS OF PHORBOL 12-MYRISTATE 13-ACETATE (PMA), FORSKOLIN (FK) AND THEIR COMBINATION ON (Fe^{2+} + ASCORBATE)-INDUCED LPO IN RAT HEART HOMOGENATES, IN VITRO.

Fig 3

Table 3

EFFECTS OF PHORBOL 12-MYRISTATE 13-ACETATE (PMA), FORSKOLIN (FK) AND THEIR COMBINATION ON (Fe^{2+} + ASCORBATE)-INDUCED LPO IN MICROSOMES, ISOLATED FROM PREINCUBATED RAT HEART HOMOGENATE.*

INCUBATION TIME, MIN	L P O I N H I B I T I O N , %		
	PMA**	FK***	PMA+FK
15	25,0±1,5	0,0	0,0
30	8,0±0,7	0,0	0,0

*PREINCUBATION FOR 10 MIN AT 30°C.
**0,1μM PMA
***1μM FK

of PMA on (Fe^{2+} + ascorbate)-dependent LPO in rat heart homogenates and microsomes isolated from homogenates preincubated in the presence of PMA was significantly less pronounced than inhibition in liver membrane preparations. These results are in agreement with expectations based on the lower PK-C activity in rat heart, and its localization exclusively in the cytosolic fraction (24,49-51). FK eliminates the inhibitory effects of PMA on LPO in heart membrane structures (Fig. 3, Table 3). In microsomal membranes LPO can be induced both non-enzymically by the (Fe^{2+} + ascorbate)- system or enzymically by the (Fe^{2+} + NADPH)- system. In microsomes isolated from homogenates preincubated with PMA enzymically induced LPO was inhibited 1.7 - 2.8 times less efficiently than non-enzymic LPO induction (Tables 1, 4), while FK had no effect on (Fe^{2+} + NADPH)-dependent LPO alone, it completely abolished the inhibitory effect of PMA.

Table 4

EFFECTS OF PHORBOL 12-MYRISTATE 13-ACETATE (PMA), FORSKOLIN (FK) AND THEIR COMBINATION ON (Fe^{2+} + NADPH)-INDUCED LPO IN MICROSOMES, ISOLATED FROM PREINCUBATED RAT LIVER HOMOGENATE.*

INCUBATION TIME, MIN	LPO INHIBITION, %		
	PMA**	FK***	PMA+FK
5	41,0±2,1	0,0	0,0
10	19,2±6.5	0,0	0,0
20	11,0±1,5	0,0	0,0

*PREINCUBATION FOR 10 MIN AT 30°C.
**0,1μM PMA
***1μM FK

These results provide further support for the concept of antagonistic interrelations between PMA and FK and they also show that whereas the inhibitory effect of PMA is sensitive to the mode of LPO induction, inhibition by FK is equally effective in both cases. The radicals generated by non-enzymically induced LPO are mainly randomly distributed in the aqueous phase, wherefrom they attack membrane phospholipids. After enzymic induction of the radicals generated by NADPH/cytochrome P-450 arise in the vicinity of the enzymes in the membrane. It may be suggested that the water-soluble radical intermediates of LPO are readily accessible to PMA-dependent inhibition. This points to the possible involvement of water-soluble antioxidant enzymes in bringing about the effect of PMA. Further experiments will show whether PK-C activation by PMA causes phosphorylation of the antioxidant enzymes.

It should be mentioned that PMA exhibits its inhibitory effect on LPO not only in vitro but also in vivo. The data in Fig. 4 show the efficiency of (Fe^{2+} + ascorbate)-induced LPO inhibition in homogenates prepared from livers and hearts of mice pretreated with PMA (intraperitoneal injection of a single dose of 10 μmol/kg body weight).

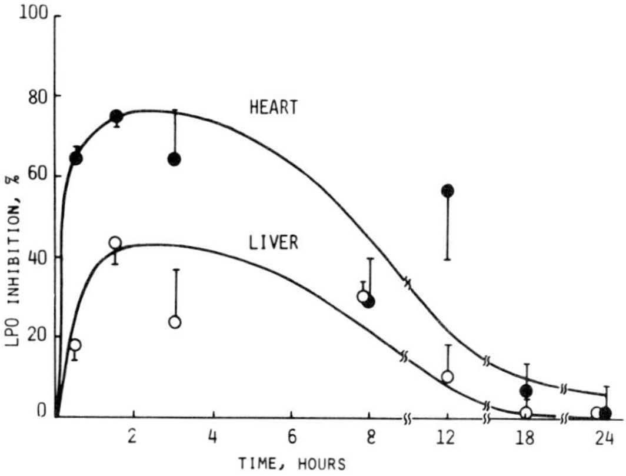

IN VITRO INHIBITION OF (Fe^{2+} + ASCORBATE)-INDUCED LPO IN LIVER AND HEART HOMOGENATES FROM MOUSE, PRETREATED WITH PHORBOL 12-MYRISTATE 13-ACETATE (10 µMOLES/KG INTRAPERITONEALLY).

Fig. 4

The pharmacodynamics of PMA are characterised by an initial rapid increase of the inhibitory effect, a period of maximal antioxidant activity persisting for several hours and a subsequent slow decline of the ability to inhibit LPO in tissues. It may be suggested that these changes are determined both by the ratio of phosphorylation and dephosphorylation rates of PMA-stimulated reactions and by the rate of PMA metabolism and excretion.

Conclutions.

In conclution the results presented show that the intensity of LPO in the cell can be regulated by PK-C- and PK-A-dependent reactions. In other words, LPO is under the control of the secondary messengers which participate in the transduction of hormonal signals across the membranes. Studies of the molecular mechanisms underlying PK-C- and PK-A-mediated LPO regulation as well as of the role of these effects in physiological conditions and in pathological states will be the aim of further studies.

References.

1. Kagan V.E., Lipid Peroxidation in Biomembranes (CRC Press, Boca Raton, Florida 1988) [in press].
2. Armstrong D., Sohal R.S., Cutler R.G., Slater T.F. Eds., Free Radicals in Molecular Biology, Aging and Disease (Raven Press, N.Y. 1984, Vol. 24) pp 1-432.
3. Corongiu F.P., Tomasi A., Vannini V. Eds., Advances in Free Radicals in Desease (Cleup Padova 1987) pp 1-133.
4. Halliwell B., Gutteridge J.M.C.: Biochem. J. 219,

(1984) 1.

5. Dormandy T.L. in: Advanced Medicine, Triger D.R. Ed. (Bailliere Tindall, London 1986) 333.

6. Autor A.P. Ed., Pathology of Oxygen (Acad. Press, N.Y. 1982) 1.

7. Halliwell B., Gutterige J.M.C., . Free Radicals in Biology and Medicine (Clarendon Press, Oxford 1985) pp 346.

8. Cornwell D.G., Morisaki N. in: Free Radicals in Biology, Pryor W. Ed. (Acad. Press, N.Y. 1986, Vol. 6) 124.

9. Kagan V.E., Serbinova E.A., Bakalova R.A., Novikov K.N., Skrypin V.I., Evstigneeva R.P., Stoytchev Ts. in: Free Radicals, Oxidant Stress and Drug Action, Rice-Evans C. Ed. (Richelieu Press, London 1987) 425

10. Pryor W. Ed., Free Radicals in Biology (Acad. Press, N.Y. 1976, Vol. 1) pp 1-239.

11. Fridovich I.: Science 201, (1978) 875.

12. Rotilio G. Ed., Superoxide and Superoxide Dismutase in Chemistry, Biology and Medicine (Elsevier Science Publishers, Amsterdam-New York-Oxford 1986) pp 1-688

13. Chance B., Sies H., Boveris A.: Physiological Rev. 59, (1979) 527.

14. Ilundain A., O'Brien J.A., Burton R.A., Sepulveda E. Biochim. Biophys. Acta 896, (1987) 113.
15. Berridge M.J.: Biochem. J. 220, (1984) 345.
16. Savov V.M., Babizhaev M.A., Kagan V.E.: Bull. Exp. Biol. Med. 101, (1986) 771.
17. Jewell S., Bellomo G., Thor H., Orrenius S., Smith M.: Science 217, (1982) 1257.
18. Cerutti P.A.: Science 227, (1985) 375.
19. Mittal C.K., Murad F.: Proc. Natl. Acad. Sci. U.S.A. 74, (1977) 4360.
20. Mulsch A., Bohme E., Busse R.: Eur. J. Pharmacol. 135, (1987) 247.
21. Nishizuka Y.: Nature 308, (1984) 693.
22. Nishizuka Y.: Science 225, (1984) 1365.
23. Kikkawa U., Takai Y., Tanaka Y., Miyake R., Nishizuka Y.: J. Biol. Chem. 258, (1983) 11442.
24. Hashimoto Y., Soderling T.R.: Arch. Biochem. Biophys. 252, (1987) 418.
25. Levitsky D.O., Aliev M.K., Kuzmin A.V., Levchenko T.S., Smirnov V.N., Chazov E.I.: Biochim. Biophys. Acta 443 (1976) 468.
26. Folch J., Lees M., Stanly G.: J. Biol. Chem. 226, (1957) 2.

27. Deamer D.W., Uster P.S. in: Liposomes, Ostro M.J. Ed. (Marcel Dekker, Inc., N.Y. 1986) 27.
28. Fleisher S., Packer L. Eds., Methods in Enzymology (Acad. Press, N.Y. 1978) pp 1-506.
29. Kagan V.E., Serbinova E., Novikov K., Ritov V., Kozlov Yu., Stoytchev Ts.: Arch. Toxicol. Suppl. 9, (1986) 302.
30. Lindsey J.A., Zhang H., Kaseki H., Morisaki N., Sato T., Cornwell D.G.: Lipids 20, (1985) 151.
31. Adam. M.A., Johnstone R.M.: Biochem. J. 242, (1987) 151.
32. Villalba M., Pajares M.A., Renart M.F., Mato J.M.: Biochem. J. 241, (1987) 911.
33. Blumberg P.M.: Crit. Rev. Toxicol. 8, (1980) 153.
34. Blumberg P.M.: Crit. Rev. Toxicol. 8, (1981) 199.
35. Stutchfield J., Jones P.M., Howell S.L.: Biochem. Biophys. Res. Commun. 136, (1986) 1001.
36. Kraft A.S., Anderson W.B., Cooper H.L., Sando J.J.: J. Biol. Chem. 257, (1982) 13193.
37. Kraft A.S., Anderson W.B.: Nature (London) 301, (1983) 621.
38. Wolfson M., McPhail L.C., Nasrallah V.N., Snyderman R.: J. Immunol. 135, (1985) 2057.

39. Jacobson K., Wenner C.E., Kemp G., Papahadjopoulos D.: Cancer Res. 35, (1975) 299.

40. Dellers M., Ruysschaert J.M., Malaisse W.J.: Chem.-Biol. Interactions 42, (1982) 271.

41. Mastro A.M. in: Limphokines, Pick E. Ed. (Acad. Press, N.Y. 1982, Vol. 6) 263.

42. Anderson W.B., Estival A., Tapiovaara H., Gopalakrishna R.: Advances in Cyclic Nucleotide and Protein Phosphorylation Research 19, (1985) 287.

43. O'Brian C.A., Arthur W.L., Weinstein I.B.: FEBS Lett. 214, (1987) 339.

44. Forder J., Scriabin A., Rasmussen N.: J. Pharm. Exp. Ther. 235, (1985) 267.

45. Trautwein W.: Triangle 24, (1985) 101.

46. Messing R.O., Carpenter C.L., Greenberg D.A.: Biochem. Biophys. Res. Commun. 136, (1986) 1049.

47. Spah F.: J. Cardiovascular Pharmacol. 6, (1984) 99.

48. Miller R.J.: TINS november/december (1986) 538.

49. Blumberg P.M., Delclos K.B., Jaken S. in: Organ and Species Specificity in Chemical Carcinogenesis, Langenbach R., Nesnow S., Rice J.M. Eds. (Plenum Publishing Corp., N.Y. 1983) 201.

50. Blumberg P.M., Dunn J.A., Jaken S., Jeng A.Y.,

Leach K.L., Sharkey N.A., Yeh E. in: Mechanisms of Tumor Promotion, Slaga T.J. Ed. (CRC Press, Boca Raton, Florida 1984, Vol. 3) 143.

51. Blumberg P.M., Jaken S., Konig B., Sharkey N.A., Leach K.L., Jeng A.Y., Yeh E.: Biochem. Pharmacol. 33, (1984) 933.

FREE RADICALS: CHEMISTRY, PATHOLOGY AND MEDICINE
edited by Catherine Rice-Evans & Thomas Dormandy, © 1988 Richelieu Press, Londo..

OXIDATION OF AMINO ACIDS WITHIN SERUM PROTEINS.

HR GRIFFITHS, J UNSWORTH*, DR BLAKE* and J LUNEC
Departments of Biochemistry and Rheumatology*
Selly Oak Hospital
Birmingham B29 6JD

ABSTRACT

The detection of free radical activity in vivo relies for the most part on the indirect measurement of oxidised biomolecules. Measurement of lipid peroxides by the thiobarbituric acid (TBA) test has been used widely but is inadequate since it also detects other lipids including the inflammatory mediators, prostaglandins (1).

Proteins are also chemically altered when exposed to a radical flux. They aggregate and display a characteristic fluorescence (Ex 360nm, Em 454nm) due to oxidation of their constituent amino acids (2). We have developed a technique to detect specific oxidised amino acids within proteins from sera and synovial fluids and present data from patients with inflammatory joint disease.

The reverse phase HPLC analyses of proteolytic digests of these proteins has shown:

(1) the presence of oxidised Trp within IgG;

(2) the levels of oxidised Trp and Cys are elevated in proteins from inflammatory synovial fluids;

(3) that an identifiable part of the fluorescence in IgG and albumin is due to oxidised Cys, Met and Trp.

These results indicate increased oxidation of proteins within the inflamed joint and support further the concept of uncontrolled free radical production during rheumatoid inflammation

----

It is widely accepted that lipid peroxidation is a major consequence of in vivo oxidative stress (3,4). However, there is now considerable evidence to indicate that proteins can undergo some form of post translational modification due to oxidation (5,6,7). This is apparent during the normal ageing process (8), but is accelerated during certain disease processes. Because of the functional diversity of proteins, from enzymes to membrane receptors, any oxidative modifications to proteins can be expected to have widespread effects.

The earliest work on the oxidation of biologically important proteins examined the effect of selected free radicals on lysozyme (9). Both thiocyanate and $OH^{\bullet}$ were found to inactivate the enzyme, implying that tryptophan residues are essential for catalytic activity, a well established fact from classical enzymology. Similarly for α-1-antitrypsin, it was found that the modification of a single methionine residue rendered the protein inactive (10). Subsequent hydrolysis and amino acid analysis revealed the presence of methionine sulphoxide. This was one of the first pieces of evidence linking amino acid oxidation to denaturation and in this case, loss of function.

Since these initial studies, three major physical changes have been described in oxidised proteins:

a.fragmentation,

b.aggregation,

c.increased susceptibility to proteolysis.

Each of these changes shall be examined in turn:

(a) Fragmentation

Oxidative attack has been shown to selectively fragment albumin (11), collagen (12) and γ globulins(13). Using steady state radiolysis to generate specific radicals in solution, the $OH^{\bullet}$ radical in the presence of molecular oxygen has been shown to degrade proteins whilst peroxy, hydro-peroxy and the superoxide radical are inert. Under normal conditions $OH^{\bullet}$ attack is random unless catalytic transition metal ions are bound to amino acids such as His or Arg. However in albumin chain scission appears to occur preferentially at proline residues, yielding fragments of defined chain length (12). This is ascribed to the greater susceptibility of Pro to oxidation. The subsequent destruction of its ring structure then results in cleavage of the peptide bond with the appearance of new amino termini.

(b) Aggregation

In addition to fragmentation, aggregates of lens proteins, caeruloplasmin and IgG are observed following oxygen radical attack (14,15,16). We have evidence to suggest that in the case of IgG, only $OH^{\bullet}$ radicals can mediate crosslink formation in the presence of oxygen. Superoxide and peroxy radicals are inert in this respect. Some, but not all, of the aggregates formed are unaffected by the reducing conditions used in SDS-PAGE and are therefore not mediated simply by a rearrangement of disulphide bonds. The amino acids involved in the formation of non reducible crosslinks remain to be elucidated. Aggregates of albumin, IgG and caeruloplasmin are found as direct multiples of the original monomer molecule suggesting that it is the native protein which is crosslinked rather than a non specific aggregation of fragments.

(c) Susceptibility to proteolysis

Oxidation causes gross conformational changes to proteins. Several groups have described that proteins denatured in this way are more susceptible to proteolysis than in the native form (17,18). This is true for both fragmented and aggregated forms and is likely to be due to structural deformation allowing greater access to proteases. These results indicate that at any site of oxygen radical production, the rate of turnover of protein would be grossly elevated in the immediate environment.

Antigenicity

Denatured proteins are not only more susceptible to proteolysis but if they come into contact with macrophages or T helper lymphocytes, an antigenic response may be induced, because of a lack of homology with self (19). In cooperation with B lymphocytes, antibodies directed towards "foreign proteins" are produced in the classical immune response. It is the production of antibodies to altered self proteins that has been proposed as the mechanism of the induction of autoimmunity. In rheumatoid arthritis, circulating autoantibodies (rheumatoid factors) directed against autologous but denatured IgG are found. We have previously suggested that oxidation may be one cause of this observed denaturation (2). A more recent finding is the presence of antibodies to albumin in the sera of patients with type II diabetes mellitus (20): this parallels the previous situation since diabetics have significantly higher levels of oxidised albumin than in age matched controls (21).

Oxidation is a normal consequence of cellular ageing. However this process seems to be accelerated in chonic inflammation, diabetes and in cataract lens formation. A method for detecting such changes in vivo could be expected to have widespread applications for monitoring disease progression.

Measurement of protein oxidation

The susceptibility of individual amino acids to $OH^{\bullet}$ attack is well documented (22). However, within a peptide backbone, the presence of bound metal ions capable of catalysing a Fenton type reaction, and the tertiary structure of the molecule are more important factors in determining relative susceptibility. Different proteins would therefore be expected to have different oxidised amino acids.

Currently, two approaches have been taken to assess oxidative damage to proteins :(a) the measurement of carbonyl groups, and (b) fluorescence emmission at 454 nm.

Carbonyl derivatives of His, Pro, Arg, Lys and Ser have been reported within different protein structures. A method has been developed by Levine and Stadtman for measuring these compounds using a simple derivitisation step with 2,4 dinitrophenyl hydrazine (23). The product has a high extinction coefficient, thereby conferring sensitivity on the assay.

However, this procedure is obviously unsuitable for chromophores absorbing at 340nm such as the haem proteins. It has been used successfully to demonstrate that protein oxidation is involved in the accumulation of altered proteins during ageing and the killing of bacteria by neutrophils (8,24).

All proteins containing aromatic amino acids exhibit a native UV fluorescence spectra. However, following attack by a mixed oxygen dependent radical system, a new characteristic fluorescence spectra is formed in the visible region (Ex_{max} 360nm, Em_{max} 454nm) (25). This has been described for albumin, γ globulins and caeruloplasmin (26). The fluorescence intensity is directly related to the flux of oxygen radicals and therefore provides an index of oxidative damage. No derivatization step is involved.

Both of these techniques provide simple assessment of overall protein oxidation, but give no information about damaged residues. Since it is well established that proteins can be protected from oxygen radical damage in the presence of the susceptible amino acid, such information might be therepeutically important.

There is considerable indirect evidence for the uncontrolled production of oxygen radicals from invading activated polymorphs and macrophages in rheumatoid arthritis (27). In particular the antibody IgG has provoked interest since it is found in a denatured form in the sera and synovial fluids of patients with this disease (28). In the following paper we describe the effects of oxygen radicals on the constituent amino acids of the major serum proteins, albumin and IgG, isolated from clinical specimens of rheumatoid patients.

Materials

Amino acids and pronase were purchased from Sigma Chemical Co., Dorset. Millipore filtered water was used for all purposes. Acetonitrile was supplied by May and Baker and all other laboratory reagents were from BDH.

γ Radiolysis

To produce oxidised standards, aerated solutions of amino acids in 40mM phosphate buffer were irradiated in a ^{60}Co-γ-source at room temperature with a dose of 1,000 gy at a dose rate of 10 gy per minute. After irradiation the solutions were kept on ice and analyzed within 24 hours.

High Performance Liquid Chromatography

An Apple IIe microprocessor controlled Gilson liquid chromatograph equipped with a variable wavelength detector and fluorimeter with ortho-phthaladehyde filter, was used. Reverse phase HPLC separations were performed on a Spherisorb 25cm column (particle size 5 μm), which was supplied by Technicol of Macclesfield, Cheshire. The mobile phase consisted of an increasing concentration of acetonitrile in phosphate buffer (0.15M, pH 5.4).

Pronase Hydrolysis

Proteins were hydrolysed in 40mM phosphate buffer pH 7.2 using 0.5 units of pronase per ml for 24 hours at $37^{o}C$.

Isolation of proteins from biological fluids.

Proteins were purified from biological fluids by selective salt precipitation using saturated ammonium sulphate. They were further purified from molecular weight contaminants by gel filtration HPLC. Finally they were extensively dialysed against PBS prior to digestion with pronase.

RESULTS

Table 1 shows the optical density of standard amino acids at 240nm before and after oxygen radical attack. Only Cys, Met, Trp, Tyr and Phe were found to absorb UV at this wavelength. A standard HPLC profile for these amino acids is shown in figure 1. The retention times varied by no more than 1% between runs.

Irradiation of individual amino acids showed a dose dependent increase of fluorescent photoproducts for Met Cys and Trp. For Trp, one of these was identified as kynurenine by spiking with a known standard (see Figure 2). The second and third major unknown peaks were assumed to be a hexahydropyrrolloindole compound and N-formyl kynurenine. (see discussion). Irradiation of cystine yielded fluorescent products which co-eluted with the native forms of cystine and cysteine. Cysteic acid was also detected. These profiles were used to standardize chromatograms of digested oxidized amino acids.

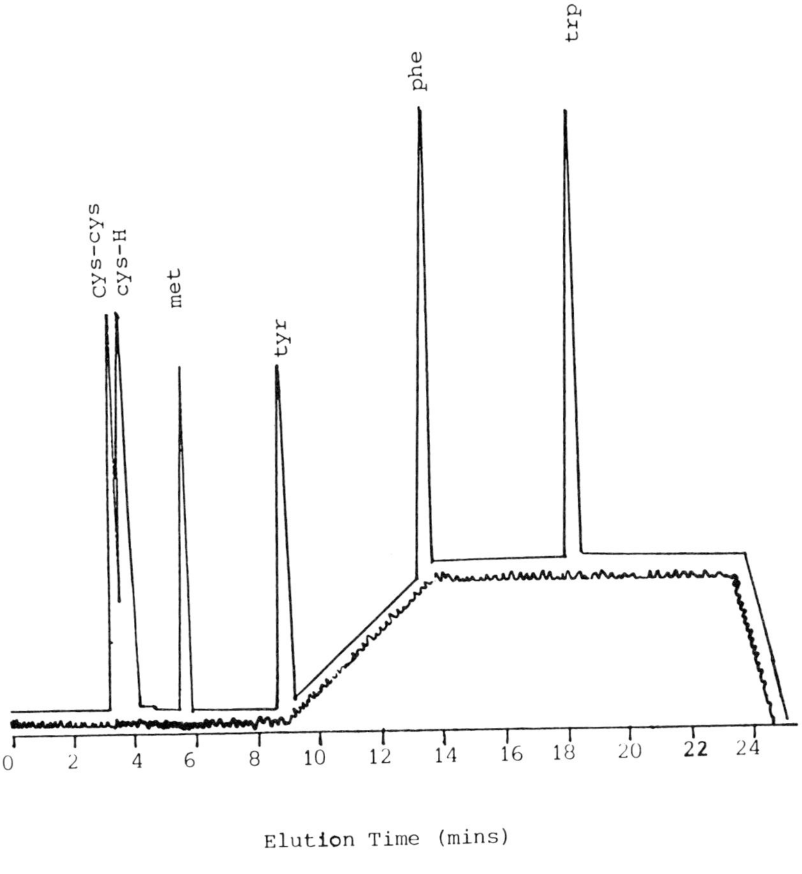

Figure 1: Elution profile of standard amino acids separated by reverse phase chromatography and monitored by UV absorbance at 240nm (-) and fluorescence ( ).

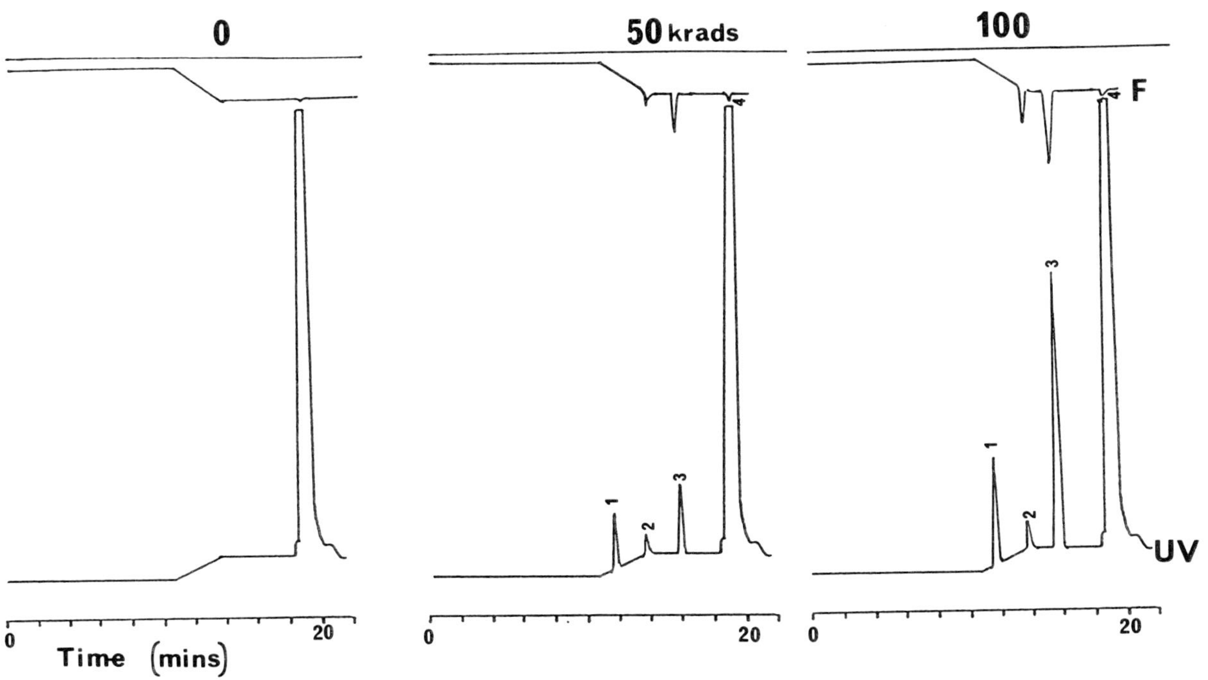

Figure 2: Dose response of tryptophan to γ radiolysis in aerated solution. Peak identities-1: HPI, 2: kynurenine, 3: N-formyl kynurenine, 4: tryptophan.

TABLE I: Optical density of amino acids before and after irradiation in aerated solution.

Amino acid 0.1 mg/ml	OD240nm 0 krad	OD240nm 200 krad
Ala	0	0
Arg	0	0
Asn	0.004	0.007
Asp	0.013	0.015
Cys	2.532	4.25
Gln	0.013	0.007
Glu	0.005	0.005
Gly	0.008	0.007
His	0.185	0.273
Ile	0.002	0.003
Leu	0.001	0.002
Lys	0.002	0.004
Met	0.471	0.549
Phe	0.272	0.351
Pro	0.023	0.022
Ser	0	0.004
Thr	0.02	0.003
Trp	18.62	20.3
Tyr	1.64	4.25
Val	0.045	0.046

In order to determine whether IgG is a selective target for oxygen free radicals released during inflammation, we studied the oxidation of amino acids of albumin and IgG in RA patients.

Examples of chromatograms of digested albumin and IgG from serum are shown in Figures 3a and b respectively. Comparison of the known amino acid sequences of these proteins with these chromatograms confirms that albumin has only 1 Trp per molecule as compared to 40 in IgG. On the other hand, there is proportionally more cystine in albumin. The oxidised amino acids have been identified by spiking with known standards and Table 2 shows the levels of oxidised cystine, methionine and tryptophan in albumin compared to IgG. The levels of the oxidised sulphur containing residues, Cys and Met, are higher in albumin. Contrary to our previous hypothesis (2), we have clearly demonstrated that oxidation products of Trp are not the major species responsible for fluorescence of oxidised proteins, but oxidised cys and met residues are also important. The unidentified fluorescence peaks are not the oxidation product of any one single amino acid.

TABLE II: The levels of oxidised amino acids within serum proteins, expressed as the percentage of total UV absorbing amino acids.

	CYS ox %	MET ox %	TRP ox %
PROTEIN			
Albumin	3.0+1.6	4.1+0.4	7.7+1.3
IgG	2.0+0.7	3.8+0.6	17.3+1.0 ***

n=14, ***=$p<0.005$

In an attempt to ascertain whether there is increased oxidative stress in the immediate environment of the inflammatory joint, we studied the oxidation of IgG isolated from synovial fluids compared to that isolated from serum.

In preliminary studies we have shown that IgG isolated from the SF of RA patients has elevated levels of the oxidised tryptophan metabolites (kynurenine) compared to the corresponding paired serum (Table 3). 5-hydroxy-tryptophan was found to be high in both isolates and did not appear to be a useful index of oxidative stress. Chromatograms representative of IgG isolated from the serum and synovial fluid of the same rheumatoid patient are shown in Figure 4.

TABLE III; Kynurenine levels in IgG isolated from serum and corresponding synovial fluids.

SOURCE/	KYNURENINE %	N-FORMYL KYNURENINE %
SERUM	11 + 1.6	6.24 + 0.5
SYNOVIAL FLUID	19.9 + 3.1 *	9.4 + 0.7 **

n=7 for each group. *=$p<0.1$, **=$p<0.05$.

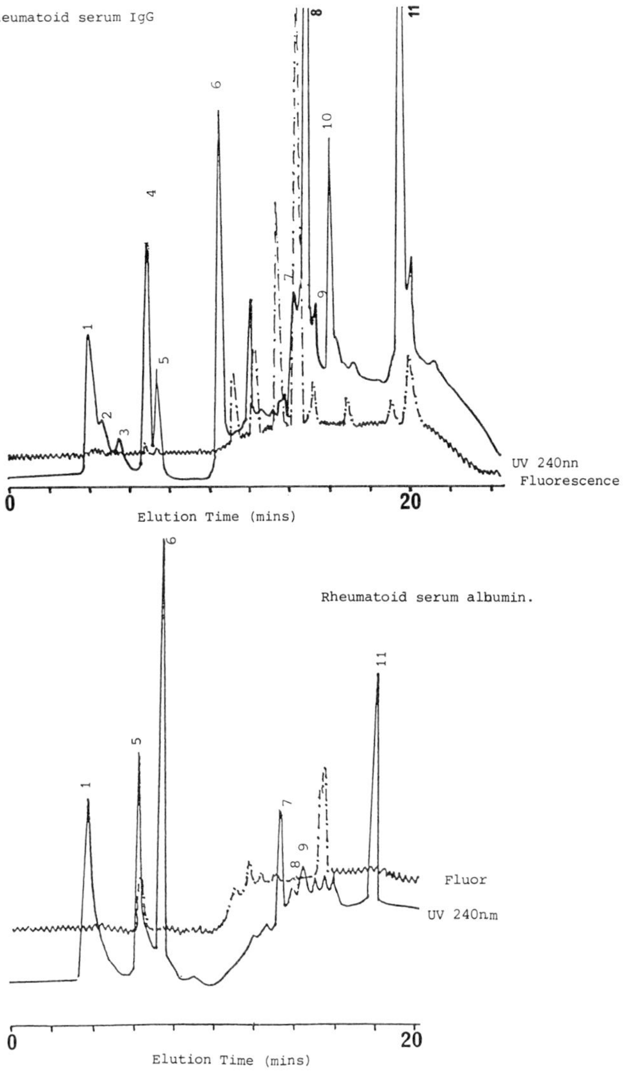

Figure 3: HPLC chromatograms illustrating the differences in oxidised and fluorescent amino acids between IgG and corresponding albumin isolated from a patient with classical rheumatoid arthritis.
Peak identities- 1: cystine, 2:cysteine 3: cysteic acid 4:oxidised met, 5: met, 6:tyr, 7: phe, 8: 5-HT, 9: kynurenine, 10: N-formyl kynurenine, 11: trp.

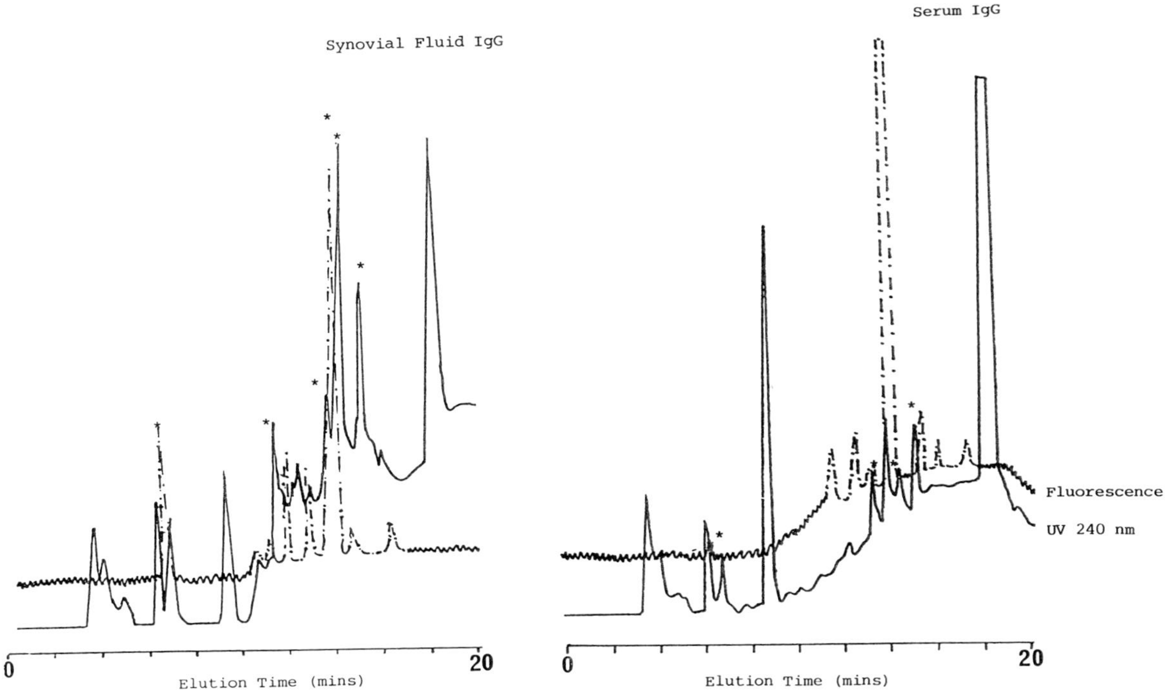

Figure 4: Comparison of the chromatograms of serum IgG and corresponding synovial fluid IgG isolated from a patient with classical rheumatoid arthritis. The fluorescent oxidised amino acids are marked by *.

DISCUSSION

Oxidised amino acids can be detected within the serum proteins of RA patients. Using a combination of UV absorbance at 240nm and fluorimetric detection, those residues which are known to be particularly susceptible to oxygen radical attack (aromatics and sulphur containing amino acids) can be detected. Figure 5 shows the established oxidation pathway of Trp(29).

Following radiolysis, the ratio of the formed products has been published as HPI 1:NFK 4 (29), and these results were used to assess the products formed in our system. The oxidation pathway of cystine is less well defined and is reviewed in ref 30.

In RA patients, the most oxidised residue in IgG was found to be Trp , whereas in albumin greater oxidation of the sulphur containing residues was detected. This suggests that the amino acids that are oxidised, may differ from protein to protein. This may possibly reflect differences in their secondary and tertiary structures or, on the other hand, may reflect attack of the protein by different radical species. It is well known that albumin is an important superoxide scavenger (31). However the work of Singh has clearly shown that the oxidation of trp to kynurenine is not a superoxide mediated event (32). Therefore the observed oxidation of Trp in IgG may reflect attack by $OH^{\bullet}$ or peroxides whereas the oxidation of sulphur containing amino acids in albumin may reflect O_2^- attack.

The presence of circulating autoantibodies to IgG in RA and not albumin suggests that the larger immunodominant residue, such as Tyr and Trp, may become antigenically important when oxidised.

ACKNOWLEDGEMENT

The authors wish to acknowledge financial support from the Arthritis and Rheumatism Council.

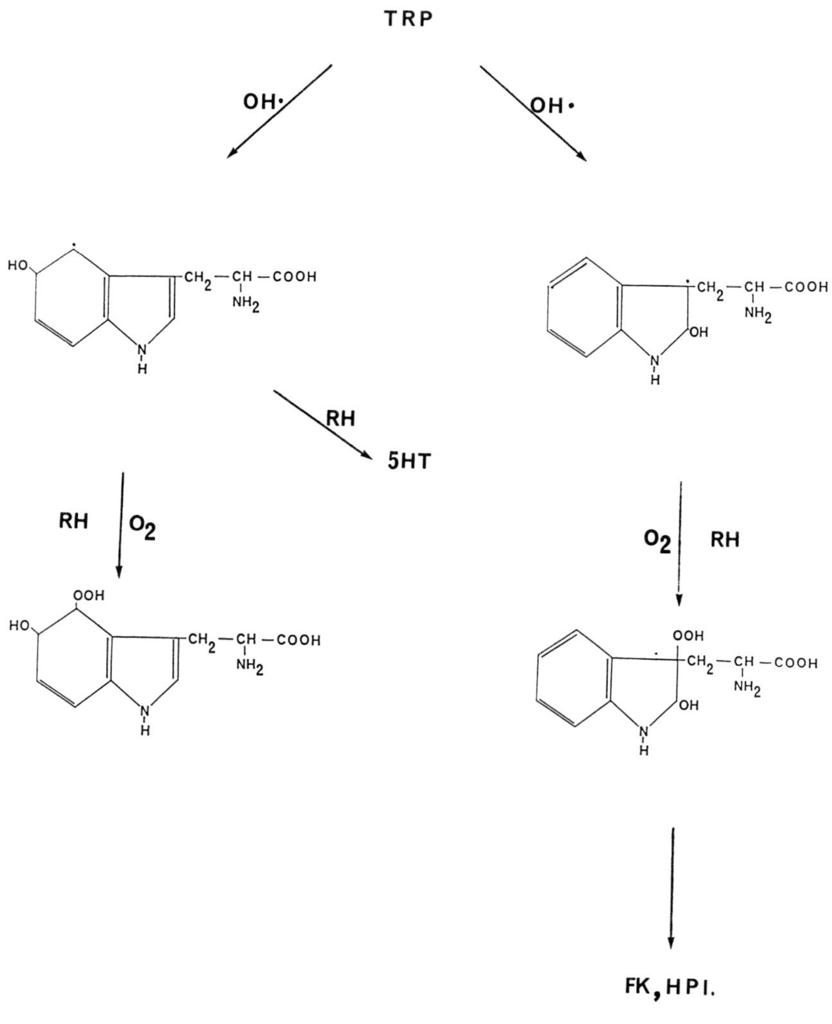

Figure 5: Schematic diagram of tryptophan oxidation by $OH^{\bullet}$. 5-HT = 5-hydroxy tryptophan, FK = formyl kynurenine, HPI = hexahydropyrrolloindole.

REFERENCES

1 Gutteridge JMC.(1986) Free Rad Res Comms.,1,(3),173-184

2 Lunec J, Blake DR, McCleary SJ, Brailsford S and Bacon PA. (1985) J Clin Invest.76(6) 2084-2090.

3 Slater TF.(1984) Biochem J. 227, 1-12.

4 Logani MK and Davies RE.(1980) Lipids. 15, 485-495

5 Davies KJA and Goldberg AL.(1987) J Biol Chem. 262 (In Press).

6 Stadtman ER.(1986) Trends in Biochem Sci. 11, 11-12.

7 Wolff SP, Garner A and Dean RT (1986) Trends in Biochem Sci. 11, 27-31

8 Oliver CN, Ahn B-W, Moerman EJ, Golstein S and Stadtman ER.(1987) J Biol Chem. 262, 5488-5491.

9 Adams GE, Willson RL, Aldrich JE, and Cundall RB (1969) Int J Radiat Biol. 16, 33

10 Carp H and Janoff, A. (1979) J Clin Invest. 65, 793-7.

11 Marx G and Chevion M. (1985) Biochem J. 236, 397-400.

12 Wolff SP and Dean RT (1986) Biochem J. 234, 399-403.

13 Phelps RA, Neet KE, Lynn LT and Putnam PD (1981) J Biol Chem. 236, 96-105.

14 Garner MH and Spector, A. (1980) Proc. Natl Acad. Sci. USA. 77, 1274-1277.

15 Winyard PG, Lunec J, Brailsford SB and Blake DR. (1984) Int J Biochem. 16(12) 1273-1278.

16 Jones AF and Lunec J.(1987) Br.J. Cancer. 55, Suppl.VIII, 60-65.

17 Khairallah EA, Bond JS and Bird JWC.(eds)(1984) Intracellular protein catabolism. Alan R Liss.pp 1-702.

18 Levine RL, Oliver CN, Fulks RM and Stadtman ER.(1981) Proc. Nat. Acad. Sci. USA. 78, 2120-2124.

19 Roitt IM, Brostoff J and Male DK.(1985) Immunology. Gower Medical Publishing, London.

20 Trioli G, Giardini E, Rinaldi A, Seddio G and Bompiani GD. (1986) J Clin Lab Immunol.21(3),113-6

21 Jones AF, Jenings PE, Wakefield A, Winkles JW, Lunec J and Barnett AH. The Lancet, Aug 23, 1986. p459.

22 Adams GE, Aldrich JE, Bisby RH, Cundall RB, Redpath JL and Willson RL. (1972) Radiat Res. 49, 298.

23 Ahn B, Rhee SG and Stadtman ER.(1987) Anal. Biochem.161, 245-257

24 Oliver CN. (1987) Arch. Biochem. Biophys. 253, 62-72.

25 Lunec J (1980) Biochem Soc Proc. 10, 21.

26 Lunec J, Griffiths HR and Blake DR.(1987) Protein fluorescence and its relationship to free radical activity. In: Free radicals cell damage and disease.Ed; C.Rice-Evans. Richelieu Press, London pp 151-167

27 Halliwell B and Gutteridge JMC (1985) In; Free Radicals in Biology and Medicine, Clarendon Press, Oxford. pp 279-286.

28 Johnson PM, Watkins J, Scopes PM and Tracey BM (1974) Ann Rheum Dis, 33, 366-370

29 Singh A, Antonsen SA, Koell GW, Kremers W and Singh H.(1983).In; ICOR III, Editors; Bors, Saran and Tait.pp491-498.

30 Creed D. Photochem Photobiol.(1984) 39,(4),pp 577-583.

31 Holt ME, Ryall MET and Campbell AK (1984) Brit J Exp Pathol. 65, 231-235

32 Singh A, Bell M-J,Koroll GW, Kremers W and Singh H. (1981) In: Oxygen and Oxy Radicals in Chemistry and Biology. pp 461-470.

FREE RADICALS: CHEMISTRY, PATHOLOGY AND MEDICINE
edited by Catherine Rice-Evans & Thomas Dormandy, © 1988 Richelieu Press, London

A Role for Calcium in Cold Ischaemic Damage to the Rabbit Kidney?

Lisa A. Cotterill, Jon D. Gower, Barry J. Fuller*, Colin J. Green

Surgical Research Group, Division of Comparative Medicine,
Clinical Research Centre, Watford Road, Harrow, Middlesex, HA1 3UJ.

*Academic Department of Surgery, Royal Free Hospital,
School of Medicine, Pond Street, Hampstead, London, NW3 2QG.

Abstract

This study describes an investigation into the possible role of calcium ions in the oxidative damage to kidneys which takes place as a result of cold storage.

Rabbit kidneys were stored at 0°C for 24 hours in isotonic saline (a poor storage medium) or for 72 hours in hypertonic citrate solution (HCA) (a good storage medium). These periods of storage resulted in significant increases in lipid peroxidation in kidney homogenates which were reduced when the calcium antagonist verapamil was added to the saline solution. Addition of calcium chloride or the calcium ionophore A23187 to the saline solution significantly increased the rate of lipid peroxidation in homogenates of the stored organs.

Addition of these substances to the HCA storage medium had no effect on oxidative damage to the kidney which may be attributable, at least in part, to the ability of citrate present in this solution to chelate calcium.

These results demonstrate a relationship between lipid

peroxidation and calcium fluxes and suggest a role for this cation in the oxidative damage resulting from cold ischaemia.

Introduction

Renal transplantation involves periods of both cold ischaemia (organ storage) and warm ischaemia (removal and replantation of the organ) followed by reperfusion with oxygenated blood. These procedures often result in poor renal function and death of tubular cells immediately after engraftment. There are likely to be several causes for this damage.

Investigations in a number of systems have linked irreversible cell injury with increases in the concentration of calcium in the cytoplasm (1). This suggests the possibility that elevated intracellular calcium levels may be the common denominator which triggers cell death and it is therefore possible that calcium ions play a role in the degenerative processes resulting from combined warm ischaemia and cold storage of kidneys.

Under normal conditions the cytoplasmic calcium concentration is 10^3-10^4 fold lower than that of the extracellular fluid. Low cytoplasmic calcium levels are maintained by a number of processes: membrane-bound ATP-dependent pumps actively remove calcium from the cell and excess calcium is sequestered by mitochondria and the endoplasmic reticulum. Dysfunction of these systems results in an increase in cytoplasmic calcium levels which lead to biochemical changes, alterations in the morphology of the cell and eventually to irreversible cell death.

There is growing evidence that oxygen-derived free radicals are responsible, at least in part, for reperfusion injury and we have shown that cold storage of rabbit kidneys results in highly elevated levels of lipid peroxidation products formed during the incubation of homogenates in vitro (2,3). Lipid peroxidation affects calcium homeostasis in a number of ways. Peroxidation renders the cell membrane more permeable to calcium ions which readily leak in from the extracellular milieu. Oxidative damage to the mitochondrion and endoplasmic reticulum membranes partially destroys their ability to sequester calcium, so leading to an increase in cytoplasmic free calcium (4,5).

Elevated cytoplasmic calcium levels result in the activation of several biochemical processes which are likely to have important consequences to ischaemic tissues. Activation of calcium dependent phospholipases results in further breakdown of cell membranes (6), the release of free fatty acids and the formation of lysophosphatides which are toxic to the cell (6). Phospholipase activity is the rate limiting step in prostaglandin synthesis and lipoxygenation (7) thus these pathways may be stimulated following ischaemia. An alteration in prostaglandin homeostasis and the formation of lipoxygenase products some of which are chemotactic and cause vasoconstriction, such as leukotriene B4 (8,9) are likely to result in vascular changes which may contribute to the "no-reflow" phenomenon.

A number of ATPases are also activated as a consequence of calcium overload and these will consume valuable ATP resources already made scarce by the ischaemic conditions (10). In addition, xanthine dehydrogenase is converted to xanthine oxidase by calcium-dependent proteolysis and this enzyme may be responsible for the production of oxygen-derived free-radicals during reperfusion (11).

During the early stages of cell injury, redistribution of intracellular calcium pools may be important. It has been shown that when kidneys are subjected to periods of warm ischaemia, mitochondrial dysfunction coupled with ATP depletion results in failure of the calcium uptake processes of this organelle (12). Renal ischaemia also decreases the ability of endoplasmic reticulum to sequester calcium (13). Thus a redistribution of intracellular calcium takes place and an increase in cytosolic calcium levels has been observed in cultured kidney cells subjected to anoxia (14). During reperfusion there is an influx of extracellular calcium as a consequence of surface membrane changes resulting in massive calcium overload. Some of this calcium accumulates in the mitochondria, probably by passive processes, leading to further mitochondrial and cellular damage (15).

Prevention of ischaemic damage before transplantation continues to pose a major challenge and there is currently considerable interest in the use of calcium antagonists, particularly slow calcium channel blockers such as verapamil. Certainly, it would appear that verapamil affords some protection to kidneys subjected to periods of cold or warm ischaemia followed by reperfusion (16-20). The beneficial effects of calcium antagonists may be due to changes in renal hemodynamics in the reperfusion period as well as direct effects on calcium fluxes.

The relationships between lipid peroxidation, calcium ions and cell death are complex and as yet not fully understood. The present experiments were designed to investigate the possible relationships between calcium ions and lipid peroxidation in kidneys stored at $0^{o}C$. The organs were either flushed and stored at $0^{o}C$ in isotonic saline solution (a poor storage medium) for 24 hours or in hypertonic citrate (HCA) solution (a good storage medium currently used in clinical transplantation) for 72 hours. The effects of adding either the calcium-channel blocking agent verapamil, calcium chloride or the calcium ionophore A23187, which evokes an increase in cytoplasmic calcium levels, to these storage solutions on the rate of lipid peroxidation in homogenates of medulla and cortex were studied. These experiments were designed to gain a better understanding of the degenerative processes which take place under these conditions with a view to improving the viability of stored organs awaiting transplantation.

Materials and Methods

Chemicals: Verapamil hydrochloride was purchased from Abbott Laboratories Ltd (Kent) and the calcium ionophore A23187 was purchased from Sigma Chemical Company Ltd (Poole, Dorset). Other reagents were obtained from standard suppliers and were of the highest quality available.

Operative Procedure: Eight groups of adult New Zealand white rabbits (average weight 3kg) were anaesthetised by intramuscular injection of 0.2ml/kg fentanyl-fluanisone (Hypnorm ®), followed by a slow intravenous injection of diazepam (valium 20 ®) at 1.0mg/kg. Frusemide (3mg/kg) and 300 i.u./kg of heparin were also administered by intravenous injection. The abdomen was opened by a mid-line incision and the kidneys were completely skeletalised and removed. The renal artery of each kidney was cannulated and the organs flushed with 30mls of sterile HCA or 30mls of sterile isotonic 0.9% sodium chloride solution (saline) already cooled to 4°C. The kidneys were placed in sterile beakers containing 60mls of the identical flush solution and stored in ice at 0° for 24 hours in saline or 72 hours in HCA.

Kidneys were treated according to the protocol outlined below. In experiments in which kidneys were stored in solutions containing verapamil (100µM), calcium chloride (1mM) or A23187 (10µM), the contralateral kidneys were stored under identical conditions without the addition of these agents thus providing controls.

Experimental Protocol:

Group (1) : Kidneys were not stored.

Group (2) : Kidneys were flushed with 30mls 0.9% w/v sodium chloride and stored for 24 hours at $0^{o}C$. Verapamil (100µM) was added to homogenates of these kidneys after storage.

Group (3) : Kidneys were flushed with 30mls HCA and stored for 72 hours at $0^{o}C$. Verapamil (100µM) was added to homogenates of these kidneys after storage.

Group (4) : Kidneys were flushed with 30mls 0.9% w/v sodium chloride solution containing 100µM verapamil and stored for 24 hours at $0^{o}C$.

Group (5) : Kidneys were flushed with 30mls HCA containing 100µM verapamil and stored for 72 hours at $0^{o}C$.

Group (6) : Kidneys were flushed with 30mls 0.9% w/v sodium chloride solution containing 1mM $CaCl_2$ and stored for 24 hours at $0^{o}C$.

Group (7) : Kidneys were flushed with 30mls HCA containing 1mM $CaCl_2$ and stored for 72 hours at $0^{o}C$.

Group (8) : Kidneys were flushed with 30mls 0.9% w/v sodium chloride solution containing 10µM A23187 and stored for 24 hours at $0^{o}C$.

Preparation of the Kidneys: Kidneys were dissected into cortex and medulla, homogenised in phosphate-saline buffer (pH 7.4) and incubated at 37°C for 60 mins in a shaking water-bath. The rate of lipid peroxidation was determined by the measurement of lipid soluble Schiff's bases and the formation of TBA-reactive material and related to the protein content of the homogenates as described previously (3).

Statistical analysis was performed using the students paired t-test.

Results

As previously reported (2,3), the cold storage of kidneys in either isotonic saline solution or HCA solution resulted in elevated levels of lipid peroxidation products in homogenates of both the cortex and medulla (Figures 1 and 2).

Addition of verapamil to homogenates of kidneys after storage (Groups 2 and 3) had no significant effect on the formation of Schiff's bases or the production of TBA-reactive material (data not shown).

When kidneys were flushed with isotonic saline solution to which verapamil had been added (Group 4), significant decreases were observed in both measures of lipid peroxidation in the cortex and medulla (Figures 1 and 2). However verapamil had no effect when added to the HCA storage solution (Group 5, data not shown).

Addition of $CaCl_2$ to the isotonic saline solution (Group 6) significantly increased the rates of lipid peroxidation in both the cortex and medulla of the stored kidney (Figure 3) but no effects were observed when $CaCl_2$ was added to the HCA storage

Figure 1

Formation of thiobarbituric acid - reactive material in rabbit kidneys flushed with saline containing verapamil (100 μM) and stored for 24 hours

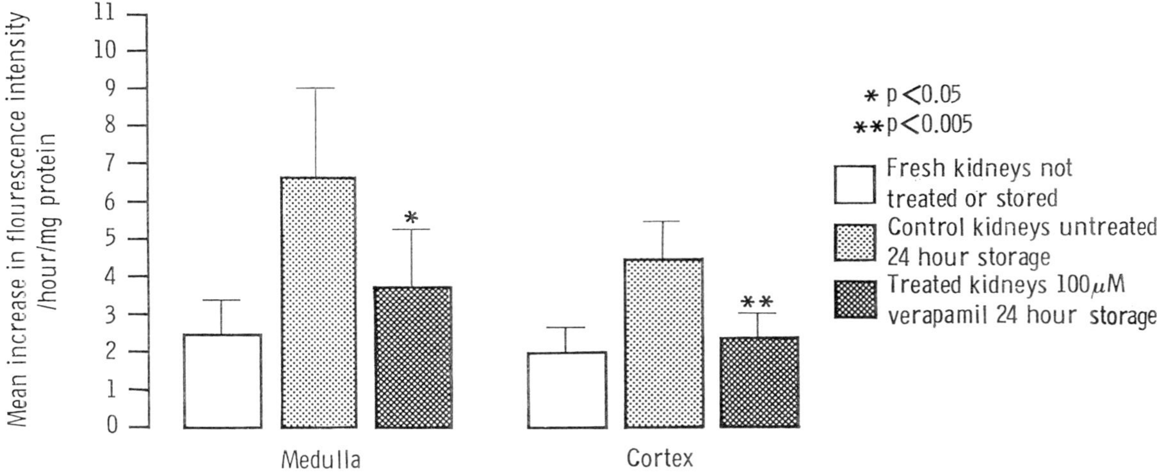

Figure 2

Formation of schiffs base in rabbit kidneys flushed with saline containing verapamil (100μM) and stored 24 hours

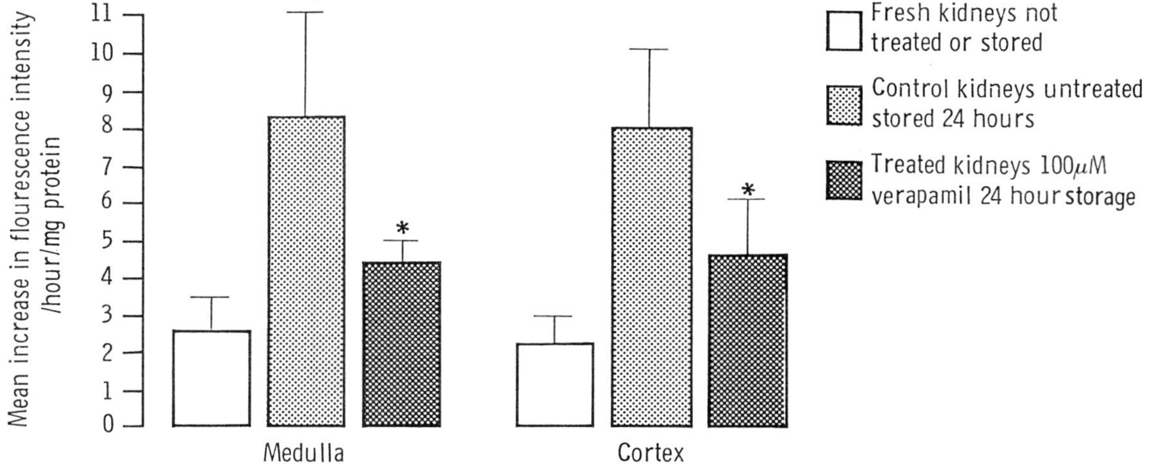

Figure 3

Formation of thiobarbituric acid-reactive material and schiffs base in rabbit kidneys flushed with saline containing $CaCl_2$ and stored for 24 hours

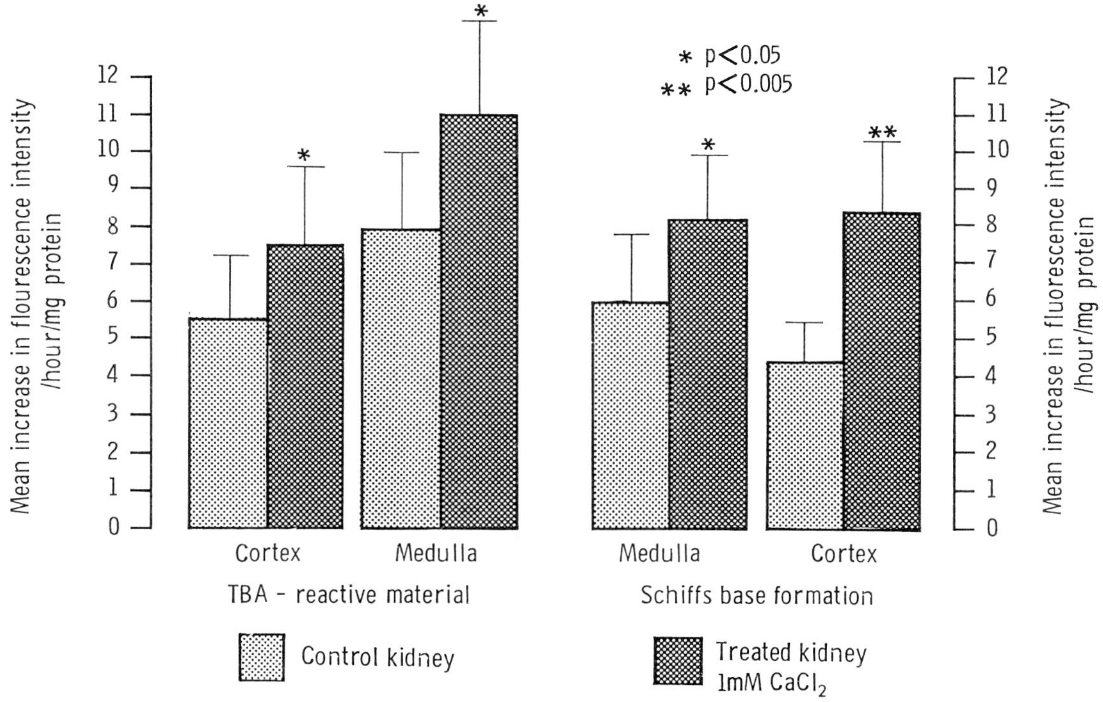

storage solution (Group 7, data not shown).

Addition of the calcium ionophore A23187 to the isotonic saline solution (Group 8) significantly increased both Schiff's base formation and the production of TBA-reactive material in the cortex and medulla of the stored kidney (Figure 4).

Discussion

These results demonstrate that increases in peroxidative damage during cold ischaemia of kidneys may be attributable in part to influx of calcium into the cytosol. Verapamil decreased lipid peroxidation in kidneys stored 24 hours in saline but had no beneficial effect when added directly to homogenates of kidneys after the storage period. It seems likely that the effect of verapamil is modulated through its ability to decrease calcium influx into the cytosol of intact kidney cells during the storage period rather than an interaction with intracellular components in the homogenised tissue.

Verapamil had no effect on lipid peroxidation in kidneys stored for 72 hours in HCA storage solution. Hence, the known value of HCA as a good storage medium compared to saline may be partly due to the ability of citrate present in the HCA solution to chelate calcium. This idea is further supported by the observation that addition of $CaCl_2$ (1mM) to the storage solutions resulted in a significant increase in peroxidative damage only to kidneys stored in saline and not in the HCA solution which contains a large excess of citrate (55mM).

Figure 4

Formation of thiobarbituric acid-reactive material and schiffs base in rabbit kidneys flushed with saline containing A23187 and stored for 24 hours

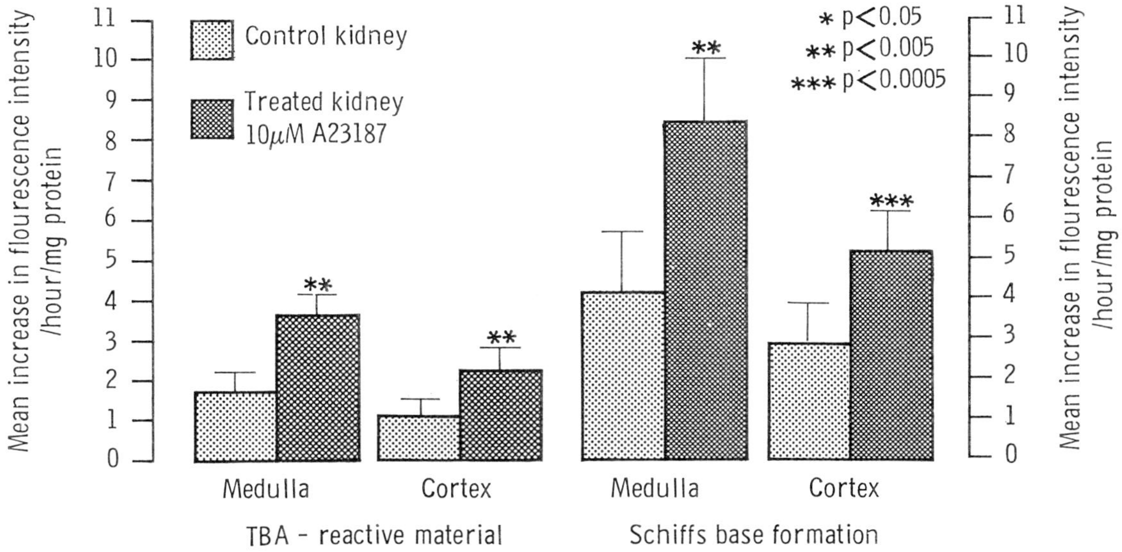

The relationship between cytosolic free calcium levels and lipid peroxidation in stored kidneys was further demonstrated by use of the ionophore A23187 which permeabilises the cell membrane to calcium and increased the rate of lipid peroxidation in kidneys stored for 24 hours in isotonic saline solution.

This study strongly suggests that prevention of increased cytosolic calcium levels is an important consideration for maintenance of the viability of organs during storage. Although the associations between calcium levels, oxidative damage and cell viability are not fully understood, it would appear that chelation of extracellular calcium during cold storage and blockage of calcium entry into the cell by agents such as verapamil may be useful approaches to protecting against ischaemic damage.

References

(1) J.L. Farber: Life Sci. 29 (1981), 1289.

(2) C.J. Green, G. Healing, J. Lunec, B.J. Fuller and S. Simpkin: Transplantation 41 (1986), 161.

(3) J. Gower, B. Fuller and C. Green: Free Rad. Res. Communs. 3 (1987), 107.

(4) A. Masini, T. Trenti, D. Ceccarelli and E. Ventura: in 'Free Radicals in Liver Injury' eds G. Poli, K.H. Cheeseman, M.U. Dianzani and T.F. Slater (IRL Press, 1986) p 135.

(5) A. Masini, T. Trenti, E. Ventura, D. Ceccarelli and U. Muscatello: Biochem. Biophys. Res. Commun. 130 (1985) 207.

(6) F.R. Ungemach: in 'Free Radicals in Liver Injury' eds G. Poli, K.H. Cheesman, M.U. Dianzani and T.F. Slater (IRL Press, 1986) p 127.

(7) P.A. Craven and F.R. DeRubertis: J. Biol. Chem. 258 (1983) 4814.

(8) A.M. Lefer: Fed. Proc. 44 (1985) 275.

(9) P.J. Piper, A.W.B. Stanton and L.J. McLeod: Prostaglandins 29 (1985) 61.

(10) C.F. Peng, M.L. Murphy and K.D. Straub: Cell Calcium 3 (1982) 227.

(11) J.M. McCord: N. Engl. J. Med. 312 (1985) 159.

(12) P.E. Arnold, U.J. VanPutten, D. Lumjergul, T.J. Burke and R.W. Schrier: Am. J. Physiol.250 (1986) F357.

(13) A. Schieppati, P.D. Wilson, T.J. Burke and R.W. Schrier: Am. J. Physiol. 249 (1985) C476.

(14) K.W. Snowdowne, C.C. Frendenrich and A.B. Borle: J. Biol. Chem. 260 (1985) 11619.

(15) C.D. Malis and J.V. Bonventre: J. Biol. Chem. 261 (1986) 14201.

(16) G.A. Gingrich, G.R. Barker, P. Lui and S.C. Stewart: J. Urol. 134 (1985) 408.

(17) D. Goldfarb, A. Iaina, I. Serban, S. Gavendo, S. Kapulek and H.E. Enanou: Proc. Soc. Exp. Biol. Med. 172 (1983) 389.

(18) J.I. Shapiro, C. Cheung, A. Itabaishi, L. Chan and R.W. Schrier: Transplantation 40 (1985) 596.

(19) T.J. Burke, P.E. Arnold, J.A. Gordon, R.E. Bulger, D.C. Dobyan and R.W. Schrier: J. Clin. Invest. 74 (1983) 1830.

(20) C.D. Malis, J.Y. Cheung, A. Leaf and J.V. Bonventre: Am. J. Physiol. 245 (1983) F735.

FREE RADICALS: CHEMISTRY, PATHOLOGY AND MEDICINE
edited by Catherine Rice-Evans & Thomas Dormandy, © 1988 Richelieu Press, London

The Effect of Oxygen Metabolites on the Zymogens of Human Pancreatic Proteases

PM Guyan, JM Braganza*, J Butler[†] (The Medical School, Royal Infirmary* and Paterson Laboratories[†], Manchester)

ABSTRACT

$O_2^{\bar{\cdot}}$, $OH^{\cdot}$ and HOCl do not activate the zymogens of human pancreatic proteases, nor destroy trypsin inhibitor.

The combination of superoxide dismutase and catalase has been shown to ameliorate acute pancreatitis(AP) in several animal models, including two in which the invasion of the pancreatic interstitium by leucocytes is a striking early feature(1). These observations suggest that the primary metabolites of oxygen somehow detonate AP and, furthermore, that the damage may be compounded by HOCl. The disease is believed to be due to premature intrapancreatic activation of trypsinogen, which then activates the other zymogens, and elastase is accepted as being the cause of pancreatic haemorrhage in AP. We have therefore investigated the effect of $O_2^{\bar{\cdot}}$, $OH^{\cdot}$ and HOCl on the zymogens and trypsin inhibitor in human pancreatic juice.

METHODS

Samples

Human pancreatic juice was obtained by endoscopic cannulation of the Ampulla of Vater or from an operative drain in the pancreatic duct. Using specific substrates and spectrophotometric or fluorometric detection, the activities of trypsin(2), chymotrypsin(3), elastase(4), and trypsin inhibitor(2) were measured before and after exposing the samples to a free radical source.

Production of Oxygen Metabolites

(a) Equal concentrations of $O_2^{\bar{\cdot}}$ and $OH^{\cdot}$ were generated, in various doses, by irradiation of air-saturated samples of pancreatic juice, using a caesium source (0.5-13 krads) and a linear accelerator (3-13 krads). (b) Xanthine/xanthine oxidase was used to generate $O_2^{\bar{\cdot}}$ in the presence of pancreatic juice samples. (c) HOCl was produced by adjusting NaOCl to pH 6.2 with dilute H_2SO_4; concentrations were determined using a molar absorptivity coefficient of 100 at 235 nm(5).

RESULTS

1) Irradiation(Table I) did not activate the zymogens of human pancreatic proteases, neither did it destroy trypsin inhibitor(TI): the potential protease activity was determined by activation with enterokinase(EK). 2) Irradiation

caused a dose-dependent loss of trypsin activity in aqueous solutions of commercial bovine trypsin, but it had no effect on trypsin in human pancreatic juice that was already activated (Table II).

TABLE I

Dose (krad)	Trypsin Fluor.units (Ex335/Em410)	Chymotrypsin Rate (ΔAbs_{400} in 1min)	Elastase Rate (ΔAbs_{410} in 5min)	TI %inhibition 10ug trypsin
0	2.1	.014	.0032	52
0.67	1.8	.015	.0035	
1.33	1.5	.013	.0034	
2.00	1.4	.014	.0039	
2.66	1.8		.0032	
3.30	2.1	.015		53
3.85	2.3		.0032	
6.61	2.1	.017		52
7.70	2.6		.0032	
11.55	2.2		.0032	
13.22	2.2	.014		50
	80(EK)	.141(EK)	.0170(EK)	

TABLE II

Dose (krad)	Commercial Fluor	Trypsin % max	Human Trypsin Fluor	% max
0	303	100	199	100
0.67	278	92	196	98
1.33	255	84	198	99
2.00	236	78	200	101

3) Xanthine/xanthine oxidase did not activate the zymogens.

4) HOCl(50-650μM) did not activate the zymogens nor destroy trypsin inhibitor.

DISCUSSION

These in-vitro studies indicate that if $O_2^{\cdot -}$ and $OH^{\cdot}$

detonate AP, as they appear to do(1), they act in ways that do not involve the activation of pancreatic protease precursors, which also resist activation by HOCl. This resistance may in part be due to the H_2O_2-dependent oxidase in human pancreatic juice(6).

References

1. JM Braganza: Proc Free Rad Soc. St Andrews(1987) In Press
2. JK MacDonald, S Ellis, TJ Reilly: J Biol Chem 241(1966) 1494
3. CJ Martin, J Golubow, AE Axelrod: J Biol Chem 234(1959) 1718
4. J Bieth, B Spiess, CG Wermuth: Biochem Med 11(1974)350
5. JC Morris: J Phys Chem 70(1966)3798
6. PM Guyan, J Butler, JM Braganza, FS Steven: Biochem Soc Trans 14(1986)980

Acknowledgements

These studies were supported by the NWRHA and the Thomas Hunt Fellowship.

Index